Ecological Studies, Vol. 88

Analysis and Synthesis

Ecological Studies

Volume 68
Stable Isotopes in Ecological Research
Edited by P.W. Rundel, J.R. Ehleringer, and K.A. Nagy (1989)

Volume 69
Vertebrates in Complex Tropical Systems
Edited by M.L. Harmelin-Vivien and F. Bourlière (1989)

Volume 70
The Northern Forest Border in Canada and Alaska
Edited by J.A. Larsen (1989)

Volume 71
Tidal Flat Estuaries: Simulation and Analysis of the Ems Estuary
Edited by J. Baretta and P. Ruardij (1988)

Volume 72
Acidic Deposition and Forest Soils
By D. Binkley, C.T. Driscoll, H.L. Allen, P. Schoeneberger, and D. McAvoy (1989)

Volume 73
Toxic Organic Chemicals in Porous Media
Edited by Z. Gerstl, Y. Chen, U. Mingelgrin, and B. Yaron (1989)

Volume 74
Inorganic Contaminants in the Vadose Zone
Edited by B. Bar-Yosef, N.J. Barnow, and J. Goldshmid (1989)

Volume 75
The Grazing Land Ecosystems of the African Sahel
Edited By H.N. Le Houérou (1989)

Volume 76
Vascular Plants as Epiphytes: Evolution and Ecophysiology
Edited by U. Lüttge (1989)

Volume 77
Air Pollution and Forest Decline: A Study of Spruce (*Picea abies*) on Acid Soils
Edited by E.-D. Schulze, O.L. Lange, and R. Oren (1989)

Volume 78
Agroecology: Researching the Ecological Basis for Sustainable Agriculture
Edited by S.R. Gliessman (1990)

Volume 79
Remote Sensing of Biosphere Functioning
Edited by R.J. Hobbs and H.A. Mooney (1990)

Volume 80
Plant Biology of the Basin and Range (1990)
Edited by B. Osmond, G.M. Hidy, and L. Pitelka (1990)

Volume 81
Nitrogen in Terrestrial Ecosystem: Questions of Productivity, Vegetational Changes, and Ecosystem Stability
C.O. Tamm (1991)

Volume 82
Quantitative Methods in Landscape Ecology: The Analysis and Interpretation of Landscape Heterogeneity
Edited by M.G. Turner and R.H. Gardner (1990)

Volume 83
The Rivers of Florida
Edited by R.J. Livingston (1990)

Volume 84
Fire in the Tropical Biota: Ecosystem Processes and Global Challenges
Edited by J.G. Goldammer (1990)

Volume 85
The Mosaic-Cycle Concept of Ecosystems
Edited by H. Remmert (1991)

Volume 86
Ecological Heterogeneity
Edited by J. Kolasa and S.T.A. Pickett (1991)

Volume 87
Horses and Grasses
Edited by P. Duncan (1991)

Volume 88
Pinnipeds and El Niño: Responses to Environmental Stress
Edited by F. Trillmich and K.A. Ono (1991)

Fritz Trillmich · Kathryn A. Ono (Eds.)

Pinnipeds and El Niño

Responses to Environmental Stress

With 81 Figures

Springer-Verlag
Berlin Heidelberg New York
London Paris Tokyo Hong Kong
Barcelona Budapest

Prof. Dr. FRITZ TRILLMICH
Universität Bielefeld
Lehrstuhl für Verhaltensforschung
Postfach 8640
4800 Bielefeld
FRG

Dr. KATHRYN A. ONO
Institute of Marine Sciences
University of California
Santa Cruz, California 95064
USA

ISBN 3-540-53634-5 Springer-Verlag Berlin Heidelberg New York
ISBN 0-387-53634-5 Springer-Verlag New York Berlin Heidelberg

Library of Congress Cataloging-in-Publication Data. Pinnipeds and El Niño / Fritz Trillmich, Kathryn A. Ono, eds. p. cm. – (Ecological studies; vol. 88) Includes index. ISBN 3-540-53634-5. – ISBN 0-387-53634-5 (U.S.) 1. Pinnipedia – Pacific Ocean – Ecology. 2. El Niño Current. I. Trillmich, Fritz, 1948– . II. Ono, Kathryn A., 1951– . III. Series: Ecological studies; v. 88. QL737.P6P56 1991 599.74'50452636–dc20 91-16442

Typesetting: International Typesetters Inc., Makati, Philippines
31/3145-543210 – Printed on acid-free paper

Preface

Work in the Galapagos upwelling ecosystem on the Galapagos fur seal (*Arctocephalus galapagoensis*) gave me a vivid impression that tropical environments are not at all as stable and predictable as is sometimes still assumed. Whenever I thought I knew the system, another year would bring further surprises. The biggest was certainly the major 1982–83 El Niño/Southern Oscillation disturbance. After this experience I no longer dared to speak of a "normal" year. However, what that event and the preceding series of different years taught me was that the life history of animals needs to evolve great flexibility to permit the continued existence of a species in such an environment. Fluctuating selection seemed to be the rule and adaptation to a constant equilibrium situation impossible. While fluctuations are certainly common in the homeland of El Niño, populations further north might live in more stable and predictable environments. Have these species evolved less flexible life histories and if so which components are different? A comparative study of El Niño effects along the north-south gradient of the east Pacific coast seemed ideal to address these questions and to get a first idea about population differences in life history flexibility. A broad overview of the effects of the major disturbance of the marine ecosystem caused by the 1982–83 El Niño would be immensely helpful to understand limits to the adaptations of diverse pinniped species to environmental stress. This, in turn, promised to provide insights into the evolution of the life history parameters of these species, in particular, growth, fertility, reproductive effort, and juvenile and adult mortality schedules.

Following the 1986 International Ornithological Congress in Ottawa I made a trip along the west coast of the USA to visit several marine mammal laboratories. In Ottawa I had listened to the lively discussions among the ornithologists about El Niño effects on seabirds. I was surprised to find that for once mammalogists could document facts in as much or even more detail than ornithologists. Due to the long-term nature of many investigations on pinnipeds, there were a surprising number of studies which could interpret data gathered during the 1982–83 El Niño against the background of several "normal" years. I therefore wanted to gather as much of the available knowledge as possible to obtain an overview of the exciting data collected by many researchers on the effects of El Niño on pinniped populations. Much of the available information was not yet published and soon we began to consider it necessary to collect the dispersed information between the covers of one book in order to fit the pieces of the puzzle together which each one of us had.

I would have despaired had I not found willing help from all the authors of this book, in particular, from my coeditor, Kathy Ono, without whose organizing and editorial skills the *Zalophus* section would never have seen the light of day. Even so, it took over 3 years, a workshop in Santa Cruz, California (which was kindly hosted by Burney J. Le Boeuf and largely organized by Carolyn Heath), and much more effort than we had expected to see this book through to publication, but it seems to us that the result is well worth the effort.

Together, we would like to thank the many people and institutions which assisted in the editing of this book. First, it is a pleasure to acknowledge all the colleagues who contributed by reviewing papers: A. Alvial, W. Arnold, N. Bonner, J. Croxall, J. Estes, P. Hammerstein, B. König, K. Kovacs, D. Lavigne, D. Martinson, G. Miller, J. Norton, D. Odell, P. Rothery, I. Stirling, J. Tarazona, and M. Zuk. D. Schmidl was immensely helpful to F.T. by preparing graphs for many papers in this book, Beth Dennis expertly drew the maps presenting our Pacific-wide study sites, and M. Horning plowed through all the references and condensed them into the final list. The director of the Max-Planck Institut für Verhaltensphysiologie, Seewiesen, W. Wickler, tolerated that F.T. delved into ecological questions for a while and supported him throughout. The Department of Biology of New Mexico kindly supported K.O.'s activities. We would both like to thank our families, Bill and Gavin Rice, and Inka and Nela Trillmich, for their patience throughout this endeavor. Special thanks are extended to the creators of the FAX machine, without which timely communications between editors would have been impossible. Last, but not least, we would like to thank the very efficient staff of Springer-Verlag and our series editor, H. Remmert, whose active interest in the role of disturbance in the dynamics of ecosystems led him to take an active interest in our project.

Bielefeld, 1991 FRITZ TRILLMICH

Contents

Part III. California Sea Lion

List of Contributors

ANTONELIS, George A.
National Marine Mammal
Laboratory
National Marine Fisheries Service
7600 Sand Point Way, N.E. Bldg. 4
Seattle, Washington 98115-0070
USA

ARNTZ, Wolf E.
Alfred Wegener Institut
für Polar- und Meeresforschung
Columbusstraße S/N
2850 Bremerhaven
FRG

AURIOLES, David G.
Centro de Investigaciones Biologicas
de Baja California, A.C.
Jalisco y Madero
Apdo. Postal 128
23060 La Paz, Baja California Sur
Mexico

BECKHAM, Claire
P.O. Box 358
Demorest, Georgia 84025
USA

BONESS, Daryl J.
National Zoological Park
Smithsonian Institution
Washington, D.C. 20008
USA

COSTA, Daniel P.
Long Marine Laboratory
Institute of Marine Science
University of California
Santa Cruz, California 95064
USA

DELLINGER, Thomas
Max-Planck Institut für
Verhaltensphysiologie
Abt. Wickler
8131 Seewiesen
FRG

DELONG, Robert L.
National Marine Mammal
Laboratory
National Marine Fisheries Service
7600 Sand Point Way, N.E. Bldg. 4
Seattle, Washington 98115-0070
USA

FAHRBACH, Eberhard
Alfred Wegener Institut für
Polar- und Meeresforschung
Columbusstraße S/N
2850 Bremerhaven
FRG

FELDKAMP, Steven D.
2095 Delridge Ave.
Roseburg, Oregon 97470
USA

FRANCIS, John
Dept. Zoological Research
National Zoological Park
Washington, D.C. 20008
USA

GENTRY, Roger L.
National Marine Mammal Laboratory
National Marine Fisheries Service
7600 Sand Point Way, N.E. Bldg. 4
Seattle, Washington 98115-0070
USA

GUERRA C., Carlos G.
Instituto de Investigaciones Oceanológicas
Universidad de Antofagasta
Casilla 170
Antofagasta
Chile

HEATH, Carolyn B.
Division of Biology
Fullerton College
321 East Chapman Ave.
Fullerton, California 92634
USA

HUBER, Harriet
National Marine Mammal Laboratory
Alaska Fisheries Science Center
7600 Sand Point Way, N.E.
Seattle, Washington 98115
USA

IVERSON, Sarah J.
(c/o R.G. Ackman)
Canadian Institute of Fisheries Technology
Technical University of Nova Scota
1360 Barrington St. P.O. Box 1000
Halifax, Nova Scota
Canada B3J 2X4

LE BOEUF, Burney J.
Department of Biology
Institute of Marine Science
University of California
Santa Cruz, California 95064
USA

LOWRY, Mark S.
Southwest Fisheries Center
National Marine Fisheries Service
P.O. Box 271
La Jolla, California 92038
USA

MAJLUF, Patricia
Paul de Beaudiez 510
Lima 27
Peru

NISBET, Jack
P.O. Box 358
Demorest, Georgia 84025
USA

OFTEDAL, Olav
U.S. National Zoological Park
Smithsonian Institution
Washington, D.C., 20008
USA

OLIVER, Charles W.
Southwest Fisheries Center
National Marine Fisheries Service
P.O. Box 271
La Jolla, California 92038
USA

ONO, Kathryn A.
Institute of Marine Sciences
University of California
Santa Cruz, California 95064
USA

PEARCY, William G.
College of Oceanography
Oregon State University
Corvallis, Oregon 97331
USA

PORTFLITT, K., George
Instituto de Investigaciones
Oceanológicas
Universidad de Antofagasta
Casilla 170
Antofagasta
Chile

REITER, Joanne
Department of Biology
Institute of Marine Science
University of California
Santa Cruz, California 95064
USA

STEWART, Brent S.
Hubbs Research Institute
1700 S. Shores Road
San Diego, California 92109
USA

TRILLMICH, Fritz
Universität Bielefeld
Lehrstuhl für Verhaltensforschung
Postfach 8640
4800 Bielefeld 1
FRG

YOCHEM, Pamela K.
Hubbs Research Institute
1700 S. Shores Road
San Diego, California 92109
USA

YORK, Anne
National Marine Mammal
Laboratory
National Marine Fisheries Service
7600 Sand Point Way, N.E. Bldg. 4
Seattle, Washington 98115-0070
USA

Part I

Introduction

K.A. Ono and F. Trillmich

The El Niño is a meteorological and oceanographic phenomenon which occurs at irregular intervals in the eastern tropical Pacific. Its most obvious characteristic is the warming of surface waters. A severe El Niño/Southern Oscillation (for a more thorough explanation of the terminology and the phenomenon, see Chap. 1) affects not only the Pacific marine environment, but also continental systems worldwide (Rasmusson 1985; Ropelewski and Halpert 1986, 1987), making it one of the greatest known natural disturbances of ecosystems. The 1982–83 El Niño was presumably the strongest in the last century (Quinn et al. 1987). During a typical El Niño, cold water currents in the eastern tropical Pacific slow down, upwelling is reduced and fisheries, particularly in Peru and Ecuador, may have severely diminished yields.

The El Niño phenomenon provides us with some insight into the short-term consequences of rapid changes in an ecosystem. Such changes are generally hard to study since it is exceedingly difficult to document and analyze the behavior of large ecosystems for which experimentation is usually untenable. Since the El Niño is a recurrent and presumably ancient event (see Chap. 1), we suspect that many animal species in the tropical marine ecosystems which have been repeatedly exposed to it may have evolved some degree of adaptation to accommodate its effects. Species which live in the temperate zone and at higher latitudes experience the disturbances associated with the tropical El Niño less frequently and at much attenuated intensity. Their response to severe El Niño's which do penetrate their environment should provide valuable clues to the impacts of a rapid change in ecosystems on large vertebrate species (such as man under conditions of global warming?) which have not evolved life history adaptations to cope with unpredictable environmental variability.

Pinnipeds are top predators in marine ecosystems and are therefore good indicator species. This book attempts to document the effects of a major environmental disturbance on a number of these species over a fairly wide geographical range. We rely on the use of the comparative method to evaluate which life-history variables of the many pinniped species respond most strongly to sudden deviations of environmental parameters from the normal.

The effect of the 1982–83 El Niño on pinnipeds appeared to have decreased with distance from the center of the anomaly. This provides excellent information about the sensitivity of different families and species to environmental perturbation, and permits us to compare the responses of closely related species to the varying strength of disturbance. We were also able to compare the effects upon animals which were in different stages of their breeding cycle during the onset and most intense phase of El Niño. This unusually intense El Niño provided us with an opportunity to observe animals which should otherwise have the ability to cope with minor El Niño occurrences under conditions of extreme stress. Our analysis of the El Niño effects may also serve as a model of the potential reaction of pinniped populations to overfishing of their main prey by man.

Fortunately, many studies of pinniped population dynamics, ecology and behavior were in progress in the eastern Pacific tropics as well as along the coast of South America, Mexico and North America during the 1982–83 El Niño. These provided a unique opportunity to bring together data from many species in many locations in order to draw a more general picture of the nature of the effect of this warming phenomenon on pinnipeds. We hope the data and analyses contained in this book will serve to document once more the value of long-term studies because only they enable a critical analysis by comparing populations before, during and after the event. This is essential to further our understanding of the influence of rapid environmental changes on ecosystems. The data in this book should also provide a useful basis for comparison with future studies as well as point out the gaps in our knowledge and show where we need further data. One topic in particular, on which in this book we can only speculate, is of obvious crucial importance for increased understanding of pinniped and other top marine predators' life histories: the abundance and distribution of food. Presently, far too little is known about this aspect of marine life.

The papers contained in this book encompass a broad spectrum of taxa and disciplines. Species, which are not represented, such as the Steller sea lion (*Eumetopias jubatus*), harbor seal (*Phoca vitulina*), the Guadaloupe and the Juan Fernandez fur seal (*Arctocephalus townsendi, A. phillippii*) and the southern elephant seal (*Mirounga leonina*) are absent due to our inability to find researchers who have data over several years encompassing the El Niño, and which was gathered in a consistent manner. We felt it was important to include only studies in which El Niño – non-El Niño year comparisons could be made using the same study areas and techniques rather than attempting to patch together data collected by several different researchers using different study sites. We tried to be as complete as possible under these restrictions. Hopefully, gaps in our data sets will show where further research needs to be done, and those with studies we overlooked will publish their work using our results for comparison.

We also limited the species and studies to those pinnipeds residing in the eastern Pacific, where the effects of the El Niño per se, and not those of the wider ranging atmospheric teleconnections or of the La Niña (Kerr 1988), were well documented. This makes the latitudinal and family/species comparisons less confusing.

Part I begins with two introductory chapters on the physical (Chap. 1) and general biological effects (stressing the effects on potential pinniped prey species; Chap. 2) of El Niño in the eastern Pacific. These chapters provide a background for those unfamiliar with the multifarious effects of El Niño. These authors concentrate on factors which are potentially important for pinnipeds, whether breeding on land or foraging at sea, and detail the time course and magnitude of the effects at different latitudes. Figure 1 illustrates the major research sites and species for the studies included in this book.

Part II is primarily concerned with several species of fur seals but also includes comparative data for co-occurring sea lion species. The South American fur seal (*Arctocephalus australis*) was studied in two locations during the El Niño: Punta San Juan, Peru (Chap. 5) and Cape Paquica to Cape Angamos, Chile (Chap. 4). Both studies also contain some data on the South American sea lion (*Otaria byronia*). Next, the Galápagos fur seal (*Arctocephalus galapagoensis*) and the Galápagos subspecies of the California sea lion (*Zalophus californianus wollebaeki*), living in the heart of the El Niño, are discussed (Chap. 6). Data for the northern fur seal (*Callorhinus ursinus*) are presented from two extreme points of its range, San Miguel Island (Channel Islands, California; Chap. 7), the southernmost breeding island, and St. George Island (Pribilofs, Alaska; Chap. 8), one of the northernmost breeding sites. Additionally, the survival of northern fur seal juveniles at sea is examined and its possible connections with El Niño is discussed (Chap 9).

Part III of the book focuses on the effects of El Niño on various aspects of the biology of the California sea lion (*Zalophus c. californianus*). Several populations in Mexico were monitored before, during and after the El Niño including animals breeding in the Gulf of California as well as on islands in the Pacific Ocean off the Baja California peninsula (Chap 11). Changes in the pattern of sea lions hauling out north of the breeding area are also discussed in this section (Chap 13). The remaining studies on breeding populations come from the California Channel Islands. Studies of this population cover the most diverse topics. Starting with a description of population dynamics (Chap. 12; Chap. 17) authors then analyze the behavior of individuals in detail. The offshore foraging behavior of lactating females (Chap. 15), foraging energetics of lactating females (Chap. 16), female cycles of presence and absence from their pups (to which we refer as "attendance pattern"; Chap. 14), pup growth (Chap 18), milk intake (Chap. 19), mother-pup behavior and pup sex ratio (Chap. 20), and the suckling behavior and sex ratio of juveniles (Chap 21) are all treated in a comparative way looking at differences between El Niño influenced years and more normal years. The completeness and multifacetedness of documentation of the way in which California sea lions dealt with changing environmental conditions during El Niño are exceptional.

The fourth section compares the influence of El Niño on northern elephant seals (*Mirounga angustirostris*) breeding on three different islands which vary in population size, exposure to wind and surf, and topography. The southernmost study area is on San Nicolas Island, Channel Islands (Chap 25). Año Nuevo (Chap. 23), the site of intensive elephant seal studies for several decades, is located approximately 470 km north of the Channel Islands. The northernmost and most recently established breeding colony on the Farallon Islands, located about 97 km north of Año Nuevo, comprises the third study site (Chap 24).

Part V consists of one chapter (Chap 26) which is a synthesis of the information contained in the previous papers. It is primarily the result of a 2-day workshop

Fig. 1. Species studied and study sites along the east Pacific coast of the Americas

attended by many of the authors. Here, the observed effects of the El Niño are summarized with respect to latitude, immediate versus long-term effects, age classes and sex. Conclusions are then drawn as to how these effects impact otariids and phocids differently due to differences in their biology. Lastly, the possible selective effects of environmental fluctuations, such as El Niño on pinnipeds, are discussed and the parameters through which selection might be mediated (maternal strategies, male size and sexual selection) are considered.

1 The Time Sequence and Magnitude of Physical Effects of El Niño in the Eastern Pacific

E. Fahrbach, F. Trillmich, and W. Arntz

1.1 Introduction

During the years 1982–84 different pinniped populations along the Pacific coasts of the Americas were heavily affected by climatic anomalies over a wide range of latitudes, extending from Alaska in the north to the south of Chile. These anomalies can be related to El Niño (EN), a climatological phenomenon characterized by anomalous conditions in the atmosphere and the ocean. It appears in a somewhat regular time sequence in the tropical Pacific, extends, however, less intensively to the mid-latitudes and beyond. During EN, anomalously warm water appears along the coast of Ecuador and Peru. As an example we show the deviation of the sea surface temperature from the long-term mean at Puerto Chicama (Fig. 1). The magnitude of this positive sea surface temperature (SST) anomaly can be used to classify the intensity of an EN.

Examination of colonial documents and shipboard records, collected since the Spanish exploration of South America, suggests that 47 strong to very strong EN events (with an SST anomaly over several months of 3 to 12 °C) have occurred during the last four and a half centuries. In addition, there were at least 32 moderate EN events with a temperature anomaly of 2 to 3 °C since 1800 (Quinn et al. 1987). Therefore, on average, an EN turns up once every 4 years, with strong to very strong events occurring once every 10 years, though the actual distribution shows significant deviations from true cyclicity (Fig. 2).

EN is related to the weather system over the entire Pacific Ocean, and in most cases, it coincides with a significant "Southern Oscillation". The Southern Oscillation describes a seesaw-type change in the air pressure distribution over the South Pacific. It is quantified by an index which is based on the air pressure difference between Tahiti and Darwin, Australia. Under normal (non-EN) conditions, pressure is high over Tahiti and low over Darwin (Fig. 3).

A reversal of this pressure distribution, indicated by a negative Southern Oscillation Index, is usually followed by an EN event (Fig. 4). The causes of these changes in the pressure distribution lie in the instability of the large-scale atmospheric and oceanic circulation systems and their mutual interaction which we are just beginning to understand at present (Cane 1986; Enfield 1987; Graham and White 1988). Despite a dense network of meteorological and oceanographic stations measuring a multitude of parameters, it is still not possible to accurately predict EN events, even 2 months before they occur (Enfield and Wyrtki, pers. comm., Guayaquil Conference 1986). This fact became most evident when the secular EN of

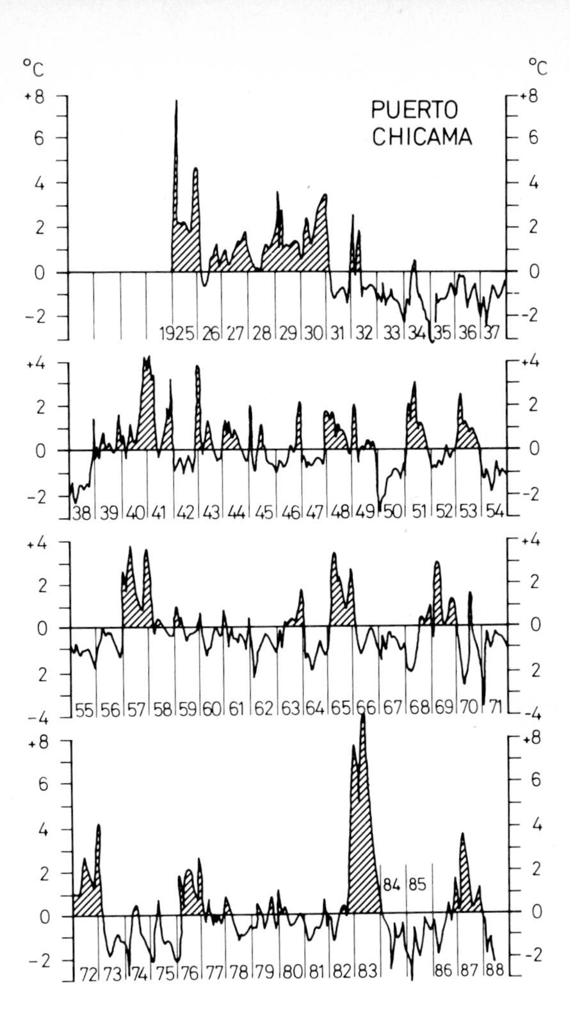

Fig. 1. Temperature anomalies (°C) at Puerto Chicama, Peru (7°42′S; 79°27′W) from 1925 to 1988 (Peruvian Navy, Callao). Positive anomalies are *shaded*

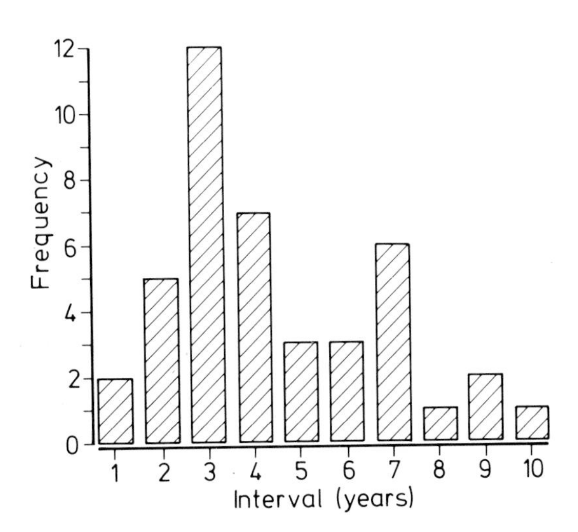

Fig. 2 Histogram of the intervals between EN events derived from data given by Quinn et al. (1987)

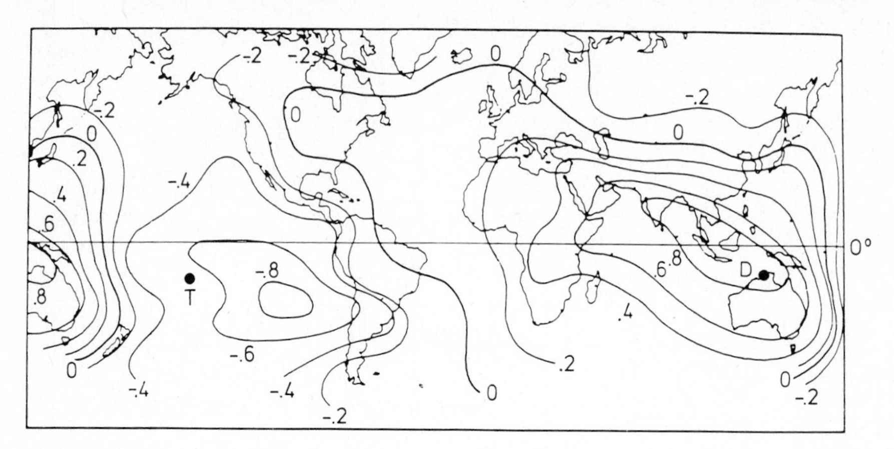

Fig. 3. The Southern Oscillation visualized by the local correlations of annual pressure anomalies with the ones at Djakarta, Indonesia, from Rasmusson (1984) after Berlage (1957). Tahiti (*T*) and Darwin (*D*) are highly correlated but with opposite sign

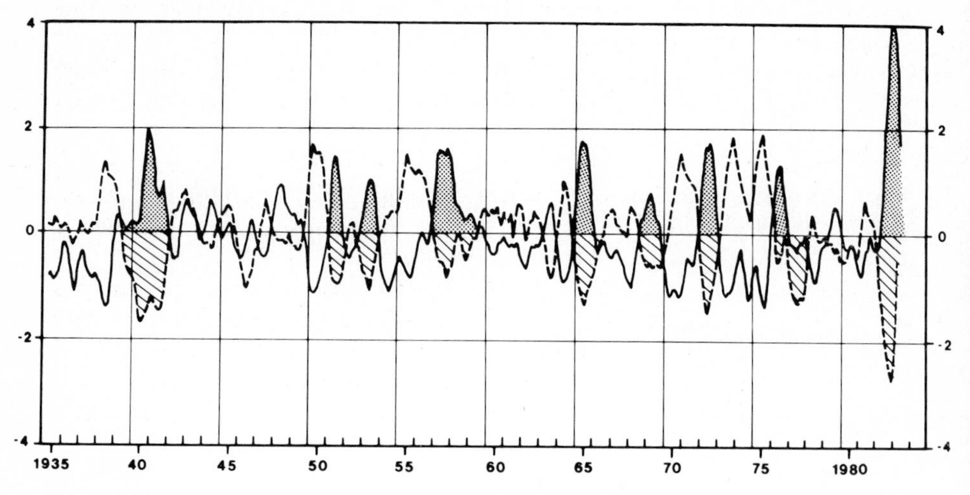

Fig. 4. Curves of the Southern Oscillation Index (SOI, *dashed line*) and sea surface temperature anomaly at Puerto Chicama (SST, *solid line*). Negative SOI coincides with positive SST anomaly (Rasmusson 1984)

1982–83 appeared completely without warning. This considerably dampened the high expectations of scientists who presumed the prediction problem had been solved through the work of Wyrtki (1982; however, see Cane et al. 1986). Wyrtki, analyzing wind and ocean conditions before and during the six EN events preceding that of 1982–83, established a number of rules which seemed to enable scientists to predict a coming event. While these rules are applicable for the six events before

1982–83 they failed in the case of EN 1982–83 due to a number of irregularities in the development of this event (Cane 1983, 1986; Rasmusson and Wallace 1983). At present, "hindcasting" is certainly more successful than forecasting; with the full amount of information at hand, the history of the more recent ENs, including the 1982–83 event, can be traced back to their origins.

We are not sure when EN-like events occurred for the first time off the American Pacific coast, and whether their appearance have since been continuous or interrupted. Evidence from sediment layers (Wells 1987), glaciers (Thompson and Mosley-Thompson 1986), and the accumulation of cadmium in corals (Shen and Boyle 1984) present proxy-indicators for EN-like events in the past. Based mainly on the study of fossil mollusks, deVries (1987) suspects that ENs may have existed for approximately 2 million years (since the Pliocene-Pleistocene boundary). An opposing view, i.e., a relatively recent "birth" of EN about 5000 years ago, has been suggested by archaeologists (Rollins et al. 1986). If deVries is correct, EN has been around for about 400 000 seal generations. At that time, the modern seal genera *Arctocephalus* and *Otaria* did not yet exist, but two Phocidae have been identified for the southern Peruvian coast (de Muizon 1981). However, according to present understanding a recent "birth" is more likely, because it is difficult to imagine that secondary variations like "El Niño" persist, if the climate is subject to dramatic changes like ice ages (Martinson, pers. comm.).

Direct and indirect changes brought about by EN events in the marine ecosystem greatly influence pinniped populations living in the eastern Pacific. This impact, both of an abiotic and biotic nature, occurred in a particularly severe and geographically widespread manner in the 1982–83 EN off South America (e.g. Barber and Chávez 1983; Cane 1983; Rasmusson and Wallace 1983; Philander 1983; Arntz 1984, 1986). Along the North American coast it was less dramatic but nevertheless the strongest event ever recorded.

First, we give a general description of some meteorological and oceanographic aspects of the EN phenomenon and then summarize information on changes of sea surface temperature, sea level, currents and temperatures at depth during the 1982–83 EN which are most likely to influence the distribution and abundance of organisms of importance to pinnipeds, either indirectly as food of their prey or directly as their own prey.

1.2 The Eastern Tropical Pacific Under Normal and EN Conditions

The eastern equatorial Pacific is unusually cold for tropical ocean areas. This is caused by equatorial and coastal upwelling and an influx of cold water with the Peru (Humboldt) Current (Fig. 5). The Peru Current is driven by SE trades which force surface waters away from the coast of Peru and Chile leading to local upwelling. This injects substantial amounts of cold water from deeper levels to the system. In contrast, the surface waters of the western Pacific are very warm (Fig. 6a). They are, as Cane (1986) states, the largest pool of very warm water in the world. It is maintained by the oceanwide trade winds, which drive currents westward (Fig. 5). In the tropics and subtropics the water is warmed on its way to the west due to the

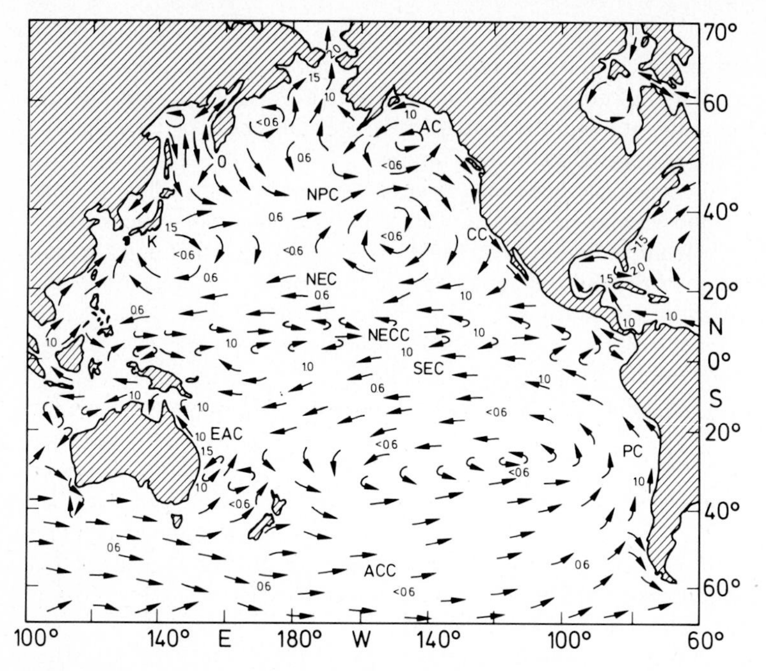

Fig. 5. Ocean Surface Currents of the Pacific Ocean in knots compiled from ship drift estimates for northern summer. (U.S. Navy 1977, 1979). Below the surface layer undercurrents can flow in the opposite direction; e.g. Peru Undercurrent and Equatorial Undercurrent. *AC* Alaska Current; *O* Oyashio; *NPC* North Pacific Current; *CC* California Current; *K* Kuroshio; *NEC* North Equatorial Current; *NECC* North Equatorial Counter Current; *SEC* South Equatorial Current; *EAC* East Australian Current; *PC* Peru (Humboldt) Current; *ACC* Antarctic Circumpolar Current

heat gain by solar radiation. Consequently, there is a significant gradient in SST along the equator.

By stirring the near-surface waters the winds generate a mixed layer. The entrainment of cold water from below induces a heat transport from the mixed layer to the deep ocean. If this heat loss is overcompensated by heat gain from the atmosphere above, the mixed layer warms up even if it deepens. Along the equator in the Pacific the mixed layer extends to about 100 m depth, becoming shallower to the east and almost disappears at the South American coast due to the penetration of cold water to the surface by upwelling. The lower boundary of the mixed layer is formed by a transition zone of increased vertical temperature gradients, the thermocline (Fig. 7a). In the tropical oceans the thermocline is comparably sharp.

The wind-induced water transport from east to west leads to an increase in sea level in the western Pacific which gives rise to a strong subsurface flow below the directly wind-driven layer from the west to the east (Fig. 7b). The Coriolis force concentrates this current on the equator. Therefore, it is generally called the Equatorial Undercurrent, however, in the Pacific, it is sometimes referred to as the Cromwell Current, after its discoverer. On its way along the equator it hits the Galápagos

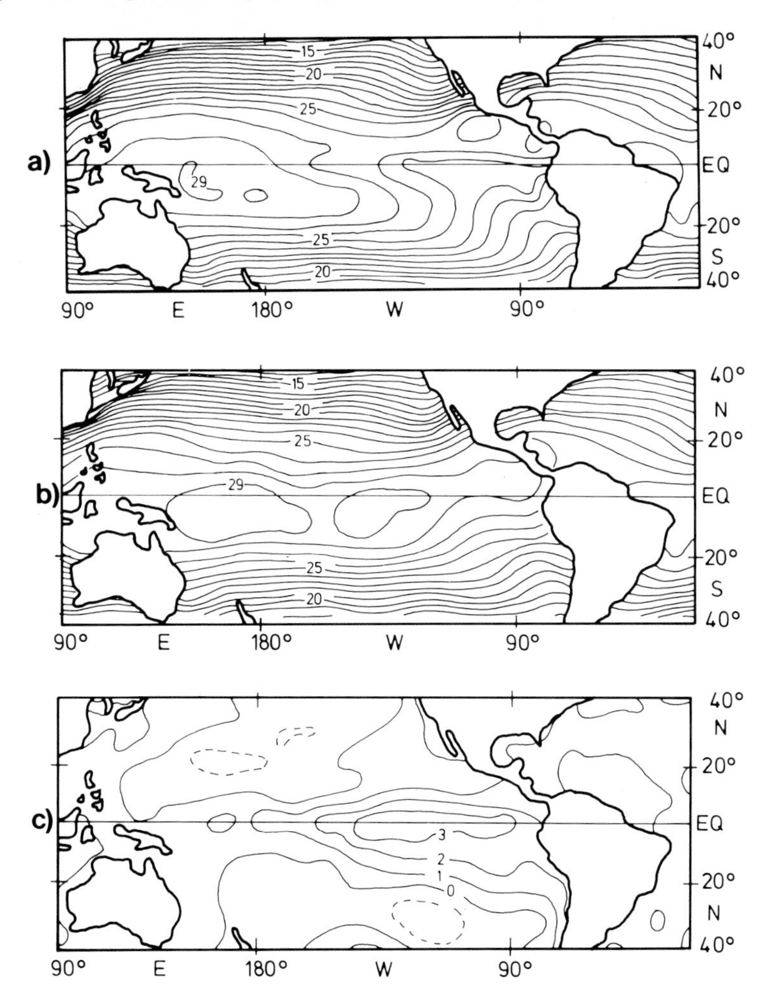

Fig. 6a-c. Sea surface temperature in °C in the tropical Pacific (after Rasmusson and Wallace 1983). **a** Climatological conditions from December to February. **b** Average conditions from December 1982 to February 1983. **c** Anomalies calculated as difference between **a** and **b**

Islands. Part of it surfaces and supplies large quantities of cold water to the upwelling ecosystem of the Galápagos. Its remainders are split into branches which surround the islands to the north and south (Houvenaghel 1984).

The SST distribution, with cold water in the east and warm water in the west, is associated with an atmospheric circulation along the equator: cold, relatively dry air flows from the eastern equatorial Pacific towards the west (Fig. 8). Over the warm water the air absorbs heat and moisture and rises. Some of this warm moist air returns to the east where it sinks, and some flows poleward at great altitudes. These patterns of air flow are related to high air pressure over the eastern tropical Pacific and lower pressure in the west.

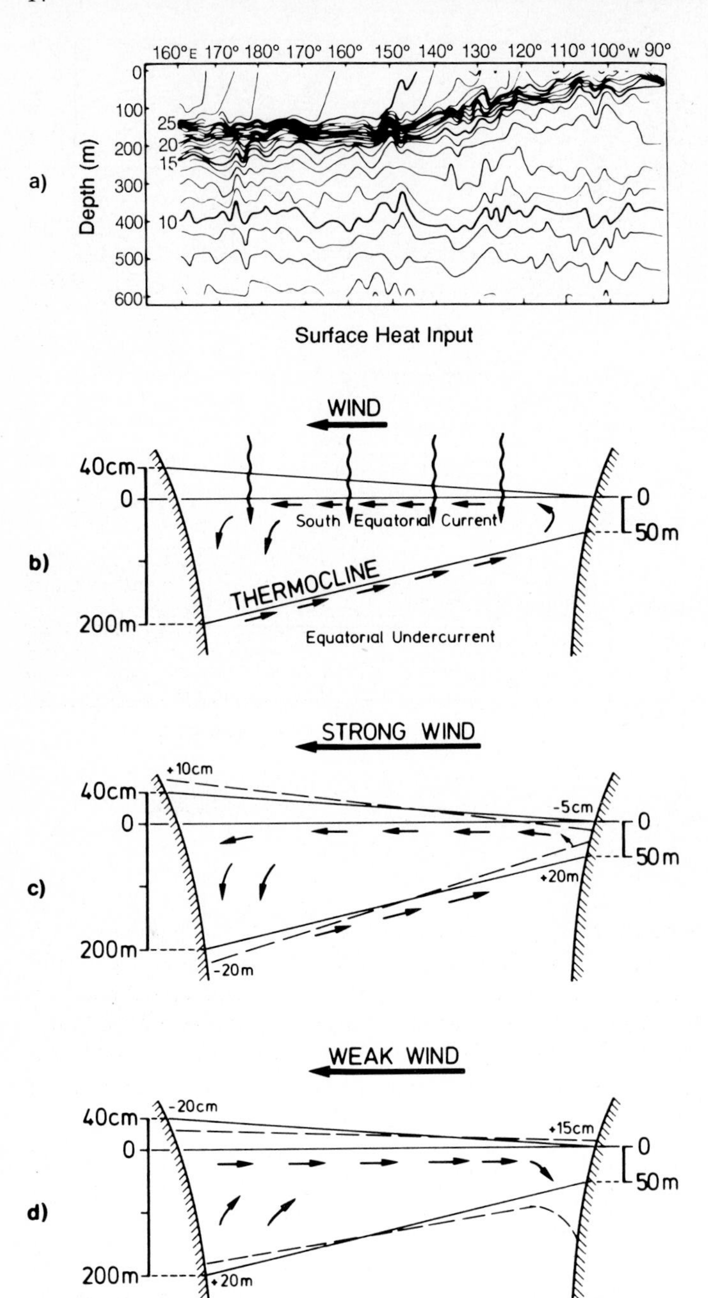

Fig. 7a-d. Thermal structure of the equatorial Pacific (after Colin et al. 1971; Wyrtki 1982). **a** Observed structure, schematic representations; **b** normal conditions; **c** strong trade winds pile up water in the west; **d** after relaxation of the wind water sloshes to the east inducing sea level rise and thermocline deepening

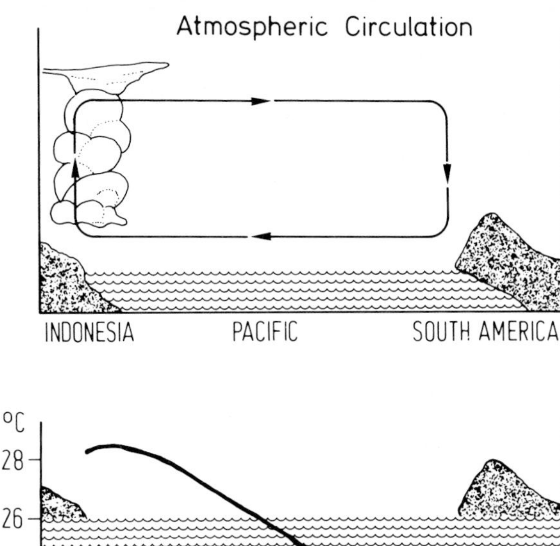

Fig. 8. Schematic representation of the Walker circulation in the atmosphere (*above*) and sea surface temperature along the equator (*below*). Warm moist air is rising over the warm western Pacific. In the east cool air is sinking (after Wyrtki 1982)

When the trade winds weaken, which is a normal event at the beginning of each southern summer, the cool Peru Current slows down and the upwelling of cold water is diminished. However, the surface heating rate is maintained, so the surface waters of the equatorial Pacific are warmed. The increasing volume of warm water depresses the thermocline from about 30 m during the cold season to roughly 50 m during the warm season. In normal years though, little warming is noticeable along the coast of Peru below 50 m depth (Chávez 1987). This is the normal situation in January (austral summer). The warm period ends towards April when the SE winds gain strength and cool conditions return. Even during this cool period occasional surface and subsurface warming occurs between April and July which may last for about 1 month (Chávez 1987).

During an EN event, instead of cooling, warming continues. This is caused by backwashing of warm waters which were accumulated in the western Pacific by anomalously strong trades during the prior year (Fig. 7). The relaxation of the trades excites wave patterns trapped along the equator which are called Kelvin waves. Due to these Kelvin waves sea level rises 10 to 30 cm and the thermocline is depressed several tens of meters. The surface current to the west is consequently reduced or even reversed. As Kelvin waves propagate with a velocity of 2 to 3 m/s they reach the South American coast 6 to 8 weeks after their excitation in the western Pacific (Fonseca 1985). When they arrive at the coast they deviate and propagate along the coast to the south and north, generating sea level rise and thermocline depression well before atmospheric anomalies can be observed. The observation that relaxing trade winds in the western Pacific induce sea level rise on the eastern coast gave birth to the notion of "teleconnection". This relationship can now be explained by propagation of Kelvin waves.

In the atmosphere anomalies occurring in the tropical Pacific excite perturbations of the extratropical weather systems by wave propagation. The atmospheric "teleconnections" are due to so-called planetary waves.

The depression of the thermocline during an EN is of considerable importance to the coastal upwelling ecosystem. The upwelled water originates from depths of 50 to 100 m which are normally below the thermocline and therefore cold and rich in nutrients. However, during EN, due to the deepening of the thermocline, upwelled water originates from the surface mixed layer and is warm and poor in nutrients. Consequently, even if the trade winds do not slacken along the coast and upwelling continues, the nutrient supply necessary to enhance primary production is interrupted (Barber and Chávez 1986).

Prior to the 1982–83 EN, several EN events were used to describe the so-called *canonical* (composite) EN (Wyrtki 1982). The 1982–83 and before it the 1940–41 ENs show a number of peculiarities which do not fit the previously established rules. Prior to the 1982–83 EN the trade wind failed to blow stronger than normal which is observed during the canonical EN and leads to the accumulation of warm water in the west (Kerr 1983). Consequently, no sea level elevation was observed there (Cane 1983). Warming on the equator started in the central Pacific and spread eastward to the South American coast instead of starting there. The 1940–41 EN was quite similar (Rasmusson and Wallace 1983), although at that time the temperature anomalies were much weaker than during the 1982–83 event. Furthermore, the 1982–83 EN lasted to mid-July 1983 instead of ending in April as EN events off South America usually do. Off North America, unusual warming even continued into 1984 (Wooster and Fluharty 1985).

1.3 Abiotic Effects of EN 1982–83 in the Eastern Tropical and Southern Pacific

In the eastern tropical and southern Pacific the effects of the 1982–83 EN were most pronounced. The mean sea surface temperature from December 1972 to February 1983 is shown in Fig. 6b, the anomaly from the climatological conditions in Fig. 6c. We attempt here to give a brief description of the sequence and extent of the effects from the Galápagos to Chile, the areas for which we have data on pinnipeds. Effects along the North American coast are presented in the next section.

1.3.1 Galápagos and Mainland Ecuador

The monthly mean sea level in Galápagos began to rise above the long-term average in October 1982, reached a first maximum of 37 cm above normal in December and a second one of 33 cm in May 1983. The absolute maximum value was 47 cm above normal on 6 January 1983 (Wyrtki 1985). Along the equator west and east of the Galápagos the usual westward flow stopped and the surface current to the west of the Galápagos was reversed from about mid-September to mid-December 1982 carrying warm water from the west into the Galápagos area. The Equatorial Undercurrent seemed to disappear or even partly reverse during the same period (Firing et al. 1983; Taft 1985; Jiménez and Intriago 1986) but returned again in January 1983. The 15 °C isotherm normally found between 50 and 100 m was deeper than 200 m in December 1982. Water warmer than 25 °C extended to 100 m and surface temperatures rose to 27 °C. In March 1983 the 15 °C isotherm ascended to about 150 m depth and the 25 °C isotherm was at 50 m. The surface water, however, continued warming and reached more than 30 °C at times. In late June 1983 the waters west of the Galápagos Archipelago cooled rapidly while the northern and eastern islands (Wolf, Darwin and San Cristobal) were bathed by warm waters for one more month. Apparently, the cool waters took much longer to reach north of the equatorial front where these three islands are situated (Hayes 1985). In October 1983 all these parameters had returned to normal, while sea level and SST were even lower than the long-term mean for Galápagos. Off mainland Ecuador, sea level was more than 10 cm above the 1975–1981 average between May and September 1982, rose sharply from October and reached a first peak of 30 cm in January 1983. After a decline in the following months, a second peak of about the same height was reached in May 1983 before a steady decline which culminated in less than 10 cm below average values in December 1983 (Cucalón 1987).

During the first peak of the event, in February 1983, SSTs off Ecuador were 4 °C higher than in 1981 which is considered a normal year. Interestingly, temperatures at 50 m were 9 °C higher than 1981, i.e., the subsurface signal of EN in this area was clearly stronger than the surface signal. The 25 °C isotherm which was found at the surface in February 1981 sank to 50 m in February 1983; at the same time, the 16 °C isotherm declined by 80 to 130 m. In previous ENs, a similar but weaker deepening of the 16 °C isotherm had been observed; in February 1972 it was at 100 m, in February 1976 at 80 to 100 m depth. By April to May 1983, SST off Ecuador continued to increase; the highest SST anomalies (up to 6.3 °C, which corresponds to a temperature of 29.5 °C) were observed from May to July 1983. Thereafter, the system gradually recovered (Cucalón 1987).

1.3.2 Peru and Chile

This description of the event largely follows Arntz (1984, 1986), Fonseca (1985), and Guillén et al. (1985). The front of the Kelvin wave packet reached the South American coast, at 10°S, on 7 October 1982, about 6 days after passing the Galápagos. A sea level rise of up to 40 cm in Ecuador and Peru was similar to that observed in Galápagos. Maxima of around 30 cm were reported as far south as Antofagasta (23°S) in Chile (Fonseca 1985). The sea level along roughly 25° of latitude peaked in January 1982 and again between April and June 1983. The southern extent of the rise in sea level is not adequately documented.

The sea level elevation led to an enhancement of poleward flow and therefore advection of warm water. The Peru Current was strongly reduced and even reversed its direction at times during April and May 1983. At 10 °S, the Peru Undercurrent increased its poleward speed from 4.2 to 25.3 cm/s (the 64-day average) with peak velocities of 35.8 cm/s. After 10 July 1983 this current returned again to its former slower speeds (R.L. Smith 1983, 1984).

In November 1982 SSTs reached 23 to 26 °C along the Peruvian coast, i.e., were 4 to 6 °C warmer than normal. The warm waters extended as far south as Arica in Chile (18°30′S). Larger cold water areas were left only off San Juan (15°20′S) and Atico (16°10′S) with SSTs less than 20 and 19 °C, respectively. During December these cooler upwelling areas shrunk drastically. Efficient upwelling resulting in nutrient supply to the near-surface waters was restricted to a few areas close to the coast. Further offshore upwelling did not reach below the thermocline. Consequently, the upwelled water was poor in nutrients. SSTs reached their first maximum in January 1983, with values predominantly between 26 and 30 °C north of Callao. In the south of Peru this peak was delayed until March attaining 23 to 27 °C. In northern Chile SST exceeding, from January to March, the long-term mean by 4.5 °C, increased to 28 °C. The warming extended southward to at least 37 °S (Kelly 1985). By June temperatures had returned to near normal in northern Chile and southern Peru where they achieved the long-term mean again in September (Fuenzalida 1985). North of 14 °S, however, warming continued until June, with a second peak during April and May 1983. After July 1983, the SST dropped drastically but had gained normal levels again by October 1983, although slight positive anomalies persisted locally north of Callao. This anomalous SST was accompanied by large-scale warming at depth. The 15 °C isotherm, normally at about 50 m depth nearshore, sank to more than 200 m in December 1982, but returned to about 130 m in February 1983 (Guillén et al. 1985). In northern Chile it dropped to 150 m depth, off Antofagasta to about 100 m (Blanco and Díaz 1985). At the peak of the event, oceanographers traced the warming down to 800 m (Guillén et al. 1985).

Dissolved oxygen was slightly reduced in the surface layers whereas at depth the warm intruding water induced an oxygen increase. A comparison of the "Humboldt" cruises 8103/04 (normal conditions) and 8212/8301 (EN) indicated a three- to seven fold increase of dissolved oxygen at the seafloor off northern and central Peru during EN (Arntz et al. 1985). An unusually well-oxygenated layer with values up to 2.5 ml/l O_2 was thus established to a depth of 150 to 200 m (Guillén et al. 1985). In contrast to the conditions off Peru, during the later phase of EN, the surface waters off northern Chile were poor in oxygen with values less than 1 ml/l (Alvial 1985; Fonseca 1985; Fuenzalida 1985).

Simultaneous with the warming of the sea, the well-known effect of a much increased rainfall in a normally arid coastal region, with both catastrophic and positive consequences, was observed.

1.4 Effects of EN Along the Coast of North America

Important populations of pinnipeds live along the coast of North America in the California and Alaska Currents. The North Pacific Current system is mainly wind-driven. Hence, its flow and position relative to the coast strongly depends upon the position of the Aleutian low and the Pacific high.

Under normal conditions, the California Current (0–200 m deep) carries water southeastward parallel to the coast throughout the year (Fig. 5). Its water comes from the Subarctic Pacific and the West Wind Drift and enters the current at about 48 °N. This water is cool, of low salinity and high in oxygen and phosphate. Volume transport is maximum in summer and minimum in winter and spring. Upwelling intensity along the coast parallels these changes in the intensity of current flow: most upwelling takes place during the spring and summer months. From October through January, the California Current is displaced about 100–150 km offshore. During this time, the California Countercurrent (which further north merges with the Davidson Current) transports warmer water nearshore northward.

Effects of the tropical EN in 1982–83 were transmitted to the coast of North and Central America through two processes, the so-called teleconnections.

1. Poleward propagation of Kelvin waves resulting in an elevation of coastal sea level, an increase in SST and stronger poleward flow.
2. Planetary waves in the atmosphere induce variations in the wind fields due to changes in the pressure systems over the central and north Pacific, in particular, intensification of the Aleutian low. They lead to the relaxation of coastal upwelling (Hamilton and Emery 1985; Norton et al. 1985).

The relative role of these two processes in the anomalies observed along the coast of North America is still not completely understood.

The sequence of the EN effects along the coast of North America varied with latitude. This was more noticeable with respect to the end of EN effects than to their onset. An increase of nearshore (0 to 200 km) SSTs was noted from Panama to Canada in October 1982 (Wooster and Fluharty 1985). In the northern Gulf of Alaska the first signs were delayed until January 1983 (Royer 1985). Peak anomalies generally occurred in the period of March to May 1983 at all latitudes from Mexico to Canada but were retarded in the Gulf of Alaska until June and July 1984 in surface waters. Warming observed in the Gulf of Alaska in 1983 and 1984 was part of a long-term warming trend and may not be an EN signal (Royer, unpubl.).

The width of the affected coastal strip, depth range and strength of the temperature signal also varied strongly with latitude. The effect of a warming event on the biological system is different from area to area. Whereas biological productivity is reduced in the California Current by the warming, it is enhanced in the Gulf of Alaska and Bering Sea (McLain 1983). In reviewing the physical effects we will proceed from the south to the north. Most of the data mentioned in the following stem from the book edited by Wooster and Fluharty (1985), where additional useful information can be found.

Off Mexico, a first maximum in sea level elevation occurred in December 1982 and a second one from March to May 1983 (Galindo et al. 1986). Warm water appeared even earlier in July 1982 near Guerrero (16° to 18 °N) (Ramos et al. 1986). SST obtained its first maximum from October to December 1982 and a second one from March to May 1983 (Galindo et al. 1986). The warm water layer extended to depths of more than 100 m (Gallegos et al. 1986). During EN more subsurface water entered the Gulf apparently because of the strengthened Costa Rica Current. This induced intense mixing with waters from the inner Gulf making more nutrients available for phytoplankton growth (Baumgartner et al. 1987).

In San Diego (southern California) oceanographers first began to notice the rise in sea level in June and July 1982 (Simpson 1984b). Along the western coasts of Baja and Alta California SSTs climbed above average in fall 1982. The rise in SST extended from the coast to about 500 to 1000 km offshore (Norton et al. 1985). Effects were strongest in winter 1982–83 and again in winter 1983–84. Further offshore (more than 1000 km) SSTs were lower than normal in winter 1982–83, but near normal in 1983–84. The warm anomalies evidently affected not only the surface but extended to a depth of more than 200 m. At 100 m depth a dramatic warming was observed in winter 1982–83. After cooling in spring and summer 1983 a second major warming occurred in late summer, fall and winter of 1983. Between 29° and 34 °N positive temperature anomalies at 100 m depth varied from 2.3 to 3 °C in 1982–83 and achieved about 1.8 °C in the following fall returning to near normal in winter 1983–84.

During winter and spring of 1983, the storm tracks associated with the prevailing westerlies moved southward because of the expanded Aleutian low and caused increased rainfall and severe floods in most of California. Extreme southerly coastal winds associated with the intensified Aleutian low suppressed coastal upwelling and caused much damage through increased swell in combination with the rise in sea level of more than 30 cm. In February and March 1983 sea level showed a monthly anomaly of 26 cm at San Francisco and about 15 cm in southern California. The northward coastal countercurrent flow increased and brought unusually warm water far north. Simultaneously, the California Current transported less subarctic water southward. These combined factors produced the strong warming event along the coast of California (Norton et al. 1985).

Further north, along the coast of Oregon, Washington and British Columbia a coastal band of warm water of about 30 km width was established in November 1982 and increased to a width of about 300 km in February and March 1983. Concurrently, the monthly average sea level rose about 30 cm above the long-term mean. SST anomalies near the coast exceeded 3 °C. The anomalies peaked around April 1983 and SSTs warmer than normal were maintained throughout summer. Most of the surface warming had disappeared by August 1983. It vanished completely along the coast of Washington and Oregon in December 1983, but continued into 1984 further north. Off British Columbia subsurface water warmed down to 800 m depth with a maximum anomaly at 200 m. Warming of up to 5 °C between 50 and 100 m depth was observed over the shelf in November 1982 and extended for about 200 km offshore in March 1983. By the end of June 1983 the subsurface layers had returned to near-normal temperatures, but along the Oregon coast anomalies were still recorded in July 1983 affecting a depth of more than 500 m (Tabata 1984, 1985; Huyer and Smith 1985).

Data from the northern Gulf of Alaska (Royer 1985) indicate that the event lasted over 1 year in Alaskan waters and invoked temperature anomalies greater than 2 °C at a depth of 100 to 200 m and of 5 °C at the surface. Warming began in early 1983 and anomalies lasted at least until June 1984. Positive temperature anomalies occurred at all depths over the shelf and had decreased again to less than 1 °C by June 1984. The Alaskan warming was clearly related to the intensification of the Aleutian low and its movement to the southeast, resulting in the flow of warmer waters from the North Pacific into the northern Gulf of Alaska. Sea level height in the Gulf increased by more than 20 cm from January to March 1983 (Norton et al. 1985) reflecting the increase in temperature of the adjacent water mass.

Warming of the Bering Sea began in November 1982 when an anomaly of 2 °C extended from the north Pacific Ocean to the Kamchatka peninsula, and northward toward the Bering Strait (NOAA 1982–84). In December, the anomaly extended eastward across the entire Bering Sea. The major warming of 3 °C in the southern Bering Sea lasted from January through April 1983, followed by a secondary peak not exceeding 1.5 °C and lasting from July 1983 through March 1984. Thereafter, it decreased below 1 °C. In comparison to the Gulf of Alaska, the warming of the Bering Sea began and ended slightly earlier, was slightly shorter, and its peak anomaly was about 2 °C less. The warming mechanism of the Bering Sea was probably an increased southerly air flow from the north Pacific Ocean associated with changes in the strength and location of the Aleutian low (Niebauer 1985). In later papers Niebauer (1988) and Niebauer and Day (1989) stated that the 1982–83 EN had little apparent effect in the Bering Sea. This is explained by the extreme eastern position of the Aleutian low and northerly winds during this period.

Acknowledgments. We are grateful to D.G. Martinson, J. Norton, W. Pearcy and R.L. Smith for their helpful suggestions which were incorporated into the manuscript. This is AWI-Contribution No. 189.

2 Biological Consequences of the 1982–83 El Niño in the Eastern Pacific

W. ARNTZ, W.G. PEARCY, and F. TRILLMICH

2.1 Introduction

The drastic changes in the abiotic environment caused by El Niño (EN) 1982–83 led to marked repercussions in the fauna and flora of the marine ecosystem, including pelagic and benthic marine subsystems and the seashore. Some of them were of a negative, others were of a positive nature. Mass mortalities of some species contrasted with enormous proliferations of others. Many species emigrated during EN from their traditional, cool upwelling areas, which had been converted to tropical or unusually warm conditions. Some moved towards deeper water, others poleward where conditions resembled those in their former habitat. Growth, body condition, reproduction and adult survival of some species were diminished. Also, changes in growth and survival of early life history stages affected recruitment of some species years later. At the same time dispersal stages of warm water species were transported south-/northward, and juveniles and adults actively invaded unusual areas. The biological effects of EN 1982–83 have been summarized in various conference volumes (Arntz et al. 1985b; CONCYTEC 1985; IFOP 1985; Robinson and del Pino 1985; Wooster and Fluharty 1985; IOC 1986; AGU 1987) and a number of reviews (Barber and Chávez 1983, 1986; Arntz 1984, 1986; McGowan 1984; Barber et al. 1985; Mysak 1986; Pearcy and Schoener 1987; Alvial 1988; Arntz and Fahrbach 1991).

Only part of the biological effects caused by EN are of major relevance to the Pacific pinniped populations, and we will concentrate on these changes and their underlying causes. Most effects were connected with changes in composition, distribution, abundance and thus availability of the prey of pinnipeds. We document these changes in order to understand the effects of the 1982/83 EN on pinniped populations along the eastern margin of the Pacific Ocean. We begin by describing the changes in the eastern tropical and subtropical Pacific where EN impacts were most severe and then document changes along the Pacific coast of North America. Documentation of EN effects along the coast of South America is less detailed and not as extensively published as for effects off North America. We therefore decided to give an overview of the effects off South America and treat the changes along the coast of North America in a species by species manner for the most important pinniped prey. In Section 2.4 we attempt to provide an overall view of the ways in which EN 1982–83 affected pinniped prey populations.

2.2 Biological Effects of EN 1982–83 off South America

2.2.1 The Planktonic Community

The Humboldt Current upwelling ecosystem, with a primary production of over 1000 g C m^{-2} yr^{-1} (Walsh 1981), is considered to be the most productive current system in the world. Within a narrow band covering only a tiny fraction of the world's ocean surface it was able to supply up to 22% of all the fish caught in the world in the 1970s (Idyll 1973). Under normal, non-EN conditions, effective year-round upwelling recycles remineralized nutrients from shallow depths (40–80 m) to the euphotic layer. There, immense populations of small diatoms make use of the nutrients and the sunlight, and provide a food base for small herbivorous zooplankton. Both phyto- and zooplankton are effectively grazed upon by small shoaling fish, mainly anchovy (*Engraulis ringens*) and sardine (*Sardinops sagax*), which are the staple food for most of the higher links in the different food chains which include predatory fish, guano birds and pinnipeds. Usually anchovy and sardine are easily accessible to predators; both species are normally found nearshore, close to the sea surface and in large shoals. Furthermore, they are small, relatively slow swimming species without dangerous defenses, like spines, and can easily be swallowed whole. Up to 1973, anchovy comprised 95% of the total pelagic fish biomass off Peru (Idyll 1973). Since then, the anchovy population has declined substantially, fluctuated markedly and temporarily been replaced by the sardine (Pauly and Tsukayama 1987).

The exceptionally strong 1982–83 EN severely disrupted this ecological system. During the peak of this EN upwelling in most areas became biologically ineffective, especially along the central and northern coast of Peru. Upwelling areas were compressed to narrow strips of 2–5 km width at sites where the cold water system normally extends several hundred kilometers offshore. Coastal winds were favorable for upwelling through March 1983 (Barber and Chávez 1986), but even though upwelling continued, the thermocline was depressed below the depth where water is entrained into the upwelling circulation. This meant that nutrients were concentrated below the depth of entrainment and were no longer transported into the euphotic layer.

EN thus enormously reduced the surface nutrient concentrations. Nitrate, phosphate and silicate values became extremely low for over 6 months (December 1982 to July 1983), and primary production was considerably diminished (Fig. 1). There was a clear N-S gradient in the intensity of the changes brought about by EN, but severe changes in the pelagic environment were registered as far south as 33°S (Muñoz 1985). In Peru, EN 1982–83 continued beyond March 1983 N of Pisco (14°S), whereas conditions returned to normal south of the Paracas peninsula in that month. However, some parts of northern Chile seem to have been affected as late as June 1983 (Avaria 1985b) and Alvial (1985) even suggested that it ended there between August and October 1983. In the Galápagos Islands, nutrient concentrations at the surface were extremely low from December 1982 to July 1983 (Kogelschatz et al. 1985). During March to May 1983, phytoplankton biomass near the center of the island group was reduced to 30% of the mean quantity present in September to

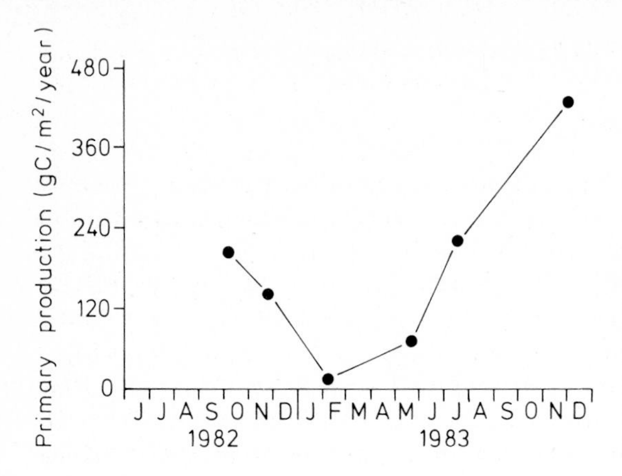

Fig. 1. Changes in primary productivity in the Peruvian upwelling system (redrawn after Chávez and Barber, unpubl.). Area weighed mean primary productivity along 85°W between the equator and 10.5°S

November 1983 (Kogelschatz et al. 1985); the reduction in the western area of the archipelago may have been even greater (Feldman 1984).

The reduction of nutrients by "nutricline depression" (Barber and Chávez 1986) and thus inefficient upwelling, or by temporary cessation of upwelling led to changes in the composition, biomass and production of phytoplankton. Small cold-water diatoms were reduced, (sub-)tropical dinoflagellates, including "red tide" species and large warm-water diatoms were transported into the area, and numerous indicators of oceanic or coastal equatorial waters appeared off Peru and Chile. At the peak of the event primary production in coastal waters off Paita was reduced by a factor of 20 (Barber and Chávez 1986), and biomass off central Peru may have been reduced even more (Rojas de Mendiola et al. 1985).

The reduced primary production also affected the herbivorous grazers in the upwelling system. Drastic changes were registered in species composition, density (reduced) and diversity (increased) of zooplankton. Oceanic copepods were found close to the shore, and large siphonophores, salps, jellyfish, euphausiids, appendicularians, pteropods and chaetognaths replaced the small copepods characteristic of the area during normal times. Biomass of small zooplankton became insignificant whereas that of the larger forms increased due to the much higher individual weight of the tropical species (Santander and Zuzunaga 1984; Carrasco and Santander 1987; Dessier and Donguy 1987).

2.2.2 *Nektonic Animals – Pinniped Prey*

Decisive changes at the lower levels of the food web, as referred to above, seem to have been the main reason for the alterations in the pelagic fish populations, which in turn influenced the food resources of their warm-blooded predators. In addition, the strong and abrupt increase in temperature and the increased oxygen concentrations near the seafloor may have been of major importance to forage fish.

Under normal conditions anchovy prefer cold water which keeps them in areas where plankton is most abundant (Barber and Chávez 1986). During the summer (January-March), when the coastal current is narrowest, they are densely concentrated in shallow waters and close to shore. At this time of the year they are readily available to fisheries, seabirds and pinnipeds (Idyll 1973; Arntz 1986). During the winter, when warm oceanic waters come closer to the shore, they disperse and remain at greater depths, below the deeper thermocline (Jordán and Chirinos de Vildoso 1965; Saetersdal et al. 1965). Spawning occurs in August and September during the winter and again on a smaller scale in January and February (Jordán and Chirinos de Vildoso 1965; Idyll 1973).

Anchovy carry out diel vertical migrations. During the day they remain in deeper waters, usually not below 50 m water depth, where they form dense shoals. Around dusk, they start dispersing and migrating upwards, until they reach the surface layers. They remain dispersed until dawn, when the whole process is reversed (IMARPE 1969). During spawning these migrations do not always occur (Jordán 1971).

Increased temperature and deteriorated feeding conditions during the 1982–83 EN affected anchovy (Fig. 2), sardine (Fig. 3), and the silverside (*Odontesthes regia*), an important item of the – nearshore – artisanal fishery. There were both behavioral and physiological responses, and both may have been of great importance for the pinnipeds feeding on pelagic fish. Silverside disappeared from shallow waters off Peru in February 1982 and returned only 2 years after EN; in the meantime no specimens were caught at all. Anchovy, sardine and silverside, while disappearing from near-surface waters, undertook active migrations to avoid high temperatures and barren food conditions in their traditional environment. These migrations seem to have been of three different kinds (cf. Valdivia 1978):

1. Some of the fish in Peru concentrated in the remnants of upwelling close to shore. Most of them were trapped there and apparently died, as indicated by occasional reports of dead fish in shallow water and the fact that seabirds or seals

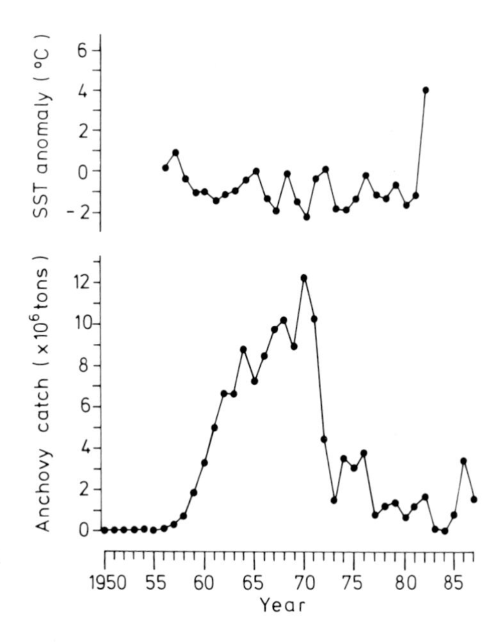

Fig. 2. 1982–1984 Temperature anomaly at Chicama (*above*) and anchovy catch off Peru (redrawn and slightly modified after Barber and Chávez 1986)

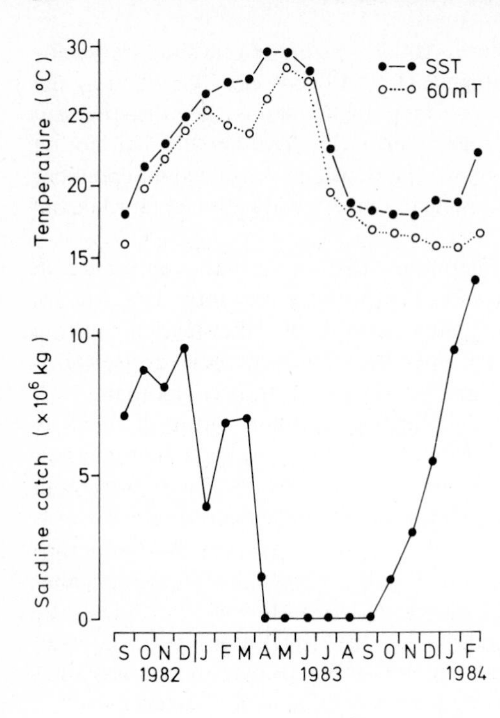

Fig. 3. 1982–1984 Temperature anomaly at Paita (*above*) and the sardine catch off Peru (redrawn after Barber and Chávez 1986)

never found any substantial fish concentrations nearshore in the later phase of EN.

2. Many fish – especially sardine – migrated south into Chilean waters where environmental changes were less drastic. A large proportion of them was caught by the Chilean purse seine fishery which landed nearly 3 Mt of sardine in 1983; few anchovy were caught, however, which means that either anchovy did not join the sardine in their southward migration, or that they were not available to the Chilean purse seines.
3. A third group of fish withdrew to deep water where temperatures were lower than near the surface, but where food conditions must have been miserable. Despite this fact, most of the fish, especially anchovy, that survived may have been those that migrated to deeper areas of the continental shelf, or even down part of the continental slope.

By the end of 1982, the anchovy had suspended vertical migrations altogether and no longer formed large shoals. They always remained below 40–50 m, sometimes close to the seafloor, and between November 1982 and January 1983 were caught at depths between 70 and 130 m by R.V. "Humboldt's" ground trawl. At the same time, the pelagic trawl caught none. During this same period, sardine disappeared altogether from coastal Ecuadoran waters (Arntz 1986; Maridueña 1986).

Anchovy caught at the beginning of summer 1983 had lost 30% of their body weight and sardines 15% of their normal weight (Dioses 1985), and their lipid content was reduced by about 56% (Romo 1985). Alamo and Bouchon (1987) registered weight losses of 10–20%, but up to 31% for Peruvian sardine in 1982–83. Sardine stomachs contained less and unusual food items compared to normal years, especially off southern Peru. As a consequence of the poor condition of the two species, anchovy spawning failed almost completely in 1982, 1983 and 1984, and sardine spawning was very weak off Peru and Chile in 1982 and 1983 (Santander and Zuzunaga 1984; Retamales and Gonzales 1985). In 1984 and 1985, sardine spawning was about normal, and in 1985 anchovy spawned in an expanded area up to 120 km offshore (Muck et al. 1987; Santander 1987).

Surprisingly, the stock of anchovy which before EN 1982–83 had declined to a very low level (Santander and Zuzunaga 1984) and showed serious depletion during the event, recovered from EN much better than the sardine. In 1985 relatively large concentrations were seen off central Peru, and in 1986, when purse seining for anchovy was opened again, considerable biomass of this species was located off Peru, Ecuador and northern Chile. A substantial proportion of 3–4-year-old anchovy were found among these fish which must have survived the 1982–83 EN, apparently at great depth.

Hake (*Merluccius gayi peruanus*), under non-EN conditions a semipelagic species restricted to the area N of Chimbote (9°S), and some other autochthonous demersal fish of the Peruvian upwelling areas responded to the 1982–83 EN by migrating to deeper waters, often onto the continental slope where they dispersed (Samamé et al 1985). At the same time they started a southward migration and disappeared from Ecuadoran waters (Herdson 1984), taking advantage of the improved O_2 and food conditions at the seafloor. If these fish were prey for sea lions and fur seals before EN, they clearly disappeared from the range of these species under EN conditions.

(Sub-)tropical and oceanic fish species that invaded the area during EN include two mackerel species (*Trachurus murphyi* and *Scomber japonicus*) which normally live further offshore and migrate towards the shore when waters become warmer, bonito (*Sarda chilensis*), dorado (*Coryphaena hippurus*), skipjack (*Katsuwonus pelamis*) and other tunas, and several sharks and rays. None of these species replaced the traditional dominants of the upwelling system as pinniped prey; their size is mostly unsuitable for smaller pinnipeds, some have spines, they never form dense shoals like anchovy and sardine, and are faster swimmers. In addition, the jack mackerel avoided the surface waters during EN, just as the anchovy did, as evidenced by acoustic surveys which showed it to be distributed mainly between 40 and 300 m (Santander and Zuzunaga 1984; Icochea 1989). The only fish species which increased during EN and might have provided suitable food for guano birds and pinnipeds were mullet species (*Lisa* sp.), but they never appeared in quantities comparable to the normal densities of anchovy and sardine.

Along the coast of Ecuador the incidental catches of small squids in the demersal fishery declined about tenfold during 1983. Limited observations in the Galápagos indicated that shallow-water fish did not spawn between October 1982 and April 1983, while in normal years January to April is the peak spawning season for most fish (Herdson 1984). No data are available for the pelagic fish around the Galápagos.

2.3 Biological Effects of EN Off North America

The influence of the 1982–83 EN along the west coast of North America was in many respects the mirror image of that found off South America. However, the biological changes that were coincident with this strong EN were most pronounced during 1983 and 1984, rather than 1982–83 as in the South Pacific. The effects of the EN lingered even though sea temperatures and sea level seemed normal in coastal waters during 1984 (Huyer and Smith 1985; McLain et al. 1985; Cole and McLain 1989).

The biological changes in phyto-zooplankton and nekton that were related to the 1982–83 EN in the Gulf of California, the California Current, the Alaska Current, the Gulf of Alaska, and in the Bering Sea are reviewed in the following.

2.3.1 *Gulf of California*

Changes in plankton and fish that occurred in the Gulf of California (or Sea of Cortez) during EN were markedly different from the rest of the west coast of North America and are therefore summarized separately. Primary production in the central gulf was very high in March 1983 before declining to pre-EN levels in November of 1984 (Baumgartner et al. 1987). This increased primary productivity was coincident with a 52% increase in the catch of sardine (*Sardinops sagax*) and a 49% increase in that of thread herring (*Opisthonema* spp.) for the 1982–83 fishing season (December-May) in comparison to 1981–82 (Mee et al. 1985). In contrast, the total fish catch in 1983–84 declined by 44%, with only 34% of the 1981–82 catch of thread herring. The high sardine catch in 1982–83 may reflect successful recruitment of sardine, whereas the 1983–84 decline in landings was presumably caused by a recruitment failure during EN (Mee et al. 1985). The thread herring migrated further north than normal into the gulf in 1982–83 due to the unusually warm waters. Under more normal thermal conditions during 1984, it penetrated much less deeply into the gulf. Thus, 1983 most likely was a year of unusually rich food resources for pinnipeds inside the Gulf of Cortez.

2.3.2 *The Planktonic Community*

Satellite imagery revealed warm sea surface temperatures (SSTs), with the greatest anomalies along the coast, weakened upwelling, changes in surface circulation, and reduced primary productivity from outer Baja California to British Columbia related to the EN event (Fiedler 1984a). In the California Current off San Diego, elevated SST and a depressed thermocline were detected in March 1983. By August 1983 the nutricline was deep, surface values of chlorophyll were very low and the subsurface chlorophyll maximum was deeper than normal (McGowan 1985). The same patterns were apparent off the west coast of Baja California (Torres-Moye and Alvarez-Borrego 1987).

Off Oregon, SSTs were also abnormally high and the deep thermocline reduced entrainment of cool, nutrient-rich water into the euphotic zone during the spring and summer of 1983. Chlorophyll concentrations were low and the non-upwelling pattern of chlorophyll distribution persisted through most of the summer of 1983, the time of the year that upwelling is usually well developed (Brodeur et al. 1985; Miller et al. 1985; Pearcy et al. 1985; Brodeur and Pearcy 1986).

Macrozooplankton standing stocks were reduced to about 50% of average values in March-May 1983 and about 10% of average in July-September 1983 off southern California (McGowan 1985), to about 30% of non-EN years off Oregon (Miller et al. 1985) and to about 50% off Vancouver Island, British Columbia (Seften et al. 1984). The species composition of zooplankton was altered during the 1983 summer off Oregon, with a persistence of southern species, which typically disappear in the summer (Miller et al. 1985).

The abundance of larval fish was several times lower off Oregon during the summer of 1983 and the species composition was unusual. Species of larvae normally found offshore were found inshore, and the inshore species were reduced in abundance. Northern anchovy larvae (*Engraulis mordax*) were abundant in April 1983, months earlier than in previous years. They were also found closer to shore, indicative of inshore spawning, or onshore advection of eggs and larvae. Osmerid larvae, usually the most common inshore larval fish, were virtually absent in 1983 (Brodeur et al. 1985).

Changes in ocean circulation and productivity and the northward shift of the Subarctic Boundary during 1983 presumably resulted in a lower biomass and smaller individual size of zooplankton in the northern portion of the California Current than found in normal years. These changes may affect the feeding effectiveness and growth rates, and possibly survival, of juvenile salmonids (Fulton and LeBrasseur 1985) and other planktivorous fish. If zooplankton standing stocks decrease in the California Current during strong ENs, they may increase in the Gulf of Alaska. Frost (1983) observed that year to year variations in the zooplankton biomass at Station "P" (50°N, 145°W) were opposite to those in the California Current system.

2.3.3 Nektonic Animals – Pinniped Prey

The following sections are devoted to fish and squids that are known to be important prey for seals, sea lions and fur seals in the California Current, Alaska Current, the Gulf of Alaska, and in the Bering Sea (Table 1). Some species were not considered that are sometimes important prey (such as capelin, sand lance, Pacific saury, gonatid and onychoteuthid squids) because little or no quantitative data are available to assess their abundances. We use a species by species approach, proceeding poleward from species that occupy the southern part of the California Current to those in the Alaska Current, Gulf of Alaska, and then the Bering Sea.

Table 1. Important prey species for pinnipeds in the North Pacific Ocean[a]

	Southern California Current	Northern California Current	Alaska Current Gulf of Alaska	Bering Sea
Northern anchovy	X	X		
Pacific sardine	?			
Pacific mackerel	X			
Jack mackerel	X			
Market squid	X	X		
Pacific whiting	X	X		
Pacific herring		X	X	X
Pacific salmon		X	X	X
Walleye pollock			X	X
Pacific cod			X	X
Atka mackerel			X	X

[a]From Antonelis and Fiscus (1980); Bigg (1985); Perez and Bigg (1986); York (1987); Kajimura and Loughlin (1987).

Northern Anchovy – *Engraulis mordax*

The northern anchovy is an important prey for pinnipeds in the California Current off California (DeLong et al., DeLong and Antonelis, this Vol.; Antonelis and Perez 1984; Perez and Bigg 1986; York 1987). During 1983 and 1984 spawning of the central stock of the northern anchovy expanded farther offshore and to the north than during previous years due to the shift in sea temperature boundaries (Fiedler 1984a; Hewitt 1985; Fiedler et al. 1986). This northward movement of anchovy probably explains the dramatic increase of this species in the diet of California sea lions and northern fur seals on the Channel Islands from 1982 to 1983 (DeLong et al., DeLong and Antonelis, this Vol.).

The total spawning biomass of the stock, which experienced a long-term decline between 1975 and 1983, was not severely affected in 1983, but it reached its lowest level in 20 years in 1984 (Methot and Lo 1987; see Table 2). Anchovy were less available in surface waters to fishermen, and perhaps to pinnipeds, as a result of the deep thermocline during 1983. Aerial spotters detected few schools (CalCoFi 1984; R. Methot pers. comm.). This northward shift in the distribution also occurred for the southern stock which apparently moved into the Southern California Bight from Baja California (R. Methot pers. comm.).

Mortality of yolk-sac larvae was high in 1983 (Fiedler et al. 1986), and the 1983 year class was weak (Methot and Lo 1987). Although the growth of larval anchovy was not affected by low availability of food during the 1982–83 EN, the growth of juvenile anchovy was retarded. Lengths of 1-, 2-, and 3-year-old anchovy were shorter than expected beginning in early 1983, and year classes experienced abrupt increases in length during the fall of 1984 (Fiedler et al. 1986; Butler 1987). Survival of the 1984 and 1985 year classes was higher than that for the 1983 year class (Methot and Lo 1987).

The reduced growth and survival of northern anchovy during EN were most likely caused by the reduced availability of zooplankton which was significantly lower in

Table 2. Summary of abundances, availability and year class success of some important species during 1983 and 1984 along the west coast of North America

Species/stocks	Availability abundance		Year-class strength	
Northern anchovy	1983 – Northward movement	Hewitt (1985) Fiedler et al. (1986)	1983 – Weak	Methot and Lo 1987)
(Central pop.)	1983 – Spawning biomass average 1984 – Spawning biomass low	Methot and Lo (1987) CalCoFi (1984), R. Methot (pers. comm.)	1984 – Average	Methot and Lo (1987)
Pacific sardine	1984 – Large schools in Monterey Bay 1983 – Catches largest in 20 years	CalCoFi (1985)	1983 – Strong	CalCoFi (1985, 1986, 1987)
Pacific mackerel	1983/84 – Northward movement 1983 – Poor catches California	MacCall et al. (1985) Pearcy et al. (1985) R. Klingbeil (pers. comm.)	1983 – Weak 1984 – Weak	MacCall et al. (1985) CalCoFi (1988)
Jack mackerel	1983/84 – Reduced availability California	CalCoFi (1984, 1985)	1983 – Weak 1984 – Strong	
Market squid	1983/84 – Both California fisheries failed	CalCoFi (1984, 1985, 1988) T. Dickerson (pers. comm.)	1983 – Weak?	
	1983/84 – Decreased catches Oregon/Washington 1983 – Increased catches in Puget Sound	Pearcy et al. (1985) Schoener and Fluharty (1985)		
Pacific whiting	1983 – Migrated farther north	R. Methot (pers. comm.) D. Ware (pers. comm.)	1983 – Average 1984 – Strong	Shaw et al. (1988) Hollowed et al. (1988)
Pacific herring	*Spawning biomass*			
N Calif. (S.F. Bay)	1983–84 Season – low	Spratt (1987a)	1983 – Strong 1984 – Strong	CalCoFi (1988), Sprat (1987b)
Oregon	1983/84 Seasons – average	J. Butler (pers. comm.)	1983 – Strong 1984 – Average	J. Butler (pers. comm.)
Washington (Puget Sound)	1983/84 Seasons – average	D. Day (pers. comm.)	1983 – Average 1984 – Average	D. Day (pers. comm.)
British Columbia				
Southern Georgia Strait	1983/84 – Average	Haist et al. (1988)	1983 – Strong 1984 – Average+	Haist et al. (1988) Haist et al. (1988)
Northern Georgia Strait	1983/84 – Low	Haist et al. (1988)	1983 – Strong 1984 – Average	Haist et al. (1988) Haist et al. (1988)

Table 2. (*continued*)

Species/stocks	Availability abundance		Year-class strength	
Southwest Vancouver Is.	1983/84 – Average	Haist et al. (1988)	1983 – Average	Haist et al. (1988), D. Ware (pers. comm.)
			1984 – Weak	Haist et al. (1988)
Northwest Vancouver Is.	1983/84 – Low	Haist et al. (1988)	1983 – Average?	Haist et al. (1988)
			1984 – Average	Haist et al. (1988)
Central B.C. Coast	1983/84 – Average	Haist et al. (1988)	1983 – Weak	Haist et al. (1988)
			1984 – Weak?	Haist et al. (1988)
Prince Rupert District	1983/84 – Average	Haist et al. (1988)	1983 – Average?	Haist et al. (1988)
			1984 – Strong	Haist et al. (1988)
Queen Charlotte Islands	1983/84 – Average	Haist et al. (1988)	1983 – Weak	Haist et al. (1988)
			1984 – Weak	Haist et al. (1988)
Alaska				
Sitka Sound	1983 – Average	Funk and Savikko (1989)	1983 – Average	Funk and Savikko (1989)
			1984 – Strong	Funk and Savikko (1989)
Prince William Sound	1983 – Average	Funk and Savikko (1989)	1983 – Weak	Funk and Savikko (1989)
	1984 – Average		1984 – Strong	Funk and Savikko (1989)
Kamishak	1983 – Low	Funk and Savikko (1989)	1983 – Strong	Funk and Savikko (1989)
	1984 – Low		1984 – Strong	Funk and Savikko (1989)
Togiak District	1983 – Average	Funk and Savikko (1989)	1983 – Average	Funk and Savikko (1989)
	1984 – Average			
Security Cove	1983 – Average	Funk and Savikko (1989)	1983 – Average	Funk and Savikko (1989)
	1984 – Average			
Goodnews Bay	1983 – Average	Funk and Savikko (1989)	1983 – Average	Funk and Savikko (1989)
	1984 – Average			
Nelson Island	—		1983 – Average	Funk and Savikko (1989)
Nunivak Island	—		1983 – Average	Funk and Savikko (1989)
Cape Romanzof	1983 – Average	Funk and Savikko (1989)	1983 – Average	Funk and Savikko (1989)
	1984 – Average			
Norton Sound	1983 – High	Funk and Savikko (1989)	1983 – Weak	Funk and Savikko (1989)
	1984 – Average			

Table 2. *(continued)*

Species/stocks	Availability abundance		Year-class strength	
Salmon	*Catch and escapement*		*Year of ocean entry*	
Chinook				
California	1983 – Low	PFMC (1984, 1985)	1983 – Average	Pearcy and Fisher (unpubl.), Pacific salmon comm. (unpubl.)
	1984 – Low		1984 – Strong	Pearcy and Fisher (unpubl.), Pacific salmon comm. (unpubl.)
Southern Oregon	1983 – Low	Nicholas and Hankin (1988)	1983 – Average	Pearcy and Fisher (unpubl.)
		Johnson (1988), PFMC (1984)	1984 – Strong	Pacific salmon comm. (unpubl.)
Northern Oregon	1983 – Average+	Johnson (1988)	1984 – Average	Pearcy and Fisher (unpubl.), Pacific salmon comm. (unpubl.)
			1984 – Average	Pearcy and Fisher (unpubl.), Pacific salmon comm. (unpubl.)
Columbia River Tule	1983 – Low	Johnson (1988)		
Coho				
Oregon prod. area	1983 – Low	PFMC (1989)	1983 – Weak	Johnson (1988)
	1984 – Low	PFMC(1989)	1984 – Weak	Fisher and Pearcy (1988)
Washington	1984 – Low	PFMC (1985)		
Carnation Creek, B.C.			1983 – Weak	B. Holtby (pers. comm.)
			1984 – Weak	B. Holtby (pers. comm.)
			1983 – Average	B. Holtby (pers. comm.)
Central British Columbia			1984 – Average	B. Holtby (pers. comm.)
Sockeye				
Barkley Sound, B.C.			1983 – Weak	K. Hyatt (pers. comm.)
			1984 – Weak	K. Hyatt (pers. comm.)

Table 2. (*continued*)

Species/stocks	Availability abundance		Year-class strength	
Central British Columbia			1983 – Average	K. Hyatt (pers. comm.)
			1984 – Average	K. Hyatt (pers. comm.)
Bristol Bay	1983 – High	Eggers and Dean (1987)	1983	D. Eggers (pers. comm.)
	1984 – Average	Eggers and Dean (1987)	1984	D. Eggers (pers. comm.)
Pink Salmon				
Southeast Alaska	1983 – High	Eggers and Dean (1987)	1983 – Strong	D. Eggers (pers. comm.)
	1984 – Average+	Eggers and Dean (1987)	1984 – Strong	D. Eggers (pers. comm.)
Walleye Pollack	*Stock biomass*			
Gulf of Alaska	1983 – Average	Megrey (1989)	1983 – Weak	Megrey (1989)
	1984 – Average	Megrey (1989)	1984 – Weak?	Megrey (1989)
Shelikof Strait	1983 – High	Megrey (1988)	1983 – Weak	Nunallee and Williamson (1988)
	1984 – Average	Nunallee and Williamson (1988)	1984 – Strong?	Nunallee and Williamson (1988)
Eastern Bering Sea	1983 – Average	Wespestad and Traynor (1990)	1983 – Weak	Bakkala et al. (1987)
	1984 – Average	Wespestad and Traynor (1990)	1984 – Average	Wespestad and Traynor (1990)
Pacific cod				
Hecate Strait, B.C.	1983 – Low	Foucher and Tyler (1988)	1983 – Weak	Foucher and Tyler (1988)
	1984 – Low	Foucher and Tyler (1988)		
Gulf of Alaska	1983 – Average	Zenger (1989)	1983 – Average?	Zenger (1989)
	1984 – Average	Zenger (1989)	1984 – Average+	Zenger (1989)
Eastern Bering Sea	1983 – Average	Thompson and Shimada (1990)	1983 – Average	Thompson and Shimada (1990)
	1984 – Average		1984 – Average	
Atka mackerel				
Aleutian Islands	1983 – Low	Kimura and Ronholt (1990)	1993 – Average	Kimura and Ronholt (1990)
	1984 – High	Kimura and Ronholt (1990)	1984 – Average	Kimura and Ronholt (1990)

1983(McGowan1985).Theinfluxofzooplanktonpredatorsfromthesouthmayhave affected survival and availability of prey for anchovy. The pelagic red crab (*Pleuroncodes planipes*), a voracious planktonic predator, increased in occurrence in net catches and strandings in 1983 and 1984 (P.E. Smith 1985) and in the diet of California sea lions, northern fur seals and northern elephant seals (Hacker 1986; Antonelis et al. 1987; DeLong et al., DeLong and Antonelis, this Vol.). Pelagic red crabs, which are normally found off the southern tip of Baja California, first appeared in California in October 1982; they reached the waters of the Channel Islands by December 1982, remained in California waters through 1983 (Stewart et al. 1984), and sometimes were caught in huge numbers in bottom trawls (CalCoFi 1984). Predation on anchovy by scombrid predators migrating into California waters from the south may also have increased during 1983 (Bernard et al. 1985).

Off Oregon, the northern stock of northern anchovy spawned earlier in 1983 than in other non-EN years and larvae were unusually abundant in the warm waters close to the coast (Brodeur et al. 1985). Large schools of anchovy that usually occur in the Columbia River plume and estuary were not observed (R. Emmett, pers. comm.).

Pacific Sardine – *Sardinops sagax caeruleus*

Trends in the abundance of the Pacific sardine were opposite those of the northern anchovy (Table 2). Catches, and presumably abundance, have increased from 1974 to 1987. The 1983 year class was strong, dominating the catches in 1984, the year in which the largest catch in 20 years occurred. Sardine, like anchovy, were more available in waters to the north in 1984 than in previous years. Schools were seen in Monterey Bay, California and a substantial portion of the total catch during that year was landed in Monterey (CalCoFi 1985, 1986, 1987).

Pacific Mackerel – *Scomber japonicus*

The total biomass of Pacific mackerel (age 1+) increased sharply in 1977, attained a peak in 1982, and then decreased substantially in 1983 and 1984 because of poor recruitment (MacCall et al. 1985). The 1983 year class was very weak and the 1984 year class was weak in subsequent landings (CalCoFi 1988; R. Klingbeil pers. comm.). This was unexpected since Sinclair et al. (1985) found that EN events and associated high sea levels favored survival of the early life history stages of this species in the California Current.

During 1983–84, this species migrated to the north in large numbers. Schools of Pacific mackerel were abundant in Monterey Bay, the Gulf of Farallon off San Francisco and off Oregon and Washington, and were reported in Puget Sound, along the west coast of Vancouver Island and around the Queen Charlotte Islands (Ashton et al. 1985; MacCall et al. 1985; Pearcy et al. 1985). Off Oregon and Washington, Pacific mackerel was the most abundant nektonic animal in purse seine catches during 1983 and 1984, although it did not even rank among the top ten species during 1979–1982 or 1985 (Brodeur and Pearcy 1986; Pearcy and Schoener 1987). Because of the impressive northward movement of Pacific mackerel, the fishery essentially collapsed in southern California during 1983 (Klingbeil, pers. comm.). This probably explains why this species decreased in importance in the diet of California sea lions on San Miguel and San Clemente Islands between 1982 and 1983 (DeLong et al., this Vol.).

Jack Mackerel – *Trachurus symmetricus*

Landings of jack mackerel declined dramatically during 1983 and 1984 because of their reduced availability in surface waters to purse seines. They were largely unavailable off southern California, but moderate landings were made off central California. Catches continued to be low between 1985 and 1987 (CalCoFi 1984, 1985, 1988). The 1983 year class of jack mackerel was weak (J. Mason, pers. comm.), the 1984 year class strong (T. Dickerson, pers. comm.).

This fish also expanded its range to the north, probably as a response to ocean warming during EN, and was abundant in purse seine catches off Oregon and Washington in 1983 and 1984 compared to other years (Pearcy et al. 1985; Brodeur and Pearcy 1986). The decrease in occurrence of jack mackerel in the diet of California sea lions of the Channel Islands off southern California (DeLong et al., this Vol.) was again likely caused by northward migration out of this area.

Market squid – *Loligo opalescens*

Two fisheries exist for market squid in California: one off southern California, the other off Monterey. Both of these fisheries failed in 1983 (Table 2). The southern fishery had its worst year since the 1960s. Catches in both fisheries were even lower in 1984, either because of reduced abundance or availability on the spawning grounds. Both fisheries rebounded in 1985 and catches have continued to increase since then in southern California with the highest catches on record reported in 1988 (CalCoFi 1984, 1985, 1986; T. Dickerson, pers. comm.). Poor catches of market squid are correlated with their decreased occurrence in the stomachs of California sea lions between 1982 and 1983 (DeLong et al., this Vol.).

Off Oregon and Washington, market squid was usually the most abundant nektonic animal in purse seine catches during the non-EN years of 1979–1982 and in 1985. However, it decreased in rank to sixth in 1983 and to fifth in 1984 (Pearcy and Schoener 1987). Landings of market squid in Puget Sound, Washington, increased substantially in 1983 and 1984 compared with other years (Schoener and Fluharty 1985). This increase was partially due to the use of purse seines and lampara nets only during these years (L. Palensky, pers. comm.).

Pacific Whiting or Hake – *Merluccius productus*

This species, one of the most abundant marine fish in the California Current System, and an important prey for sea lions and fur seals, spawns during the winter off Baja California and juveniles and adults migrate northward during the spring as far north as British Columbia before returning to their winter spawning grounds (Bailey et al. 1982). Neither catches nor spawner biomass of this relatively long-lived fish was depressed during 1983–84. The 1983 year class was weak, but the 1984 year class was strong (Hollowed et al. 1988; Shaw et al. 1988).

Obvious changes in the depth distribution of whiting were not noted during 1983 or 1984. All age groups of Pacific whiting migrated farther to the north in

1983 than in other years (Table 2). Ware (pers. comm.) found that Pacific whiting were farther northward and seaward in 1983 than in the 4 years after 1983. This northward, farther offshore distribution may explain the decreased occurrence of this species in the diet of California sea lions in the Channel Islands off southern California (DeLong et al., this Vol.).

Pacific Herring – *Clupea harengus pallasi*

Because of the many separate spawning stocks of Pacific herring from central California to the Bering Sea, this species, like species of Pacific salmon, can provide unique information on the latitudinal effects of major EN events. The Pacific herring is also an important prey for pinnipeds throughout its range (e.g. Bigg 1985; Perez and Bigg 1986; York, this Vol.). Catches of the herring fishery in San Francisco Bay declined by 40% between 1982 and 1983, and the following winter season of 1983–84 was the poorest since the fishery began in 1973 (CalCoFi 1984, 1985). Spratt (1987a,b) believed that the EN caused unusually high natural mortality of 2–5-year-old San Francisco Bay herring and decreased the abundance of Tomales Bay herring because of altered migration patterns during EN. Growth of herring was also poor during the 1983 EN year. In 1983 there was no weight gain by San Francisco Bay herring (Spratt 1987c). Growth was good during 1984 and the San Francisco fishery recovered from the transitory effects of the EN during the 1984–85 season (CalCoFi 1986; Spratt 1987c).

Despite the impact of EN on the availability and survival of spawners, the San Francisco population of herring produced a series of four consecutive relatively strong year classes (1982, 1983, 1984, 1985) (Spratt 1987b; CalCoFi 1988). Considering that California is the southern limit of major herring spawning concentrations along the west coast of North America, it is surprising that stocks were not affected more by the strong 1982–83 EN (Spratt 1985).

The spawner biomass of herring in Yaquina Bay, Oregon, was about average during the 1983 and 1984 seasons. A strong year class was produced in 1983 which has contributed substantially to the 1986–1988 catches. The 1984 year class was not exceptional (J. Butler, pers. comm.).

The stock size and catch of herring in the Strait of Georgia (Washington) has experienced long-term decline since the early 1970s, and no unusual trends are apparent in either the spawner biomass or year class strength for 1983 and 1984 (D. Day 1987, pers. comm.).

In British Columbia, the estimated spawning stock biomass of herring was about average during 1983 and 1984 for most of the seven districts, but was low during the 1983–84 season in the northern Georgia Strait and northwest Vancouver Island. Year-class strength did not vary consistently among the regions. D. Ware (pers. comm.) concluded that 1983 year-class strength was average and that no striking decrease in size-at-age occurred in 1983 for herring off the west coast of Vancouver Island. Ware and McFarlane (1986) and D. Ware (pers. comm.) reported a strong negative correlation between year-class strength of herring along the west coast of Vancouver Island and SST, which they suspected resulted from the extended northward movement of piscivorous Pacific whiting during warm years. The average year class of herring produced during the warm year of 1983 is an exception to this trend that may be explained by the unusual offshore distribution of whiting during 1983 (D. Ware, pers. comm.; see above).

In Alaskan waters, the stock biomass of herring from Sitka Sound and Prince William Sound was about average during 1983 and 1984 compared to other years. Data are available which compare the spawner biomass in 1983 and 1984 with other years for five of the seven herring fishing areas in the Bering Sea. All stock biomasses were about average during 1983 and 1984 except for higher than average biomass in Norton Sound (Funk and Savikko 1989).

Over the west coast range of herring, these results suggest that the 1982–83 EN only had drastic effects on the availability and survival of adult herring in the southern part of its range in 1983 and 1984. However, these years were correlated with good survival of the larvae that hatched early in 1983 for stocks of herring from San Francisco, Oregon and Georgia Strait. The 1984 year class of several stocks along the Gulf of Alaska also enjoyed good survival.

Strong year classes of herring were correlated with previous ENs and warm-water conditions in the northern part of their range (Pearcy 1983; Bailey and Incze 1985; Mysak 1986; York, this Vol.), perhaps because high sea levels and onshore convergence during EN reduce offshore transport of larvae (Taylor and Wickett 1967; Pearcy 1983). The strong 1958–59 EN was associated with unusually strong year classes in the northern part of the range and poor year classes in central California (Bailey and Incze 1985). This appears to be the opposite of the 1982–83 EN where strong year classes were most pronounced in southern British Columbia and southward.

Pacific Salmon – *Oncorhynchus* spp.

Both otariids and phocids are known to feed on salmon in the California Current, Gulf of Alaska and the Bering Sea (Antonelis and Fiscus 1980; Perez and Bigg 1986; York 1987; Kajimura and Loughlin 1988). Pacific salmon, like Pacific herring, have many discrete stocks ranging from central California to the Arctic Ocean. Stocks in the southern part of the range were severely impacted by the 1982–83 EN. Runs farther to the north were not unusually depressed, and in Alaska good production was recorded during these years.

The 1982–83 EN may have altered migration patterns of salmonids in the ocean. The migration of Fraser River sockeye around Vancouver Island is the classic example of migratory pathways that are related to ocean temperatures. During the recent years of high sea level and warm sea temperatures, a high percentage of the run has returned through the northern passage on the east side rather than along the west side of Vancouver Island (Groot and Quinn 1987). The highest diversion rate was recorded in 1983, when the effects of the EN were most pronounced.

Walleye Pollock – *Theragra chalcogramma*

The pollock fishery currently ranks as one of the most productive fisheries in the world. Walleye pollock are especially abundant in the Gulf of Alaska and the Bering Sea. Pollock is an important prey for Steller sea lions (*Eumetopias jubatus*), northern fur seals (*Callorhinus ursinus*) and harbor seals (*Phoca vitulina*) and other phocids in the Gulf of Alaska and the Bering Sea (Kajimura and Fowler 1984; Hacker and Antonelis 1986; Perez and Bigg 1986; Lowry et al. in press). Based on trawl surveys, the stock biomass of walleye pollock in the Gulf of Alaska and the eastern Bering Sea was about average during 1983 and 1984, except that high biomass was reported in Shelikof Strait in 1983. The strength of the 1983 and 1984 year classes was below average in the western Gulf of Alaska based on Fishery data (Megrey 1988), but the 1984 year-class strength was thought to be strong in Shelikof Strait based on acoustic surveys (Nunallee and Williamson 1988). The 1983 year class was weak and the 1984 year class was above the 1980–1987 average in the eastern Bering Sea (Bakkala et al. 1987; Wespestad and Traynor 1988).

Pacific Cod – *Gadus macrocephalus*
Pacific cod is listed as a preferred prey of harbor seals, Steller sea lions and northern fur seals in the Gulf of Alaska (Kajimura and Loughlin 1988). Landings and catch per effort of Pacific cod in Hecate Strait, British Columbia, decreased since 1979 and this trend continued into the years 1983–84 when catches were also low. Based on the numbers of age-3 cod, the 1983 year class in British Columbia was the lowest since 1961 (Foucher and Tyler 1988). Farther north in the Gulf of Alaska and in the Bering Sea (Thompson 1988), both the abundance and the year classes of 1983 and 1984 were about average.

Atka Mackerel – *Pleurogrammus monopterygius*
This species, which is sometimes important prey for northern fur seals in western Alaska and the Bering sea (York 1987), produced low catches in 1983 and high catches in 1984 around the Aleutian Islands. Numbers of age-3 fish indicated that the 1983–84 year classes were not exceptional, stronger than the 1978–1981 year classes, but weaker than the strong 1975 and 1977 year classes (Kimura and Ronholt 1988).

2.3.4 Discussion of Effects on EN Along the Coast of North America

Obvious effects of this EN seemed to be confined to the California Current System from Vancouver Island to the south as evidenced by catches, spawner biomass and survival of Pacific herring and Pacific salmon, two groups of fish that spawn in both the California Current and farther north in the Gulf of Alaska and the Bering Sea. The availability of market squid and herring were severely affected on their traditional spawning grounds in California. However, the EN years produced surprisingly strong year classes of San Francisco Bay herring. Farther to the north unprecedented mortality of coho salmon (*Oncorhynchus Kisutch*) occurred during 1983 off Oregon and California. Returning adult coho and chinook (*O. tshawytscha*) salmon were extremely small, and coho smolts migrating to sea during both 1983 and 1984 survived poorly.

Bailey and Incze (1985) concluded that the strong 1958–59 EN appeared to be beneficial for some fish that spawned at the northern end of their ranges, including subtropical stocks of Pacific sardine, northern anchovy and jack mackerel, which had strong 1958 and/or 1959 year classes off California. This also applied to some temperate stocks at the northern end of their ranges, such as the Pacific herring stocks off Canada, in the Gulf of Alaska and in the Bering Sea, which had unusually strong year classes in 1958 or 1959. They noted that strong ENs may have disastrous effects on recruitment of stocks that inhabit the southern end of their ranges, e.g., Pacific cod off British Columbia and Pacific herring off California, both of which produced weak year classes in 1958.

Bailey and Incze's predictions that strong ENs and abnormally warm years will have severe impacts on temperate stocks living at the equatorial ends of their geographic ranges and more favorable effects on subtropical stocks spawning towards the poleward end of their ranges were not always substantiated during 1983–84 off North America. Herring stocks in the southern part of their range often realized better than average year classes in 1983, and particularly strong year classes of herring were not common in 1983–84 north of Vancouver Island. Among the more subtropical species at the northern end of their ranges, only the Pacific sardine had a strong year class in 1983. Jack mackerel, Pacific mackerel and northern anchovy produced weak 1983 year classes. Since we have included only several of the 58 stocks that Bailey and Incze (1985) listed, for which recruitment data will eventually become available for the 1983–84 year classes, our conclusions are preliminary. Nevertheless, it is obvious that the 1982–83 EN had effects on pelagic species of the northeastern Pacific that were different from the 1958–1959 EN, the only other big northern EN for which considerable data are available. On the other hand, as Sharp (1980) and Bailey and Incze (1985) predict, those species in the California Current that experienced the most significant effects during 1983 and 1984 were species that home to specific localized spawning sites, such as herring, salmon and squid. Nomadic or migratory species were apparently able to find suitable conditions for spawning outside their normal ranges.

Ocean climate may be one of the reasons why ENs of similar magnitude have different impacts. The 1982–83 EN was unusual in timing and hydrographic changes (McGowan 1985; Fahrbach et al., this Vol.). It is also well documented that EN in 1982–83 influenced the Bering Sea much less than previous EN events since the position of the Aleutian low moved unusually far eastward (Niebauer 1988; Fahrbach et al., this Vol.).

Furthermore, the strong 1982–83 EN produced a signal that was embedded in longer-term, climatic events in the Pacific ocean. Along the west coast of North America, for example, a warming trend commenced years before the 1982–83 EN. Elevated sea levels and high SSTs occurred along the coast after 1976 (Norton et al. 1985; Cole and McLain 1989). This climatic change in 1976 is correlated with low survival of coho salmon off Oregon (Pearcy 1988), increased availability and northward expansion of Pacific mackerel, increased catches of Pacific sardine, and decreased catches of northern anchovy in California (CalCoFi 1984, 1985, 1986, 1987, 1988), and decreased catch per effort of Pacific cod in Canada (Foucher and Tyler 1988). These long-term changes in the pelagic ecosystems, which are independent of EN events, may have effects on the population biology of pinnipeds as important as isolated EN events. The long-term decline in the population of the Steller sea lion on the extreme southern end of its breeding range on the Channel Islands off California (Bartholomew 1967) may be the result of such a change in ocean climate.

A final caveat: it must be emphasized that Eastern Boundary Current Systems are inherently variable, and that it is difficult to ascribe changes observed during an EN event unequivocally to that event. We have correlations and can only try to interpret which is cause and which effect (Paine 1986; Pearcy and Schoener 1987).

2.4 Conclusions

Major similarities in the response of pelagic fish to the EN in both the southern and the northern hemisphere were the strong poleward migrations, the intrusion of oceanic and subtropical species toward the coasts, and the poor conditions for feeding, growth and survival of some species. Differences included the earlier impacts off South America than off North America and the return to normal conditions earlier in the South Pacific. Effects on fish populations were obvious farther from the equator in the North Pacific (50°N) than in the South Pacific (33°S), but this may be partly a consequence of missing observations in the far south. Those species in the California Current that experienced the most significant effects during 1983 and 1984 were species that home to specific localized spawning sites, such as herring, salmon and squid. Nomadic or migratory species were apparently able to find suitable conditions for spawning outside their normal ranges.

The EN had both short- and long-term effects on the prey populations of pinnipeds. The most obvious short-term responses were changes in the availability of prey caused by poleward, inshore-offshore, or bathymetric movements observed for anchovy, sardine, mackerel and hake in both hemispheres. Actual changes in the abundance and age structure of populations were related to higher mortality rates of adult, juvenile or larval stages. Body condition, growth rates and reproductive capacities were also reduced. The influx of new predators, such as subtropical scombrids, into an area modified the abundance and availability of the usual prey populations. All of these changes could affect the availability, vulnerability and nutritive value of prey populations to marine mammals and therefore the quantity, quality and the ease of capture of prey.

Delayed responses to a major EN perturbation are less obvious than immediate effects because they may affect recruitment years after the event. Elevated temperatures can alter the time of spawning, fecundity and viability of eggs, rates of development and metabolism, as well as the species composition and seasonal timing of prey and predator populations; changes in the circulation of water can affect the dispersal or retention of eggs and larvae (Bailey and Incze 1985). All of these factors can influence the survival of early life history stages and the strength of year classes. Hence, weak year classes resulting from EN could alter the availability of pinniped prey years after the EN event.

Pinnipeds (and seabirds) find themselves confronted with a number of difficulties during and shortly after major EN events which seriously interfere with normal feeding:

1. Their food is not where it used to be; fish emigrate to cooler (deeper, more poleward, or remaining upwelling) areas;
2. Food can no longer be found near the surface where it is easily accessible at low energetic cost;
3. Even if food can be reached (e.g., by deep-diving pinnipeds), it is of poor quality and occurs at lower density; e.g., Peruvian anchovy had decreased fat content;
4. Invading species such as mackerel or tropical immigrants are unsuitable as pinniped food due to their size, strong defenses, and high swimming speeds; some of them also live in deeper water and do not form dense shoals like anchovy and sardine;
5. In addition, these invading species are often predators of early life history stages of the normal pinniped prey;
6. There may be major time lags in the recruitment of anchovy, sardine, pollock, capelin and other species caused by EN which render the feeding situation insecure in the years following the event;
7. Physical manifestations of EN are strongest near the equator and become attenuated towards higher latitude; biological effects are pronounced near the equator and on species that are near the equatorial ends of their ranges and have localized spawning grounds. At high latitudes (e.g., Chile or Alaska) the resulting warming may even have been beneficial.

Acknowledgments. Completion of this work would not have been possible without the generous and timely cooperation of many friends and colleagues. D. Aurioles (Centro de Investigaciones Biologicas de Baja California Sur); R. Klingbeil, T. Dickerson, J. Spratt, and P. Wolf (California Department of Fish and Game); R.T. Barber (Monterey Bay Aquarium Research Institute); J. Mason (Pacific Fisheries Environmental Group, National Marine Fisheries Service); J. Butler (Oregon Department of Fish and Wildlife); D. Day and L. Palensky (Washington Department of Fisheries); R. Emmett, K. Bailey, R. Brodeur, R. Methot, E. Sinclair, G. Antonelis, R. DeLong, R. Gentry, A. Hollowed, A. York, V. Wespestad, and H. Huber (Northwest and Alaska Fisheries Center, National Marine Fisheries Service); A. Tyler, K. Hyatt, D. Blackbourn, D. Ware, and B. Holtby (Canadian Department of Fisheries and Oceans, Pacific Biological Station); D. Eggers, H. Savikko (Alaska Department of Fish and Game); and D. Amend (Southern Southeast Regional Aquaculture Association). We also thank A. Alvial, A. Schoener, E. Sinclair, and G. Norton for helpful comments on the manuscript. D. Schmidl cheerfully prepared several versions of the graphs; we thank him for his patience. F. Trillmich was supported through W. Wickler, Max-Planck Institut, Seewiesen; W. Pearcy was supported by the Northwest and Alaska Fisheries Center (NA 88 AB-N-00043).

Part II
Otariid Pinnipeds, General

3 Introductory Remarks on the Natural History of Fur Seals

F. TRILLMICH

Fur seals are distributed worldwide. The genus *Arctocephalus* consists of eight species, four of which occur in the eastern Pacific. These are from south to north, the South American fur seal (*A. australis*), the Juan Fernandez fur seal (*A. philippii*), the Galapagos fur seal (*A. galapagoensis*), and the Guadaloupe fur seal (*A. townsendi*), which is the only *Arctocephalus* occurring north of the equator. The northern fur seal, *Callorhinus ursinus*, is the only species of its genus.

We have data on two *Arctocephalus* species, the South American and the Galapagos fur seal. The South American fur seal is distributed from Uruguay in the east to Peru in the west all around the southern tip of South America, and also occurs on the Falkland Islands. Its distribution over this wide range is very spotty. The Galapagos fur seal (at 0° lat.) is most abundant along the coasts of the western islands of the archipelago. The northern fur seal breeds in the Bering Sea and the Sea of Okhotsk (between 47° and 57°N) as well as on San Miguel Island off the coast of California (at 34°N).

All fur seal females, of the species described in this book, have a body mass of 30–50 kg, although bigger individuals may occasionally be observed. Females grow throughout life, but growth slows down very much once they reach sexual maturity at an age of 2–5 years. Males are much bigger than females. They grow faster than females and begin to partake in reproductive activities as territorial males when 7–12 years old. The size dimorphism is most pronounced in the northern fur seals where males may be five times heavier than females (York 1987). No data are available for the size dimorphism in South American fur seals, but it seems to be intermediate between that of the northern and the Galapagos fur seal, where males may be only twice as heavy as females (Trillmich 1987).

The social organization and maternal behaviour of all fur seals are very similar and differences between species, which will be described in more detail in the following chapters, are variations on a general theme. Females are gregarious on land, either during the breeding season only, as in the migratory northern fur seal, or almost year-round, as in the nonmigratory Galapagos fur seal. During the breeding season, when females give birth and enter estrus, males establish territories in female aggregations which they defend vigorously against each other. Females can give birth to a single pup every year. After parturition they remain on land with their newborns for about 1 week. They then enter estrus and copulate, usually with the territorial male. Following copulation females make their first postpartum trip to sea. This is usually a very short trip. For the remainder of the nursing period they alternate between foraging at sea and nursing the pup ashore. We refer to the pattern of stays ashore as the "attendance pattern". Most fur seals feed offshore over deep water predominantly during the night when organisms (fish, squids, and crustacea) of the deep scattering layer migrate closer to the surface (Gentry and Kooyman 1986a). The alternation between hunting at sea and suckling the pup onshore continues for about 4 months in the northern fur seal, but may extend over 1–3 years in the South American and Galapagos fur seal.

After weaning, mothers and pups of the northern fur seal live pelagically until the next breeding season. The South American fur seal in Peru is much less migratory. Mother-young pairs can be observed year-round in the breeding colony. Some animals, however, may migrate and this may have been particularly pronounced during the 1982–83 EN (Guerra and Portflitt, this Vol.). The Galapagos fur seal appears to be entirely nonmigratory.

In the following section, Guerra and Portflitt describe the immigration and establishment of the South American fur seal in northern Chile in the course of the 1982–83 El Niño (EN). Majluf has followed the fate of a population of this species in Peru during EN in 1982–83 and was able to compare these data with observations of the same population during the much weaker EN in 1986–87. Finally, Trillmich and Dellinger observed the effects of the 1982–83 EN on the Galapagos fur seal.

Gentry observed northern fur seals from the Pribilof Islands during the 1982–83 breeding season and York reports on EN effects on survival of juveniles during the first 2 years of pelagic life. DeLong and Antonelis describe the effects of EN on the southern population of northern fur seals on San Miguel Island.

In the three papers on southern fur seals some additional information is also given for the sympatric sea lions, the South American sea lion (*Otaria byronia*) in Chile and Peru and the Galapagos sea lion (*Zalophus californianus wollebaeki*) in the Galapagos. More information on the natural history of sea lions of the genus *Zalophus* is given by Ono (Chap. 10, this Vol.).

4 El Niño Effects on Pinnipeds in Northern Chile

C.G. GUERRA C. and G. PORTFLITT K.

4.1 Introduction

The impact of EN on pinnipeds in northern Chile was quite different from that in Peru (Majluf, this Vol.) and in the Galapagos (Trillmich and Dellinger, this Vol.), thus demonstrating that the response of populations varied strongly with latitude.

The South American sea lion (*Otaria byronia*, hereafter referred to as "sea lion") inhabits the SE Pacific coast without major distributional gaps from Cape Horn to Zorritos, Peru (04°00′S), whereas the South American fur seal (*Arctocephalus australis*, hereafter "fur seal") is distributed patchily with no colonies between Arica (Chile, 18°27′S) to Maiquillahue (39°27′S) (Aguayo and Maturana 1973; Torres et al. 1983; Guerra and Torres 1987). The fur seals thus live in two, probably isolated, populations along the coast of South-America: one from Cape Horn to Chiloe Island (43°54′S) and another mainly inhabiting southern Peru (Majluf and Trillmich 1981; Guerra and Torres 1987).

We here document the changes in abundance and distribution of the sea lion inhabiting the northern Chilean coast. We were also able to observe an invasion of the fur seal to northern Chile where it established colonies between 21° and 23°S during and after the 1982–83 EN.

The South American fur seal is known to compete aggressively with sea lions for space and we describe how the invasion of fur seals changed the distribution of sea lions in one study colony.

4.2 Methods

Annual surveys were made from the southern summer 1982–83 to 1987–88. They covered the area from Cape Paquica (21°54′S) to Cape Angamos (23°05′S), approximately 145 km of the north Chilean coast. Each summer sea lions and fur seals were censused in February or March for 3–4 days, and habitat use and behavior were observed. Telescopes (60x zoom), binoculars (7.5 × 50, 10 × 50) and photographs were used to identify and count animals. We used 4WD vehicles and boats to get close to the seal rookeries.

Next to Cape Angamos there is a rookery on Abtao Islet about 500 m offshore (Fig. 1). Since 1981–82 boat surveys were made there every summer to record and photograph the distribution of fur seals and sea lions on an outlying rock and the main islet (Fig. 2).

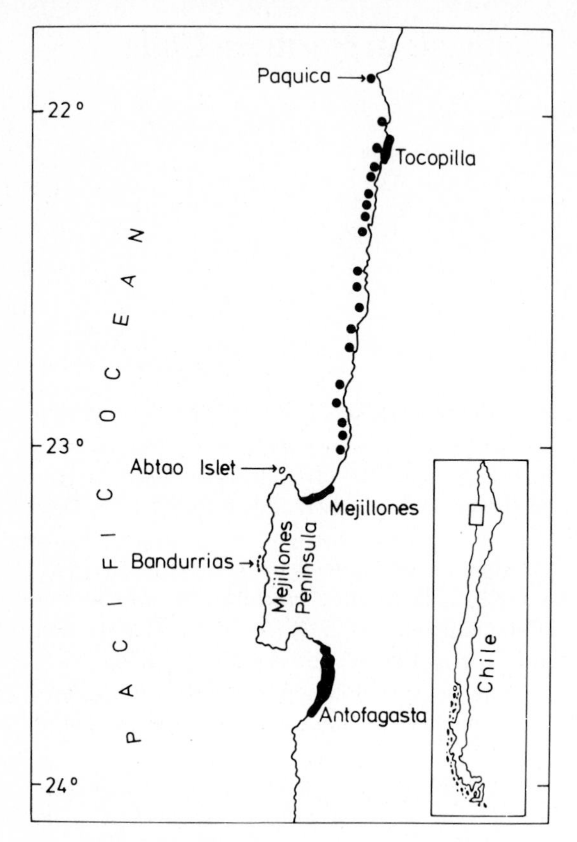

Fig. 1. The study coast and its position in Chile (*inset*). *Dots* indicate localities where seals were seen. Place names in Table 1 refer to these dots from N to S

The largest sea lion colony (Bandurrias) is situated in the midwest section of the Mejillones peninsula (23°18′S, Fig. 1). In our censuses there we distinguished the following animal categories:

1. Pups;
2. Adult females;
3. Active territorial males, and
4. Peripheral males.

The counts were made during (November to January) or shortly after every breeding season (until May when females and pups have not yet left the breeding rookeries; Table 2).

Oceanographic data were obtained from the scientific information published by researchers from Peru and Chile. Commercial fish landings in the Antofagasta area (second region of Chile) were used as an index of food availability.

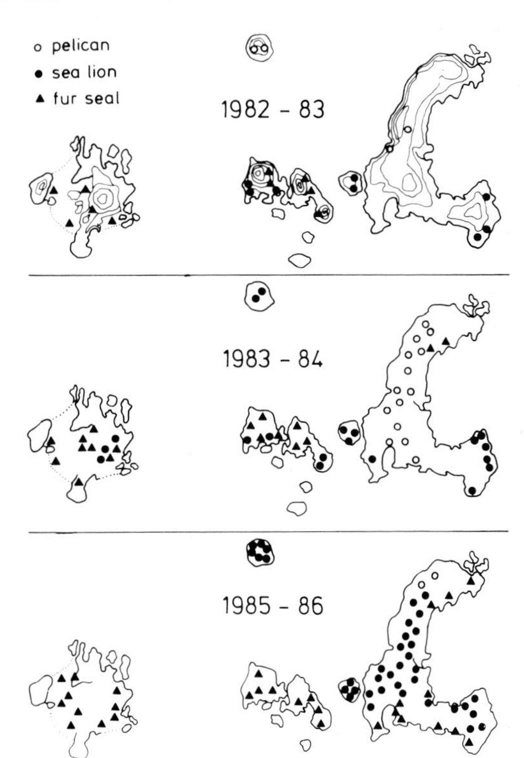

Fig. 2. Distribution of sea lions, fur seals, and pelicans (*Pelecanus occidentalis*) on Abtao islets for the years 1982–83 (60 sea lions and 100 fur seals), 1983–84 (180 sea lions and 180 fur seals), and 1985–86 (371 sea lions and 294 fur seals). North is towards the bottom of the graph

4.3 Results

4.3.1 The Fur Seals

Before EN only one young male South American fur seal was seen in the coastal section under study. This individual was observed in February 1982 at Abtao Islet (Torres et al. 1983; Guerra and Torres 1987). In January 1983, the first peak of the 1982–83 EN event (Fahrbach et al., this Vol.), 228 fur seals were counted along the Paquica-Abtao coastal section. During this census, fur seals were observed at 12 sites, with the two greatest congregations at Paquica and Abtao (44% of all fur seals). Three other sites (Algodonales Islet, Cape Cobija, and Cape Thames) each hosted about 10% of all individuals censused (Table 1). All individuals observed appeared to be subadults similar in size to one male which was captured and mea- sured at Paquica (tip of nose to tip of tail, straight length 118 cm).

In the 1983–84 census (2–6 March) when marine conditions had returned to near normal (Martínez et al. 1984; Cañon 1985), the distribution of fur seals along the study coast was changed. Except for three animals at Algodonales Islet all

Table 1. Fur seal and sea lion distribution between Cabo Paquica and Abtao Islet during the southern summers 1982–83 to 1987–88 (SL = sea lion; FS = fur seal)

Census date	12–17 Feb. 1983		2–6 March 1984		5–8 Jan. 1985		22–26 Feb. 1986		27–29 Feb. 1988	
Site	SL	FS	SL	FS	SL	FS	SL	FS	SL	FS
Paquica	321	40	647	46	252	19	121	48	51	44
Paso malo					21					
Ana	9		4		7					
Algodonales	75	22	186	3	35		3		<10	
Blanca	1									
Aqua dulce	7	1								
Alala	46	8	3		1		2		3	
Band. N	40		22		2		1		30	
Los Chinos	13		20		9				2	
Grande	35		112		73		2		10	
Cobija	1	22	1		1		7			
Guasilla	53	1	21		14					
Tamira	11									
Thames	86	26	137		261		65			
Huaque	8	1	47							
G.-guala	120	4	76		92		8		50	
Yayes	15									
Hornos	3									
Itata	17	1	66		141				4	
Chacaya	3	2								
Abtao	82	100	180	180	189	144	371	294	261	218
Total	946	228	1522	229	1198	163	613	342	411	262

fur seals were now found at Cape Paquica (20% of the total) and at Abtao Islet (75%). These two sites remained stable colonies in the following years (Table 1).

Most animals in these fur seal colonies were males but a few young females were also seen. In 1983–84 most of the males were subadults, but a few were the size of a large male found dead at Mejillones from which we have only a measure of penis bone (111 mm) and condylobasal skull length (245 mm). In Dec. 1984 males at those two colonies appeared larger than in 1983–84 and all of them were in good physical condition. Unfortunately, we were not able to approach the rookery at Abtao close enough to observe pups; but the presence of turkey vultures (*Cathartes aura*) could be considered an indication of births in this colony.

Over the next 2 years fur seal numbers at both colonies remained stable or even increased (Table 1). At Abtao some pups and juvenile fur seals were observed during the 1985–86 census, finally proving that reproduction occurred in this colony. In 1987–88 some very young fur seals were also observed at Paquica.

4.3.2 Sea Lions

In 1982–83 sea lions were distributed more evenly along the coast than fur seals. At 6 sites more than 50 individuals were counted (Table 1). Paquica was clearly the largest colony. Of the 20 sites used by sea lions, 12 were also used by fur seals in that year. In 1983–84 far more sea lions were censused along the study coast than in January 1983 (Table 1). In March 1984 sea lions formed dense, nonreproductive colonies along the censused coast. The only small breeding colony was Cape Paquica. In 1985–86 (22–26 February 1986) sea lions were mostly concentrated at Cape Paquica (20% of the total censused; Table 1) and Abtao Islet (61%), as were the fur seals, and this distribution remained very similar in the 1987–88 census (27–29 February 1988).

Further south a colony of sea lions existed at Cape Bandurrias (23°18′S), on the Mejillones peninsula. This colony was much less populated in December 1984 (only 132 individuals) than in all other years when counting took place in the month of October (in October 1980 we counted 1985 individuals) or February. Table 2 shows that pup production in this colony decreased dramatically in 1983 during EN, remained low in 1984, 1985 and returned to normal in 1986.

However, this colony had been subject to pup culling from 1978 to 1980 (pup harvest per year: mean = 2068, SD = 360). Its normal pup production was thus decreased by previous mismanagement (Guerra et al. 1987) and EN hit just at the time when the population began to recover. Furthermore, the geomorphological structure of the colony site may have led to high pup mortality during EN. The breeding area is on narrow platforms at the base of cliffs which are 30–50 m high. These platforms may have been completely awash during the extremely high tides which occurred in the study area during EN when the highest sea level in 40 years was recorded (Fonseca 1985).

Wave action, in addition to being a cause of pup mortality, may also have induced adults and juveniles to abandon the usual breeding areas. During the 1982–83 warm event, sea lions from the Bandurrias colony apparently dispersed along the

Table 2. Changes in the structure and density of the sea lion (*Otaria byronia*) colony at Cape Bandurrias (23°18′S) (after Guerra et al. 1987)

	Feb. 1981	Feb. 1982	Jan. 1983	Dec. 1984	June 1985	April 1986
Females and juveniles	1778	1682	1256	56	1279	1677
Territorial males	218	251	230	22	64	64
Nonterritorial adult males	315	482	630	44	113	380
Pups	1110	1220	695	9	479	1373
Adults in the water	33	71	14	1	343	104
Total	3454	3706	2825	132	2278	3598

coast. Juvenile individuals settled inside the Mejillones bay (23°05′S) on a public access beach, which was never used by sea lions before or since.

4.3.3 Species Interactions

Before EN 1982–83, the Abtao Islets were a brown pelican (*Pelecanus occidentalis thagus*) colony, with less than 50 sea lions (estimated from photographs) using the site. During EN, in 1982–83, pelicans did not breed (due to competition for space with guanay cormorants (*Phalacrocorax bougainvillii*) and Peruvian boobies (*Sula variegata*). The density of both of the latter species was increased by immigration of birds, presumably from Peru. In 1982–83 sea lions used the lower platform of the main islet, the southern rock and shared the central islet with the fur seals (Fig. 2). In 1984–85 pelicans bred again on the highest platform of the main islet (Fig. 2). As pinniped numbers increased (1985–86) sea lions invaded this platform and ousted the pelicans (Fig. 2). Sea lions used sites on the main islet and two rocks, all of which had flat, nearly horizontal surfaces. The eastern islet which is surrounded by small pebble beaches had the highest fur seal density. It was on this site that pups were observed in January and March of 1985 and 1987, respectively. The central islet was also claimed by fur seals in 1985–86. This area consists of steep rocks without extensive horizontal surfaces. Aggressive encounters between sea lions and fur seals in this area were invariably won by fur seals.

4.4 Discussion

Fur seals immigrated only recently to northern Chile. Before EN, only single animals were seen, presumably as a consequence of the increase of the Peruvian fur seal population (Majluf and Trillmich 1981); but a real invasion into northern Chilean waters was precipitated by the 1982–83 EN (Guerra and Torres 1987).

Our study area is characterized by its intense upwelling processes. During EN 1982–83, this system was negatively affected locally (Arntz et al. this Vol.). On the other hand, EN caused southward migration of sardines (*Sardinops sagax*) from Peruvian waters to the northern Chilean coast, greatly increasing the landings of the fishing fleet operating in the zone (Cañon 1985). The northern area of the Mejillones peninsula (Fig. 1) was particularly rich in fish during EN, especially in January and February 1983 (Martínez et al. 1984; Cañon 1985). This was probably due to narrowly circumscribed local upwelling as described for Arica and Punta Gruesa during the same period (Kelly and Blanco 1984).

We compiled a species list of the diet of sea lions and fur seals based on reports in the literature for the Pacific area (Table 3). Only three items in the diet of these two pinniped species overlap. Table 4 shows the landings of the four most important items in the diet of sea lions and fur seals. According to this information only anchovy (*Engraulis ringens*) decreased during the EN years. In contrast, landings of jack mackerel (*Trachurus symmetricus murphy*) increased almost twofold over those in 1981, and landings of sardine increased, particularly in 1984. Among ceph-

Table 3. Comparison of the diet of sea lions and fur seals from data taken from the literature

Species	Sea lion	Fur seal	References[a]
Trachurus spp.	X	X	1, 5, 6
Engraulis ringens		X	5, 6
Clupea bentinckii	X		2
Genypterus spp.	X		1
Merluccius gayi	X		2, 3
Sardinops sagax	X	X	1, 3, 4, 7
Callorhynchus callorhynchus		X	2
Paralabrax humeralis	X		3
Scomber japonicus		X	5
Stromateus stellatus	X		1
Todarodes sp.	X		1
Cephalopoda (cuttlefish, squid, octopus)	X	X	1, 5
Elasmobranchia	X		4
Isacia conceptionis	X		4

[a]1, George-Nascimento et al. (1985); central Chile. 2, Secretaria Regional de Planificación y Coordinación (1981); séptima región. 3, Aguayo and Maturana (1973); Valparaiso. 4, Guerra et al. (unpubl. data); Antofagasta. 5, Tovar and Fuentes (1984); Peru. 6, Majluf (1987); Peru.

Table 4. Annual landings of selected species in the study area. All these species are food items for sea lions and fur seals (numbers given are metric tons; source: Servicio Nacional de Pesca, Annual Reports)

Species	1980	1981	1982	1983	1984
Trachurus murphy	25 778	17 332	32 769	37 842	122 034
Sardinops sagax	173 324	229 089	230 812	265 667	363 110
Engraulis ringens	7 902	80 525	3 301	23	4 261
Octopus vulgaris	1	6	7	267	672

alopods, captures of octopus (*Octopus vulgaris*) were augmented during and after the warm event, which was clearly unrelated to changing fishing effort.

Fur seals in Peru consumed mostly anchovy, sardine and jack mackerel during EN (Majluf 1985, and this Vol.). According to the literature, anchovy apparently is not a part of the sea lions' diet, but direct observations by C.G. from fishing vessels indicated that in northern Chile sea lions occasionally feed on anchovy. Nevertheless, EN 1982–83 did decrease food availability at least from September to December 1982 (Cañon 1985) when local fish populations migrated south or withdrew to deeper waters. This might have caused the slight mortality of sea lions observed within the study area. This contrasts strongly with the high mortality observed in Peru (Majluf, this Vol.) and Galapagos (Trillmich and Limberger 1985; Trillmich and Dellinger, this Vol.). Within our study area in Chile less than 20 dead sea lions were recorded, accounting for only 0.4% of the average population (mean = 5139, SD = 1080, n = 8 years; Guerra et al. 1987). Underweight animals were not generally observed and notably, some fresh dead animals did not show clear evidence of starvation. According to Cañon's data (1985), the area received immigrant fish stocks from early 1983.

The highly dispersed occurrence of sea lions in 1982–83 may indicate a wide-ranging search for food or a strategy to minimize intraspecific food competition. The newly immigrant fur seals, slightly later followed by sea lions, concentrated in areas where fish catches were higher than average. This was particularly clear in the area of Cape Angamos where Abtao Islet is located.

There are few suitable sites for hauling or reproduction of pinnipeds along the coast of the second region of Chile. It seems that fur seals require places with less human disturbance than is tolerated by sea lions. This fact, and the short distances to areas of local upwelling with the concomitant high food availability may explain why fur seals concentrated at only two sites.

The decrease in sea lion abundance at a site taken over by fur seals was similarly observed by Vaz-Ferreira and Ponce de Leon (1984) in Uruguay. Fur seals thus are able to displace the much larger sea lions from suitable fur seal habitats. Abtao Islet allowed us to observe these distributional changes. Obviously, sea lions there preferred areas with more or less horizontal surfaces, whilst fur seals inhabited areas with steep rocks. Similar substrate preferences are known from Uruguayan (Vaz-Ferreira and Ponce de Leon 1984), and from Peruvian colonies (Trillmich and Majluf 1981).

5 El Niño Effects on Pinnipeds in Peru

P. MAJLUF

5.1 Introduction

The South American fur seal (*Arctocephalus australis*) and the South American sea lion (*Otaria byronia*) are the only two species of pinniped exploiting the rich Peruvian upwelling system. Both, like most other pelagic feeding vertebrates in Peru, feed largely on anchovy (*Engraulis ringens*), a small clupeid fish which previous to 1973 occurred in abundance in this system (Idyll 1973). Because of the economic importance of anchovy to the Peruvian fishing industry, the biology of anchovy and the effects of El Niño (EN) on its distribution and abundance have been relatively well documented (Glantz and Thompson 1981; Cushing 1982; Pauly and Tsukayama 1987) and thus it is possible to understand how EN events affect the seals' prey availability. Here, I shall compare the foraging behaviour, diet, pup growth and mortality of the fur seals and sea lions in Peru under EN and non-EN conditions and examine whether the differences can be explained by the changes in anchovy availability known to take place during ENs. Greater emphasis will be placed on the fur seals because there is only limited information on sea lions in Peru.

5.1.1 Punta San Juan: 1983–1988

The study was carried out at Punta San Juan (PSJ, 15°22'S, 75°12'W, Fig. 1), which holds up to 50% of the fur seal and about 30% of the sea lion populations in Peru (Majluf and Trillmich 1981). PSJ is the area of coldest waters, most intense upwelling (Zuta et al. 1978) and highest primary productivity (Cushing 1982) off the coast of Peru. Because of these factors, combined with its southerly location (15°S) and the very intense local winds and extremely narrow continental shelf, upwelling tends to persist longer around this area even when surrounding areas are affected by EN conditions (Villanueva et al. 1969).

During this study, there were two EN events; one in 1982/83 considered to be the most intense EN in the last 400 years (Quinn et al. 1987) and a second, milder event in 1986/87. Figure 2 shows the monthly sea surface temperature (SST) anomalies recorded at PSJ throughout the periods covered in this study. In 1982/83 SST rose to 8 °C above the normal temperature for the time of the year, and the thermocline was depressed from the usual 50 m to about 100 m (Fig. 3). In 1986/87, SSTs rose only to 2 °C above the monthly mean, but little is known yet of the effects of this event. The reproductive seasons of the fur seals in 1984,

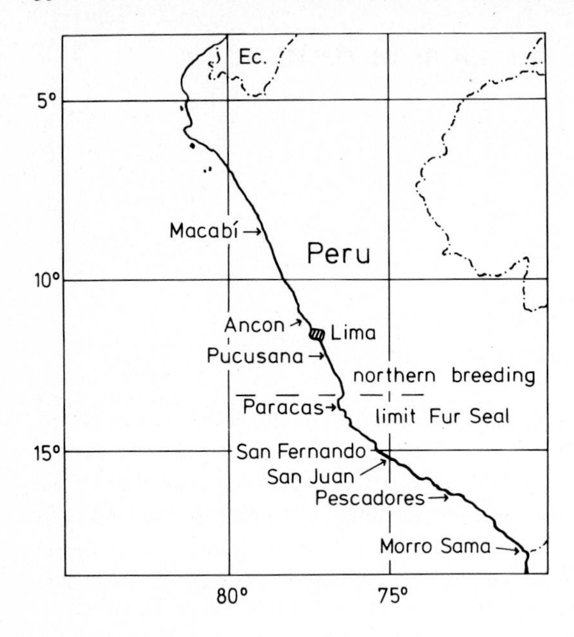

Fig. 1. Coastline of Peru showing location of seal colonies mentioned in the text

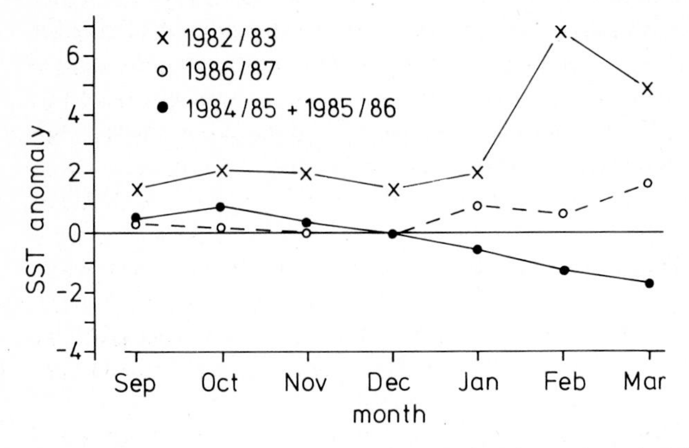

Fig. 2. SST anomalies at Punta San Juan during the study period (September-March 1982–87). Average monthly SSTs (1958–74) taken from Zuta et al. (1978). SST data from Punta San Juan (1982–87) courtesy of Cmdt. Hector Soldi, Oceanography Division, Peruvian Navy

1985 and 1987 are considered to represent non-EN conditions with 1985 being exceptionally favourable.

5.2 Methods and Study Site

Data on the fur seals were collected between January 1983 and August 1988 as part of a long-term study of the breeding ecology of female fur seals at PSJ (Majluf 1987).

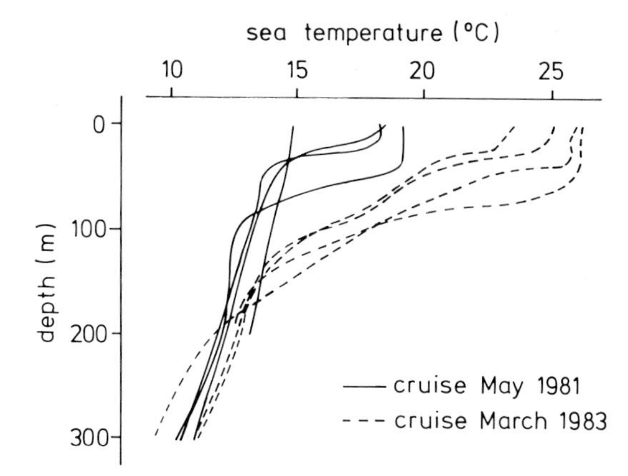

Fig. 3. Temperature distribution with water depth off Punta San Juan under normal conditions (*solid lines*) and during El Niño 1982/83 (*dashed lines*) (data courtesy of Dr. Wolf Arntz, PROCOPA/IMARPE – Callao)

Most observations were made at two sites, called S3 and S6 (for a map of PSJ, see Majluf 1985). Data on the sea lions were collected only between January and August 1983. Methods used to study the fur seals are described in detail in Majluf (1987) and for the sea lions in Limberger et al. (1983).

From 1984 onwards, adult females and yearlings were captured using a hoopnet, and individually weighed and tagged. Some of these animals were further marked by clipping the guard hairs on their backs and bleaching the underfur using hair bleach. Throughout the study pups were caught by hand, weighed, and marked.

Attendance behaviour of female fur seals was studied between January and August in 1983, and between September and February in 1984/85 and 1985/86. Female-offspring pairs in which both the female and her young were marked or where only the pups were marked were checked at least once and up to four times each day to assess the duration of the females' presence at the colony. Arrivals or departures were assumed to have taken place in the middle of the interval between the last check when they were present (or absent) and the following observation (i.e. if a female was last seen in the 0600 check and the next check was at noon, she was assumed to have left at 0900). Nocturnal arrivals or departures were assumed to have occurred at midnight.

Diving behaviour of adult female fur seals was studied using time-depth-recorders (TDRs). The recorders used in this study are described in Gentry and Kooyman (1986b) and details of the analysis of the four records obtained are given in Majluf (1987).

Fur seal diet was studied from the analysis of hard parts (fish otoliths and squid beaks) found in fresh scats.

In January/February 1983 pup mortality under EN conditions was estimated from daily counts of dead pups on one beach at PSJ. The details are described in Trillmich et al. (1986). Survival between February and March 1983 was estimated from resightings of pups tagged on another beach (n = 14) between 7–10 February and resighted from 7 March.

From 1984/85 onwards, each year at least 200 pups (mean age = 30 days) were tagged in late December on the main fur seal breeding beach at PSJ. Minimal survival (because it included tag loss and dispersal) up to 1 year of age was estimated from resightings of these tagged pups throughout the year. The beach and nearby areas were searched daily for tagged animals for periods of at least 15 consecutive days at varying intervals throughout the year. Minimal survival up to any given month was estimated as the proportion of the total number of pups tagged in December which were seen alive after a given date.

5.3 Results

5.3.1 The Fur Seals

1. Changes in Foraging Behaviour

Attendance Patterns. Attendance behaviour under EN conditions was studied between January and August 1983. In January/February, when EN conditions were most severe, females nursing 1-3-month-old pups showed very erratic behaviour. During EN, foraging trip duration varied widely between 1–10 days ($\bar{x}$ = 4.61, SE = 0.52 days, C.V. (coefficient of variation) = 54%, n = 23) while in 1985/86 the range was smaller (1–6 days, $\bar{x}$ = 3.12, SE = 0.13 days, C.V. = 37%, n = 84, Fig. 4). Median trip duration in 1983 (4 days) was significantly greater than in 1984/85 and 1985/86 (3 days, $\chi^2 = 4.71, p < 0.05$, df = 1, median test corrected for continuity, Siegel 1956). Visits ashore during EN ($\bar{x}$ = 1.3, SE = 0.13 days, n = 26) appeared to be shorter than the visits of females in 1984/85 and 1985/86 ($\bar{x}$ = 1.52, SE 0.05 days, n = 92) but this difference was not significant.

Diving Patterns. There are four complete dive records for the fur seals in Peru. One was obtained in January 1983, under severe EN conditions and the other three in December 1984 (1) and 1985 (2), under "normal conditions". In total these records cover 558 h at sea (23.3 days), 3482 dives and each includes at least two and up to

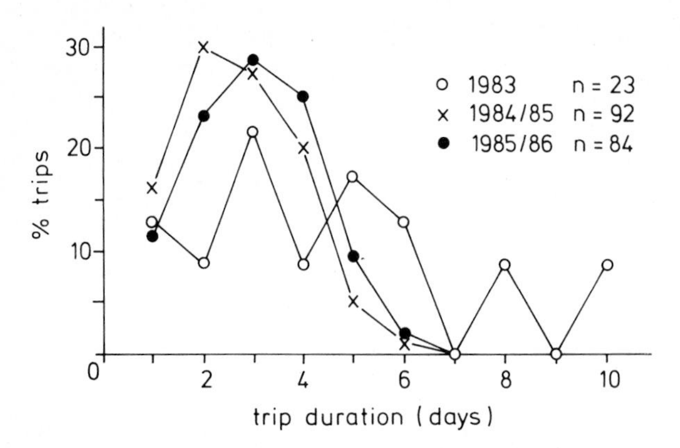

Fig. 4. Frequency distribution of trips to sea of female fur seals at Punta San Juan in 1983 (EN), 1984/85 and 1985/86. Data includes trips of females suckling 2–3-month-old pups only

four complete foraging trips ranging between 0.3–7.0 days in duration. Time at sea comprises between 43–63% of the total recording time.

In all four records most dives occurred at night. Diurnal dives comprised less than 20% of all dives recorded. Maximum dive depth was 161 m and the longest dive took 9.3 min. However, the modal dive depths were between 10–40 m (65.4%) and 89% of all dives were less than 3 min long (Table 1).

The three-dimensional frequency distribution of dive depth with time of day for the four records is given in Fig. 5. The three seals studied in 1984 and 1985, under non-EN conditions showed consistent dive patterns (Fig. 5): they made the deepest dives around dusk and dawn and throughout the night they made only shallow (<40 m deep) dives. For these animals, dives exceeding 40 m in depth comprised fewer than 20% of all dives 8–17%). In contrast, during EN, the female kept diving to a more or less constant depth range throughout the night (Fig. 5). A higher proportion (59%) of the total dives recorded for this female were over 40 m deep.

Table 1. Mean dive depth and duration for each of four instrumented females

Date	Dive Depth (m)			Dive Duration (min)		
	Mean	SD	Max.	Mean	SD	Max.
January 83	46	17.6	160.8	1.9	0.76	6.3
December 84	22	17.9	126.7	2.0	0.83	4.7
December 85	22	14.8	116.5	1.1	0.77	8.0
December 85	29	16.3	114.5	2.0	1.21	9.3

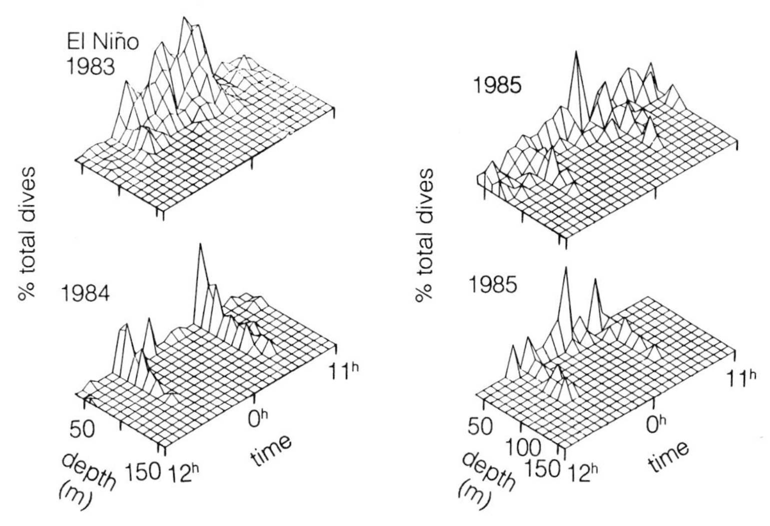

Fig. 5. Three-dimensional plots of dive depth, time of day, and percentage of dives for each instrumented female. Only dives deeper than 40 m were included to show peaks of deep dives in each record. 1983: n = 583, TD = 996. 1984: n = 119, TD = 1125. 1985 *upper*: n = 136, TD = 782. 1985 *lower*: n = 50 TD = 579. *n* refers to dives deeper than 40 m, *TD* to all dives recorded by the instrument

Diet. Only limited data on the fur seals' diet during EN is available. From a sample of 35 scats collected between January and April 1983, it appears as though fur seals at PSJ were still mainly feeding on adult (2–3 years old) anchovy (81 out of 103 were anchovy otoliths). Later in the year, however, between June and August, anchovy were missing from the fur seals' scats and they appeared to be feeding mainly on sardine (*Sardinops sagax*) and jack mackerel (*Trachurus symmetricus*).

In 1984/85, even though oceanographic conditions appeared to be "normal", fur seals took a wider range of prey and fed mainly on small mesopelagic fish (bathylagids and myctophids) and juvenile anchovy. Adult anchovy did not appear in significant numbers in the fur seal scats until 1985, when they accounted for more than 80% of the otoliths found in the scats (Fig. 6).

Signs of Dispersal in Relation to a Decrease in Food Abundance. Before the 1982/83 EN, most fur seals in Peru (80–90%) were found at only three sites: the Paracas peninsula (13°50′S), San Fernando (15°09′S) and PSJ. These sites corre-spond to centers of intense upwelling and Paracas and PSJ are protected from poaching (Majluf and Trillmich 1981). Only rarely were fur seals observed north of Paracas. Between January and August 1983, groups of three to ten individuals, mostly subadult males, were observed at Pucusana (12°29′S), Ancon (11°46′S; P. Majluf, unpubl. data) and in Macabi (8°47′S; C. Hayes, pers. comm.; see Fig. 1). Also at this time, a large decrease in numbers at PSJ and San Fernando and the pres- ence of large numbers of fur seals in previously empty colonies further south were observed (Table 2). Unfortunately, the extent to which these changes were due to mortality and/or migration cannot be assessed.

2. Effects on Patterns of Reproduction

Interannual Differences in Pup Growth. Pup weights during the 1982/83 EN were only collected in February 1983 and growth rates could not be calculated. However, February weights were used as an index of early pup growth for the year. Pup weights in February 1983 and 1987, during EN, were significantly lower than in 1985, 1986 and 1988 when conditions at sea were favourable. For these 5 years

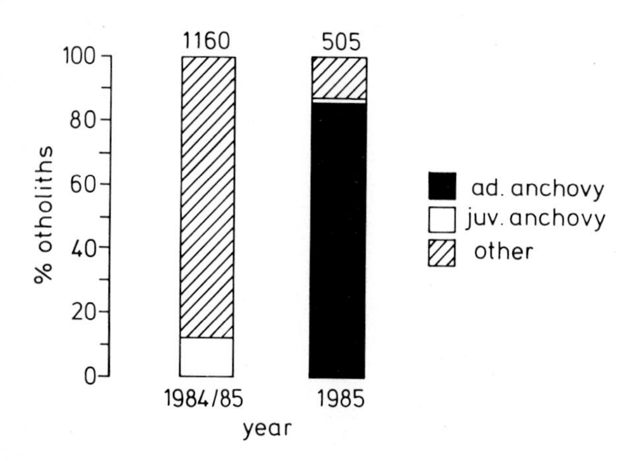

Fig. 6. Composition of fur seal scats in 1984/85 (Oc-tober-Jan-uary) and in 1985 (September-De-cember). *Columns* show percentage of total number of otoliths (*above columns*) recov-ered from scats

Table 2. Numbers of fur seals in some colonies in southern Peru before and during EN (January-August 1983)

Colony	Lat. S	1979[a]	1983[b]	Source[c]
San Fernando	15°08′	4500	300	CH
San Juan	15°22′	9644	2000	PM
Pescadores	16°24′	0	4000	CH
Morro Sama	18°00′	0	400	CH

[a]From Majluf and Trillmich (1981).
[b]Direct counts taken in late April 1983.
[c]CH, Copelia Hays (pers. comm.); PM, P. Majluf (pers. obser.).

there was a significant negative correlation between mean SST and weight of pups in February ($r = -0.99$, $n = 5$, $p < 0.01$; Fig. 7).

Weaning Age. Young fur seals in Peru may be weaned between 9–36 months of age. In 1985, 54% of the yearlings resighted (n = 36) continued to suck into a second year. In September-October 1987, of the pups tagged in December 1985, 121 were resighted as 2-year-old and of these, 51 (42%) were observed still sucking regularly at the beginning of the breeding season (September-October). Only two 3-year-old immatures were observed sucking in 1987.

3. Mortality

There is no information on pup mortality during the breeding season (October-December) under EN conditions. However, between January and March 1983, during EN, pup mortality was extremely high. Of about 70 pups on beach S6, at least 29 (41%, not counting those washed out to sea) were seen dead on land between 22 January and 12 February (22 days) (Trillmich et al. 1986). Afterwards, of 14 pups tagged between 7–10 February on beach S3, only three (21%) had been resighted by 7 March. Most pups

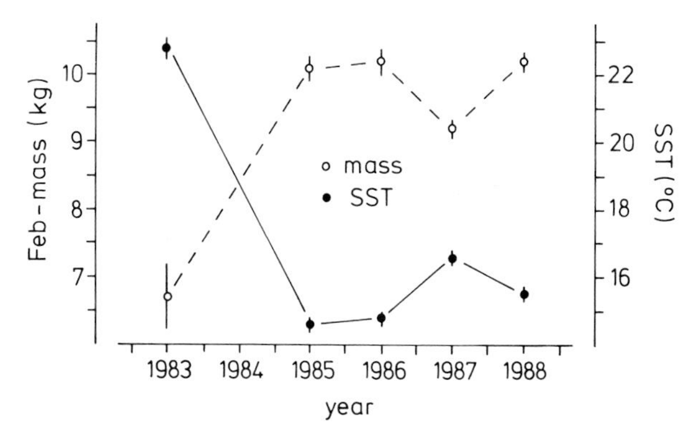

Fig. 7. Annual variation in mean pup weight (male and female pups combined) in February (90 days old) with mean SST in February. Years refer to the date of weighing, i.e. pups born in 1982 were weighed as 90-day-old in 1983

were in very poor condition and apparently most died of starvation after their mothers repeatedly spent long periods at sea foraging. Only one of the three pups surviving in March 1983 was resighted during the breeding season in 1984. It was the only pup that weighed over 9 kg when captured in February. Weights for the pups that died or disappeared ranged between 4.9–7.9 kg, well below the average for pups in February in all other years (11 kg). In these years pup mortality between January and February did not exceed 20% (Majluf 1987). There is no information available on mortality occurring later in the year during EN.

In subsequent years, within any one year there was a relationship between survival outside the breeding season (>30 days old) and early weights. Animals which were resighted as yearlings were significantly heavier when 1-month-old than those which were not resighted (Fig. 8; $F_{1,679} = 30.92$, $p = 0.0001$). Differences in pup

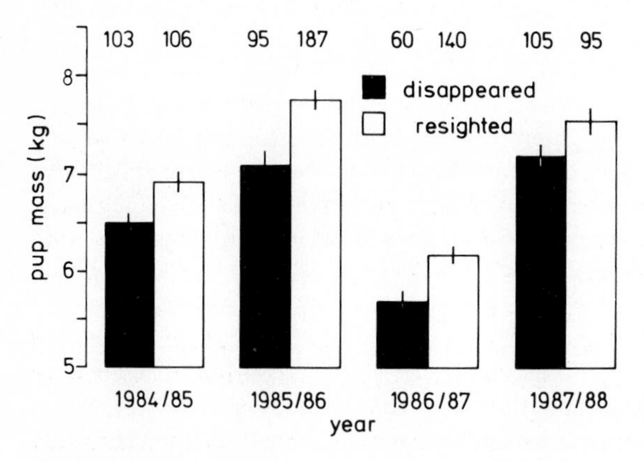

Fig. 8. Comparison between mean weights (± SE) at 30 days of age of pups resighted as yearlings and those not resighted and assumed dead in 1984/85, 1985/86, 1986/87 and 1987/88. Figures *above columns* show sample sizes

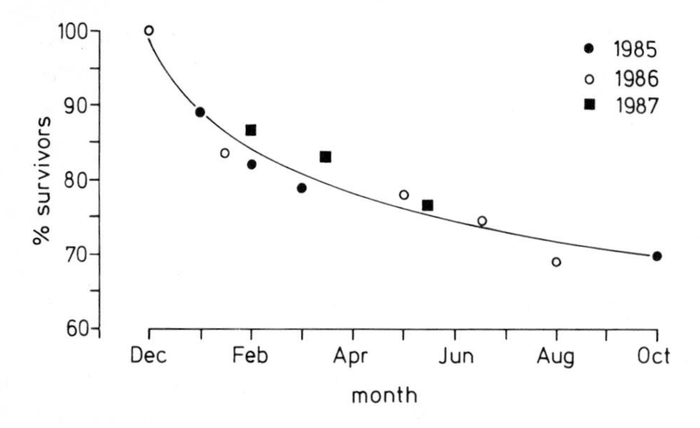

Fig. 9. Minimal survival to 1 year of age (calculated from tag resights) of pups born in 1984, 1985 and 1986 and tagged in December at a mean age of 30 days. Mortality between November-December was equal (ca. 40%) for all 3 years

weights between years, however, did not appear to have an effect on survival. Even under varying environmental conditions, pup survival rates in 1985, 1986 and 1987 (between 30–360 days old; Fig. 9) were high and did not vary. In all years about 70% of all tagged pups survived to 1 year of age.

Adult and juvenile mortality is difficult to assess because of juvenile dispersal at weaning and dispersal of adults outside the breeding season. During the 1982/83 EN juvenile and adult fur seal carcasses were regularly found washed ashore along the perimeter of the peninsula at PSJ, a rare occurrence in favourable years. No adult female weights under EN conditions are available. However, females observed throughout January and February in 1983 appeared emaciated and many were found dead ashore.

Adult territorial males may have been particularly affected by EN due to their fasting during the breeding season (October-December). Unfortunately, no information on adult sex ratios previous to EN are available to estimate the impact of EN on the territorial male population.

5.4 The Sea Lions

South American sea lions were studied at PSJ between January and August in 1983 and no information is available for this species in "normal" years.

In January 1983, mean foraging trip duration of female sea lions with marked/recognizable pups was 6.5 days (SE = 0.76 days, n = 6) and visits ashore ranged between 1–2 days. At this time of the year, sea lions also fed mainly on anchovy (21 out of 25 otoliths found in the scats were anchovy) but from March onwards switched to feeding on jack mackerel and sardines. No female weights were obtained but many appeared emaciated.

Based on counts of dead pups observed on beach S6, a minimum pup mortality of 46% (n = 24) over a 22-day period was calculated. This mortality appeared to be due mainly to starvation caused by the mothers' absences. Pups weighed on beach N5 when they were between 1–3 weeks old had hardly grown beyond birth weight (males: 8.25–10.15 kg, n = 4; females: 7.8–9.65 kg, n = 3). A male newborn weighed 8.75 kg.

No quantitative information on weaning age is available but yearlings were frequently observed sucking. As in the fur seals, many yearling and adult sea lion carcasses were found washed ashore during EN; a rare event under normal conditions.

5.5 Discussion

The changes in behaviour and condition of the seals at PSJ during the 1982/83 EN appear to be a reaction to low food availability at sea. Fur seal mothers had to dive more frequently to greater depths and spent longer times at sea foraging. Females under these abnormal conditions were still not able to obtain enough food since they and their young were in poor condition and suffered higher mortality. The sea lions were similarly affected.

Under normal conditions during the summer, females nursing young pups made very short and regular trips. At the same time of the year, under EN conditions, females made long trips more often and trip duration varied widely. This suggests a dispersed but patchy distribution of prey. Some females may have been feeding close to shore, perhaps on the anchovy patches trapped by the warm waters offshore. Thus, they could probably return to shore sooner. Females that did not encounter one of these patches were possibly making longer foraging trips. These females either had to travel longer to get to a patch, or because prey density was so low, spent more time foraging, trying to adequately fulfill their own and the energy demands of their young.

The different dive pattern observed during EN appears to be related to an altered prey distribution. The female did not dive during the day, but made deep dives throughout the night, and a much higher proportion of dives exceeded the depth of 40 m. During EN, anchovy populations were not only restricted to great depths, but were found in lower densities than normal because part of the population either died or migrated south (see Arntz et al., this Vol.). If the sample record does represent the dive pattern of females foraging under EN conditions, it indicates that females had to make a larger number of deep dives and therefore spent more energy in order to obtain their prey. The California sea lion is the only other otariid for which dive records were obtained during both EN and normal conditions. Animals during EN made longer dive bouts and fewer shallow dives than animals in normal years (Feldkamp 1985; Feldkamp et al., this Vol.).

Increased foraging effort was not compensated for by increased foraging success. When the recorder was recovered, the female was noticeably emaciated (Trillmich et al. 1986). Other females having to forage under these conditions were making longer trips to sea and not only were their pups dying of starvation, but also the females were severely emaciated. This apparently low foraging success suggests that the different dive pattern of this female is due to the changes in vertical distribution of prey rather than to individual differences in dive patterns. All three individuals recorded in 1984 and 1985, under normal conditions at sea, were similar in their diving behaviour and were in good physical condition when recaptured. The northern fur seal is the species showing the greatest individual variation in dive patterns. For this species, three different dive patterns have been described (deep, mixed and shallow divers) but all are of roughly equal efficiency (Gentry et al. 1986a). However, decreased foraging success resulting in the observed poor physical condition of females at PSJ during EN could also be due to a decreased calorific value of their prey. Fat content of anchovy varies as a function of SSTs which is an index of primary production (see Arntz et al., this Vol.). During EN, when sea temperatures are higher, the condition of anchovy is reduced (Pauly and Tsukayama 1987) and thus, even if fur seals were able to catch the same number of fish, they obtained less energy than during normal years. More information is needed on the individual variation of dive patterns in this species before the effects of EN on these patterns can be assessed.

Seals at PSJ may be in an advantageous position, relative to those in other colonies in Peru, because of the longer persistence of cold waters and hence, concentration of pelagic fish in the area during ENs (Villanueva et al. 1969; Vilchez et al. 1988). In March, early on during the 1983 EN, a survey of the Peruvian coast carried out by the Peruvian Marine Research Institute (IMARPE) showed that the *only* anchovy patches were found close to shore near PSJ (IMARPE, unpubl. data). This presence was apparent in the scats of both the fur seals and sea lions. Both species, however, switched to sardine and jack mackerel later in the year. By then, the few anchovy schools that had been present in January/February were either dead or had migrated into deeper water or further south (Arntz et al., this Vol.). However, during the winter, even in "normal" years, anchovy become less available as the warm water front moves away from the coast. With the information available, it is not possible to separate the EN effects from the normal winter dispersion.

Pup mortality in this population was high only in 1982/83 under extreme EN conditions. Pups died of starvation as a result of the long absences of their mothers foraging at sea (Majluf 1985; Trillmich et al. 1986). In subsequent years, even under varying environmental conditions, survival up to 1 year of age did not vary significantly. This might be due to the fact that in milder ENs like the one in 1986/87, the persistence of cold waters and pelagic fish around PSJ may allow female fur seals to forage close to shore and return to their pups often enough to avoid starvation and death in the first year of life.

Extreme variability between years is a characteristic of the Peruvian upwelling system. In Peru, seals have to endure unpredictable fluctuations in their food supply from year to year ranging from having available one of the largest pelagic fish biomasses in the world (Idyll 1973) to the severe conditions of the 1982/83 EN. Possibly as an adaptation to this extreme variability, fur seals themselves may have evolved highly flexible patterns of behaviour. Gentry et al. (1986a) have discussed in detail the relationship between EN and variations in attendance and diving behaviour in fur seals. Of particular interest in this population is the wide variation in duration of lactation. The weaning process is still poorly understood in this population; variations in food supply at sea, early growth rates of the young and the mothers' reproductive status during the breeding season have been suggested as possible factors affecting weaning age (Majluf 1987). A fixed duration of lactation would probably result in decreased reproductive success; females weaning their young early in bad years are likely to lose two young (the yearling and the new pup), instead of just one (the pup). Females are likely to be affected by at least one, and up to three ENs during their reproductive lives. Since they breed only once a year, losing two to six young would mean losing a high proportion of their lifetime reproductive output. Fur seals in Peru may be maximizing their lifetime reproductive success by being able to vary weaning age according to conditions at sea.

Acknowledgments. Many thanks to Rob Harcourt and Fritz Trillmich for helpful comments on the manuscript. The diving studies were done in collaboration with Mike Goebel, Gerry Kooyman and Fritz Trillmich. Pup weights and tag resight data in 1986 and early 1987 were collected by Pedro Vasquez. Assistance in the field was provided by Pedro Llerena, Daniel Gateño, Johanne Ouellette, Jorge Pejoves, Frances Weick, Maria Ines Kuroiwa, Cesar Carcamo and Antonio Tovar. Logistic support was kindly provided by Pesca-Peru, Hierro-Peru and the Peruvian Navy. I am most grateful to B. Knauer for typing the final version of the manuscript and to D. Schmidl for drawing the figures. This study was funded by Wildlife Conservation International (a division of the N.Y. Zoological Society) while I was a Ph.D. student at the Dept. of Zoology, Large Animal Research Group, University of Cambridge, supervised by Dr. T.H. Clutton-Brock.

6 The Effects of El Niño on Galapagos Pinnipeds

F. TRILLMICH and T. DELLINGER

6.1 Introduction

The effects of the 1982–83 El Niño (EN) were most severe in the eastern tropical Pacific (see Fahrbach et al. and Arntz et al., this Vol.) and the seal populations in the area of Galapagos and Peru suffered most from the impact of this event. The ocean around the Galapagos warmed rapidly in September 1982 and remained warm until July 1983 with a short, somewhat cooler period in February/March 1983.

The climate of the Galapagos islands is seasonal despite the islands' position at the equator. During the cold season, from about May to December, waters around the archipelago are cool and upwelling along the western islands is strong, whilst during the warm season, from about January to April, water temperature increases and upwelling is diminished. This seasonal cycle comes about by the movement of the intertropical convergence zone which lies N of the islands in the cool season, but moves S during the warm season bringing the northeasternmost islands under the influence of warmer waters. A coincident drop in the strength of the southeastern trade winds weakens the Humboldt current which leads to reduced influx of cold water and this in turn slows the flow of the Cromwell countercurrent which upwells on the coasts of the western islands (Fahrbach et al., this Vol.).

Here, we describe the influence of EN on the Galapagos populations of fur seals (*Arctocephalus galapagoensis*) and sea lions (Zalophus californianus wollebaeki). Fur seals and sea lions reproduce predominantly during the cold season. The 1982–83 EN most severely affected these species during the 1982 breeding season. We summarize previously published material (Limberger et al. 1983; Limberger 1985; Trillmich 1985; Trillmich and Limberger 1985) and then address the long-term effects of EN on the two populations. Place names are shown in Fig. 1.

6.1.1 Background Information on the Natural History of the Galapagos Species

The Galapagos fur seal is nonmigratory and stays close to its breeding colonies year-round. Adult females of this species weigh about 30 kg and adult, territorial males between 60 and 70 kg (Trillmich 1987). This species thus shows the least sexual size dimorphism of any fur seal. Reproduction takes place between August and November, the local cold season. Females alternate between 1–2 day periods

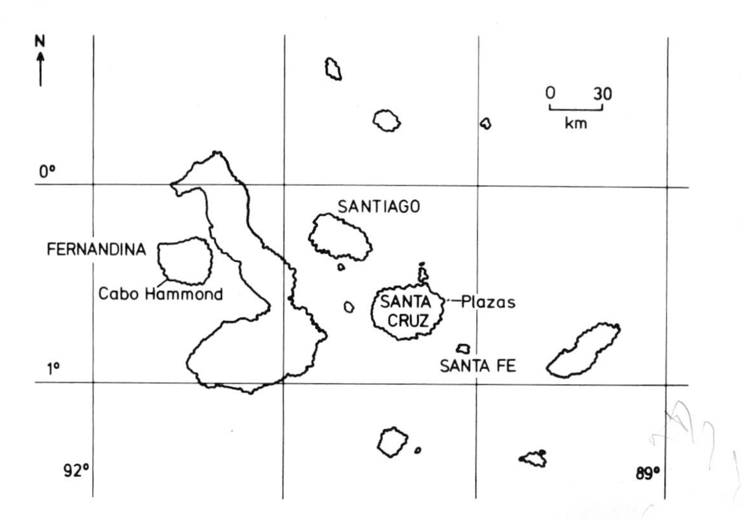

Fig. 1. Map of the Galapagos sites mentioned in the text

ashore, nursing the pup, and 1–4 day periods away, foraging at sea and resting elsewhere. Galapagos fur seals feed during the night on organisms which ascend to the ocean's surface during the dark hours from the deep scattering layer (Dellinger 1987). In most years young are weaned when about 2 years old (Trillmich 1986a).

Galapagos sea lion females weigh about 80 kg. No adult male weights are available. Mothers usually forage during the daytime (Trillmich 1986c) and are absent on foraging trips for about 1 day. Sardines are their staple food (Dellinger 1987). Galapagos sea lions, just like the fur seals, reproduce mainly during the cold season with the peak of births occurring between October and November. However, the pupping season of the sea lion lasts much longer, in the extreme case of the Plaza Islands (in the center of the archipelago, Fig. 1) from about June of one year to March of the next. On Santiago, in 1977, pups were born from June to September; on Santa Fé and Fernandina sea lions pup from about October to January. Young are suckled for about 1 year, but if a mother does not pup after 1 year, suckling may be continued for up to 3 years.

6.2 Methods

Galapagos fur seals were studied from 1976 to 1985, and in March/April 1986 at Cabo Hammond (0°28′S, 91°37′W) on Fernandina Island, the westernmost island of the Galapagos. Observations were made every year, except for 1978, during the reproductive season from approximately September-November and during the warm seasons of 1978 (February) and 1986 (March and April). The fur seal study site extended over a section of coastline of about 800 m of which 180 m were censused regularly (for further methodological details, see Trillmich and Limberger 1985).

During these field seasons much less systematic observations were gathered on a nearby sea lion colony which comprised about 150 animals living on a flat rocky shelf with a number of tide pools.

The diet of these species was determined from scats collected during the 1983–85 cold seasons (Dellinger 1987).

6.3 Results

6.3.1 The Galapagos Fur Seal

Pup production in the study area at Cabo Hammond fluctuated at around 200 pups per year from 1979–1982 (Fig. 2). For the years 1981 and 1982 no exact pup counts are available, but the observers' impression was that these were years of normal pup production. In contrast, pup production in September-October 1983, right after the end of EN, was only 11% of normal (Fig. 2). In 1984 a large number of pups were born, but in 1985 pup production was about half of the pre-EN level (Fig. 2). This effect was caused by the synchronizing effect of EN on the reproductive status of fur seal females: in normal years about half of all females suckle yearlings or 2-year-olds and, as a consequence of the energetic stress of lactation, the majority of them do not produce a pup (Trillmich 1986b). Thus, before EN, only about half of the reproductive females produced a pup in any given year. During EN, in 1982, all females lost their pups, due to starvation, and very few produced one in 1983 so that nearly all of them lived without the energetic drain of lactation for all of 1983 and most of 1984, up to the reproductive season in September-October. Nearly 100% of the adult females pupped in 1984, but then many of them did not pup again in 1985 due to the lactation effort. Therefore, 1985 pup production is more comparable to production before EN than production in 1984. Thus, the 1985 pup number can be

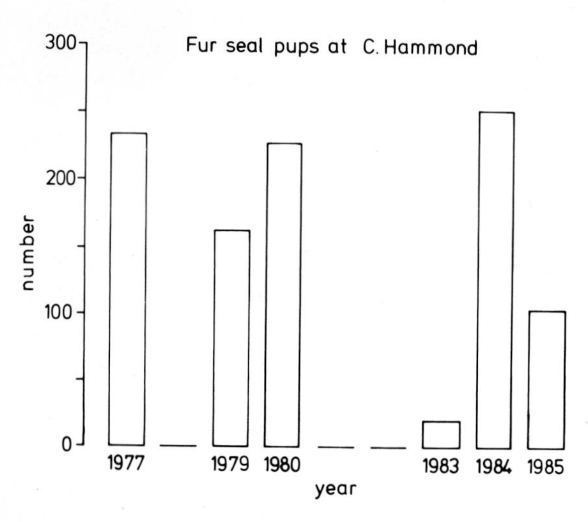

Fig. 2. Pup numbers of the Galapagos fur seal at the Cabo Hammond study site on Fernandina. No data available for the years 1978, 1981, and 1982

used as another index of the decrease in the number of reproductive females due to EN. This would indicate a ca. 50% decrease of adult females (Fig. 2).

The mean mass of newborn pups produced in 1982 and 1983 was about 10% lower than in normal years (Trillmich and Limberger 1985). Most of the pups born in 1982 lost weight rapidly before starving to death. The pups of the 1983 cohort, born about 2 months after the end of EN, grew normally.

Thirty-three percent of the pups born in 1982 died during their first month of life and 100% were dead after 5 months. In contrast, 90% of the pups born shortly after the end of EN in 1983 survived the first month as observed in other "normal" years. During the period of September 1982 to March 1983, both 1- and 2-year-olds, from the 1980 and 1981 year classes, of which many were tagged, suffered 100% mortality and about 70% of the 1979 year class (3-year-olds) also perished (Trillmich and Limberger 1985). Figure 3 (upper panel) roughly illustrates the fate of the 1979–1982 cohorts.

The reason for starvation of pups and dependent juveniles during EN in 1982–83 lies in the unusually long (foraging) absences of mothers. Absence duration of females during the cold (breeding) season from August-November varied little between years; females rarely stayed away for more than 4 days (Fig. 4a,b). During EN 1982, at a time of year which is normally the Galapagos cold season, mean absence duration lengthened significantly (Mann-Whitney U-test; $p < 0.01$) and variance in absence duration increased drastically (Fig. 4c). Interestingly, mean absence duration during the 1986 warm (nonbreeding) season was even slightly longer than during EN, but the variance of the warm season data was much smaller (Fig. 4d; F-test; $F_{54,27} = 8.3$, $p < 0.001$).

Between 1979 and 1981 the maximal number of adult females censused ashore in the study site averaged 132. In 1983 this number declined to about 70% of its former value (Trillmich and Limberger 1985). By 1984 numbers returned to pre-EN

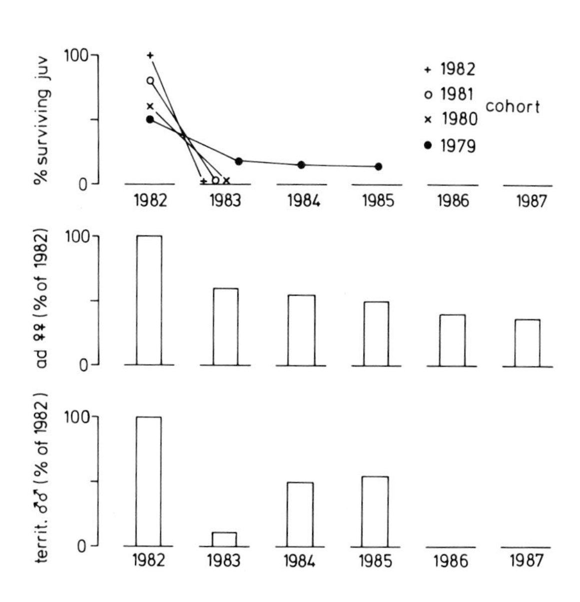

Fig. 3. A graphic summary of EN effects. *Upper panel*: the percentage of juveniles surviving from the cohorts 1979–1982. No data for 1986 and 1987. Estimated survival of a given cohort to 1982 is taken into account; therefore, the various cohorts begin at different percentages of juveniles surviving at the beginning of EN. *Middle panel*: effects on the number of mature females relative to the numbers in 1982. *Lower panel*: effects on the number of territorial males in the study area relative to those in 1982. No data for 1986 and 1987

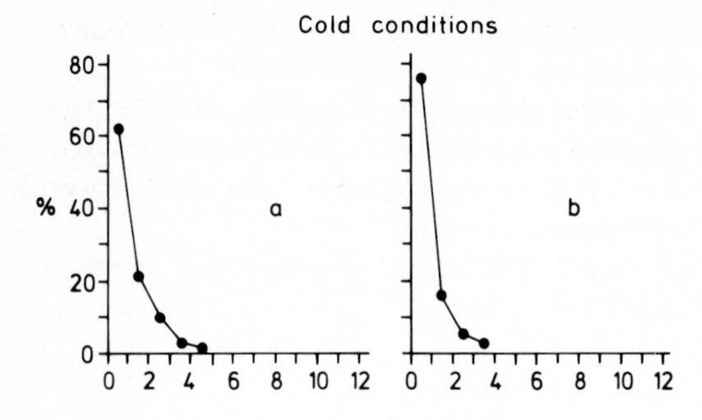

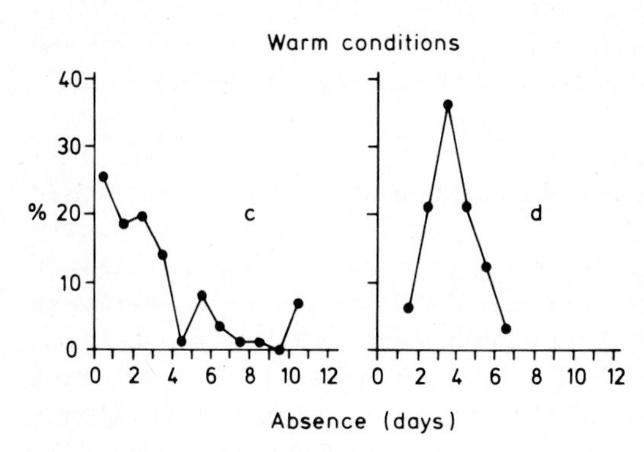

Fig. 4a-d. Comparison of the distributions of the duration of maternal absence under cold and warm conditions. **a** (*Upper left*) values from the months Sept.-Nov. 1979, 1980, 1981, 1984, 1985 (n = 345 trips). **b** (*Upper right*) values from Sept.-Nov. 1983, immediately after the end of EN effects (n = 38). **c** (*Lower left*) values during EN-related warm conditions in Sept.-Dec. 1982 (n = 86). **d** (*Lower right*) values from February 1978 and March-April 1986 (n = 33), normal warm sea sons

levels in the study area: the maximal number counted was 155 females; but large surrounding areas of fur seal habitat, which had been used by fur seals in former years, remained empty. Apparently, the reduced population had concentrated into the most suitable habitat. If we estimate the reduction in female numbers from the 1985 pup production (see above), 50% of the adult female population may have been lost during EN. Figure 3 (middle panel) shows these estimates and indicates that due to the loss of the 1980–1983 cohorts, the number of adult females must have kept declining at least until 1987 when the first females of the 1984 cohort may have matured.

In contrast to adult females, territorial males suffered close to 100% mortality during EN. In 1980 and 1981 we recorded 31 to 32 males, respectively, holding territories of about 200 m^2 each in our study area (Trillmich 1984). Males fast for up to 6 weeks while holding territories during the reproductive season between August and November. In 1982 about the same number of territorial males appeared to be present. At the end of their fast these males returned to a warm ocean which was apparently providing very sparse food supplies, as also evidenced by the unusually long foraging absences of mothers with pups (see above; Fig. 4). None of these males were observed again in the 1983 breeding season, immediately after the end of EN; they all must have starved to death. Instead, only five smaller males, of about 80% of normal breeding-male size (which in other years would have been categorized as subadult males) established dominance status over huge areas of about 800 m^2, and males of female size, i.e., less than half the mass of normal breeding males, were able to sneak many copulations during this season (Trillmich and Limberger 1985). By 1984, 15 bigger males were in the study area and the territorial system, except for unusually large territory sizes, was established again. This trend continued into 1984 (see Fig. 3, lowest panel). Two of the males observed in 1984 had been tagged as subadults before EN and had obviously grown at an increased rate between the 1983 and the 1984 reproductive season.

6.3.2 The Galapagos Sea Lion

The most complete data set on pup production and mortality during EN stems from Santa Fé (A. Laurie, pers. comm.). About 145 pups were born there in 1981 and 1982 (Table 1). Although the sea surface temperature began rising in September 1982, this had no noticeable impact on pup production in 1982. However, pup production in 1983 was about 30% lower than in previous years (Table 1). The breeding season also began later, as we only counted 26 pups on November 2, 1983 (Trillmich and Limberger 1985), whereas in previous years about 50% of the pups were already born by that date. The number of pups born in 1984 was the same as before EN, but in 1985 it declined to roughly 74% of the pre-EN value (Table 1). Presumably, a synchronization effect, as described above for the fur seals, influenced sea lion pupping on Santa Fé as well.

Pup production in 1983, after the end of EN, varied widely between colonies on different islands with a highest value of 65% of pre-EN values on Santa Fé, an island where sea lions produce their pups late in the year, and a low of only 3% of the pre-EN value of Santiago. Only six pups were born in the Fernandina colony where in former years more than 100 had been born. Similar to the fur seals, sea lions on Fernandina produced unusually many pups (166) in 1984, but only 97 were born in 1985.

The survival of pups born in 1982 was much lower than usual. Normal pup mortality during the first 6 months of life is about 5% (Trillmich and Limberger 1985). On Santa Fé, mothers abandoned their pups in March-May 1983, when oceanographic EN effects reached a second peak (Fahrbach et al., this Vol.). Only 19 (14%) of the 140 pups counted in December 1982 were still alive by April 1983. During the 1982–83 EN, pups on Fernandina began to show signs of starvation in February 1983, shortly after the first peak of EN and only about 4% of the pups born in 1982 survived to the age of 1 year.

Table 1. Sea lion pup counts on Santa Fé (data courtesy of A. Laurie)

Date	Pup number
Dec. 29, 1981	149
Dec. 26, 1982	140
Dec. 29, 1983	94
Dec. 29, 1984	140
Dec. 31, 1985	108

High yearling mortality apparently occurred throughout the Galapagos during EN, in late 1982 and early 1983. Counts in October 1983 on various islands showed that the 1982 yearling cohort was reduced to between 5 and 20% of the size of previous cohorts. The 20% value most likely is an overestimate of actual survival rate, since sea lions moved, just like fur seals, from less preferred into better habitats after EN.

Adult numbers clearly decreased during EN, but no censuses are available. Assuming that the number of pups born in 1985 of Santa Fé (Table 1) can serve as an index to the number of adult females left after EN, it seems likely that about 20% of the adult sea lion females died. Robinson (1985) had the impression that the so-called sea lion pox disease (the infectious agent responsible for the disease is unknown) became much more prevalent during 1982–83 and killed many animals. Inhabitants of the Galapagos also saw many dead adult animals, and their reports suggest that the mortality of territorial males was particularly severe.

6.4 Discussion

During EN decreased upwelling, increased sea surface temperatures and a depressed thermocline all contributed to a rapid depletion of nutrients for phytoplankton growth (Kogelschatz et al. 1985). This in turn led to reductions in zooplankton. Finally, fish starved, died, and emigrated or migrated to greater depths. There are no data on pelagic fish for the Galapagos area. However, Feldman et al. (1984) have shown that the phytoplankton concentration around the western islands was still high (>1 mg m^{-3}) in February 1983, but had decreased dramatically by March 1983. This is exactly the time when the first signs of starvation became noticeable in Galapagos sea lion pups on Fernandina (Limberger 1985), suggesting that sardine availability decreased around this time. Barber and Chávez (1986) reported a decrease of sardine catches to almost zero in mainland Ecuadoran waters. Local Galapagos fishermen claimed that benthic fish disappeared from their usual depth but returned immediately after EN, indicating that many fish indeed migrated to greater depths during the warm water phase.

Fish living in the deep scattering layer are the main food resource of the fur seals which hunt for this prey when vertical migration brings it near the surface at night. In normal cold seasons Galapagos fur seals feed mostly on myctophid and, to a lesser extent, on bathylagid fish (Table 2; Dellinger 1987). The starvation of pups and adults alike suggests that these fish either died or moved to a depth where they could not be reached easily or in sufficient quantity. After EN, in late 1983, fur seals fed on unusual prey such as sardine (*Sardinops sagax*; 6% of otoliths) and *Selene declivifrons* (5%), suggesting that myctophids – although representing the majority of the prey with 81% (Table 2) – were still not as abundant or energy-dense as in other years. Apparently, the recruitment of myctophids failed during 1983 leading to reduced availability of this prey type in 1984. Bathylagids then became a more dominant part of the fur seals' prey (Table 2). But, in both 1984 and 1985, fur seals were again feeding almost exclusively on organisms from the deep scattering layer.

Table 2. Numerical composition of Galapagos fur seal and Galapagos sea lion food during the cold seasons 1983–85. Data are given as percent of total numbers of otoliths found in scats (lowest line). Only the most frequent prey items are listed for a minimum of 90% (numerically) of the diet

	Fur seal			Sea lion		
	1983	1984	1985	1983	1984	1985
Myctophidae	81.0	41.7	77.4	0.2	1.0	0.6
Bathylagidae	4.7	56.8	21.6	–	0.6	–
Sardinops sagax	6.2	0.4	0.2	70.7	74.8	84.8
Chlorophthalmus sp.	–	–	0.1	23.1	21.1	5.0
No identified "type 34"	–		–	–	–	4.5
Total otoliths (n)	4545	3136	4766	416	659	1450

The diet of Galapagos sea lions is always dominated by sardines which contribute over 70% by number of otoliths to the diet (Dellinger 1987). However, *Chlorophthalmus* sp. (presumably *agassizii*; T. Hecht, pers. comm.) contributed 23% and 21% of the diet in 1983 and 1984, but only 5% in 1985 when the proportion of sardine otoliths reached 85% (Table 2). Apparently, as a consequence of EN, there was some shift in sea lion food resources as well. In October 1983, shortly after the end of EN, sea lions regurgitated more frequently than observed in previous or later years, actually vomiting whole stomach-fulls of partly digested sardines. Only at that time did we find about as many spewings as scats on land, an observation which may indicate illness or overeating of the previously starved sea lions.

In both species reduced food availability during EN made it far more difficult for mothers to gather sufficient food for their own and their pup's metabolic needs. Fur seals were probably affected more severely than sea lions since the former appear to forage farther away from the coast and can spend less time under water. Judging from the time they need from leaving the coast until foraging dive bouts begin, sea lions forage roughly 4–5 km from the colony (at least near Fernandina), while fur seal mothers travel about 16 km before they begin to forage (Kooyman and Trillmich 1986a,b). During EN, the cost of foraging became so high that fur seal mothers reduced the numbers of round trips and stayed at sea for very prolonged periods. The mothers' absence times exceeded their pups' fasting abilities. Females still returned and nursed their pups, but it seems like that milk transfer was minimal. In the otariid rearing strategy the effect of food shortage thus becomes immediately noticeable during pup rearing since mothers cannot rely on fat reserves gathered prior to the pup rearing period as phocids do (see Trillmich et al., this Vol.).

Larger pinnipeds can stay under water longer than small ones (Gentry et al. 1986a). If fur seal and sea lion prey stayed at greater depth and/or decreased in abundance during El Niño, then yearlings of both species would have had more difficulty catching prey than conspecific adults. This could explain the much higher mortality rates of juveniles, and perhaps, the more dramatic mortality observed in the smaller species, the fur seal.

Extrapolating from our study site, the 1982–83 EN has caused a loss of three cohorts in fur seals and a reduction in the size of two cohorts in sea lions. This extrapolation is probably an underestimate of the population reduction, as the ocean area around Fernandina is the most productive within the Galapagos archipelago. Loss of whole cohorts through exceptionally strong ENs is a rare and random event. If such an event does not reduce the pinniped populations to a very small size (on the order of 100 or less productive females), sampling drift is unlikely to significantly influence the evolution of the species. However, a reduction in the size of cohorts could po- tentially be a very strong selective agent. For example, slow growing young may survive better during food stress due to lower energy and material requirements.

The differential mortality of most large, territorial males in 1982, as observed for fur seals and suspected for sea lions, may influence selection for sexual size dimorphism. In otariid pinnipeds sexual selection through male-male conflict favors males large in size, because only large males can successfully fight for and maintain territories during the reproductive season. Repeated natural (survival) selection against exactly these largest males during periods of EN-induced food shortages will oppose sexual selection for ever-increasing male size. This could explain why the Galapagos fur seal, and most likely the Galapagos sea lion as well, have much less sexual dimorphism in body size than their close relatives in cooler environments.

Lastly, EN events occur unpredictably over the lifetime of a female. Aperiodic unpredictable reductions in food availability could select females to maintain larger body reserves at the cost of a reduction in pup growth and survival since the survival of the mother is far more important to her lifetime reproductive success than the survival and rapid growth of a given pup. This could explain the evolution of an extreme phenotypic flexibility of pup growth rate and time to weaning as found in the Galapagos fur seal.

Acknowledgments. We would like to thank A. Laurie for providing data on the Santa Fé sea lions. The permit for this study was granted by the Galapagos National Park authorities and we are most grateful for the support of the Galapagos National Park through its intendants M. Cifuentes and H. Ochoa as well as by the Charles Darwin Research Station through its directors C. McFarland, H. Hoeck, F. Köster, and G. Reck.

B.J. Le Boeuf, J. Croxall, and K. Ono provided much appreciated constructive criticism of the manuscript. Tom Hecht was immensely helpful by tracing an unknown otolith to the genus *Chlorophthalmus*. We are also most grateful to W. Wickler who continuously supported this study and to B. Knauer and D. Schmidl for preparing the graphs. This is contribution No. 434 of the Charles Darwin Foundation.

7 Impact of the 1982–1983 El Niño on the Northern Fur Seal Population at San Miguel Island, California

R.L. DELONG and G.A. ANTONELIS

7.1 Introduction

Northern fur seals reestablished a breeding colony at San Miguel Island (at 34°02′N, 120°26′W) during the late 1950s or early 1960s after having been absent from the island for over 100 years (DeLong 1982). Fur seal remains in Indian middens indicate that both northern fur seals and Guadalupe fur seals (*Arctocephalus townsendi*) were abundant on the island prehistorically (Repenning et al. 1971; Walker and Craig 1979). Unregulated commercial sealing caused the destruction of both species on the California islands by 1835.

Between 1969 and 1978, the northern fur seal colony at Adams Cove on San Miguel grew rapidly, averaging a 46% annual increase in the number of pups born. Between 1972 and 1978, annual pup births on Castle Rock increased by 34% (DeLong 1982). These high rates of increase exceed the species' innate capacity for population growth and were made possible by immigration of adult females from the breeding populations in the Pribilof Islands, Robben Island, and the Commander Islands (DeLong 1982; Antonelis and DeLong 1985; Gearin et al. 1986).

The presence of warm sea-surface water which marked the onset of the El Niño condition in the Southern California Bight was first noted in October 1982 (McGowan 1985). By mid-December the cold water plume below Point Conception and around San Miguel Island, which originates from the California Current and coastal upwelling, was 3 °C warmer than the plume of the previous year (Fiedler 1984a). By late December 1982, large schools of pelagic red crabs (*Pleuroncodes planipes*) first appeared in the waters around San Nicolas Island, indicating that northern displacement of plankton and nekton had begun (Stewart et al.1984).

Northern fur seal pups born in June and July 1982 were weaned between October and late November. These pups and pregnant fur seal females began the pelagic phase of their annual cycle at the onset of physical and biological changes from the El Niño in the California Current ecosystem. We present data on long-term fur seal population growth on San Miguel Island and observations on changes in northern fur seal pup production, pup growth, feeding cycles and food habits of adult females, cohort survival, and possible adult mortality, as well as data on the long-term impacts of the 1982–83 El Niño event.

7.2 Methods

We conducted studies of the northern fur seal population at San Miguel Island during each breeding season from 1969 through 1987 from at least mid-June through mid-August. Daily censuses were conducted to record the number of births and their temporal distribution and population composition data. Information on the weight and sex of the pups was collected when they were about 3 months of age, in late September. Female feeding cycles were quantified by observing naturally marked parturient females and recording their presence and absence on the rookery on a daily basis. Data for female time at sea for 1985 were obtained from Stewart (1985). Scats were gathered during July and August of 1982 and 1983 from rookery areas occupied only by northern fur seals. Since adult males do not feed during territorial tenure and juvenile males are generally excluded from reproductive territories, most scats originate from females which are actively feeding and returning to land to nurse pups. Identifiable hard parts of prey were recovered from scats by washing materials through nested sieves of various sizes (Antonelis et al. 1984). Prey species occurrence was recorded for each scat and the frequency of occurrence data represents the percent of total scats containing individual prey species. Each September since 1976, a sample of about 20% of the live pups has been tagged. Data on cohort survival were obtained by resighting tags on live seals between 1976 and 1988. Data used to construct a survival index were restricted to resightings of live animals for 5 years after their birth. Recovery of tags on animals found dead on beaches provides most of the information on animal distribution away from the rookery island. The rate of increase of pup births was calculated by taking the slope of a regression line fitted to the natural log of numbers of pups born each year. The slope provides an instantaneous rate of increase (r) which is converted to a finite annual rate of increase by raising e (the base of the natural logarithms) to the r power (Krebs 1972).

7.3 Results

After 14 years of increasing production, northern fur seal pup births declined at San Miguel Island in 1983 (Fig. 1). In Adams Cove, the number of pups born decreased 60%, from 1029 to 408 between 1982 and 1983 (Table 1). On Castle Rock, pup births fell 64%, from 680 in 1982 to 245 in 1983. The number of pups born increased at annual rates of 15 and 21% following the decline in 1983, but had not recovered to the 1982 level by 1987 (Fig. 1). In 1983, the first pup birth occurred on 10 June and the mean birth date was 1 July. Both dates are later than those recorded for any other year. In most years, the first birth occurred during the last week of May, and 25 June was the estimated mean birth date. More rigorous analyses of possible differences in the mean pupping dates are precluded, however, by inconsistent methods of data collection between years.

The mean weights of pups at about 3 months of age were significantly less in 1983 and 1984 than in years before or after the El Niño (one-way ANOVA, $F = 45.91$; $P < 0.01$; Fig. 2). In 1983, 3-month-old pups were only about 2 kg heavier than pups at birth observed at Adams Cove in 1981 (DeLong et al. 1982).

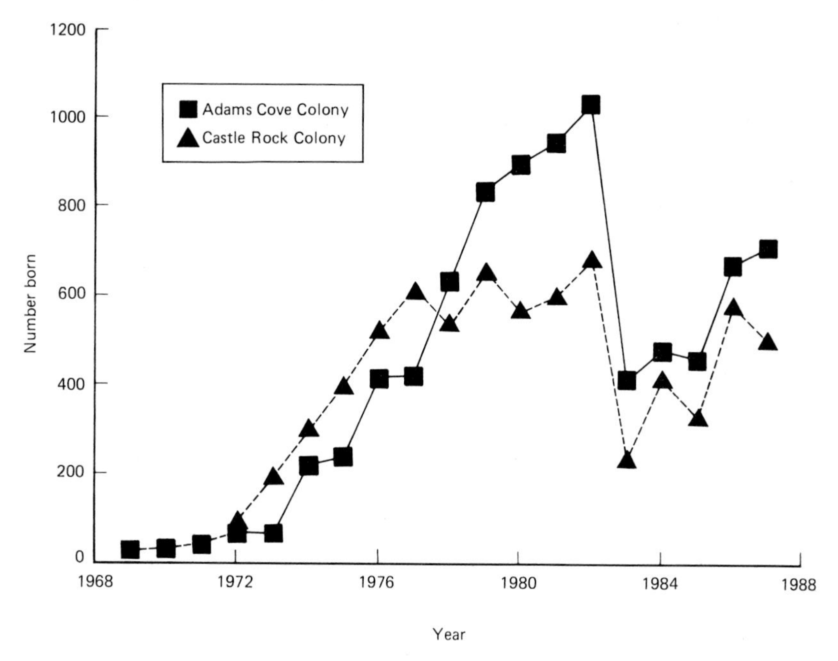

Fig. 1. Number of northern fur seal pups born at San Miguel Island, California, 1969–1987

Table 1. Adams Cove, San Miguel Island, northern fur seal population composition

Year	Pups	Maximum counts: Adult females	Subadult males	Adult males
1981	941	717	95	21
1982	1029	628	88	52
1983	408	377	37	61
1984	478	333	49	44
1985	458	315	54	41
1986	670	422	9	66
1987	709	399	32[a]	74

[a]Partial survey.

In 1983, the duration of foraging trips for parturient adult females increased significantly (one-way ANOVA, F = 3.59; $P = 0.02$) over the length of trips for 1982 and 1985, but it was not different from 1984 (Fisher's least significant difference test, $P > 0.05$; Fig. 3).

From 1982 to 1983 prey consumed by lactating females during the summer did not differ in composition, although changes in the frequency of occurrence of most prey were recorded (Fig. 4). Northern anchovy (*Engraulis mordax*) showed the

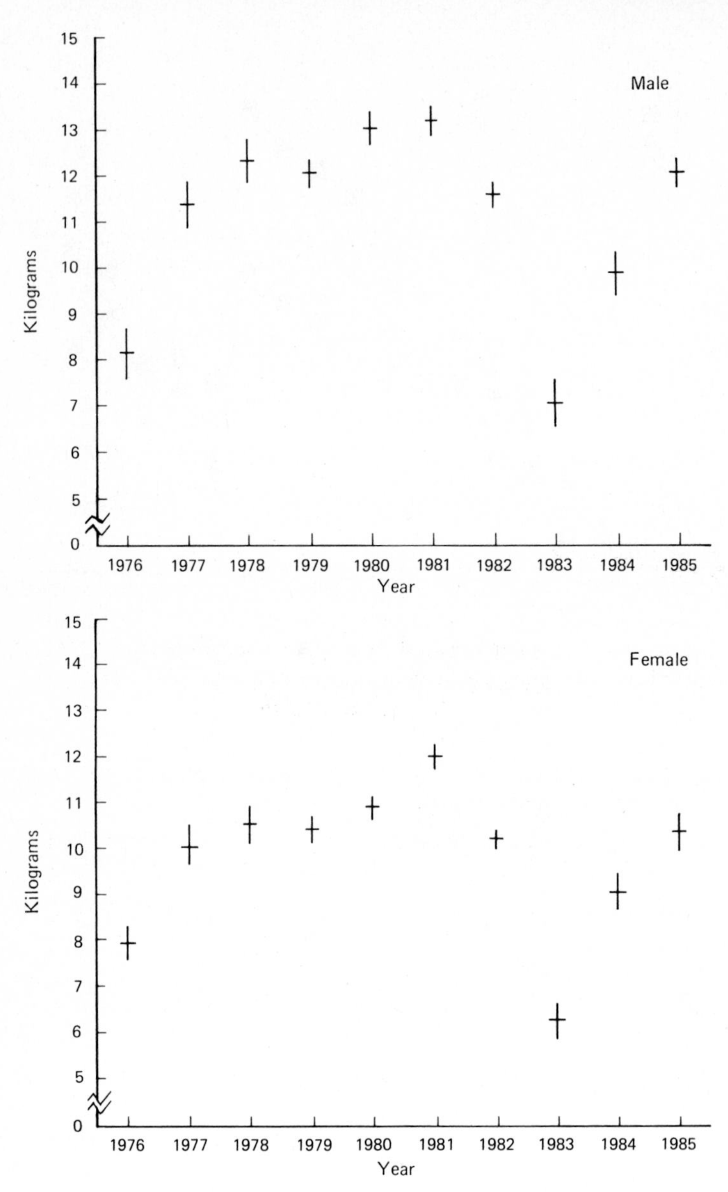

Fig. 2. Weights (means) of approximately 3-month-old northern fur seal pups in Adams Cove, San Miguel Island, 1976–1985 (*vertical bars* are 95% confidence intervals, i.e., ± approx. 2 SD errors)

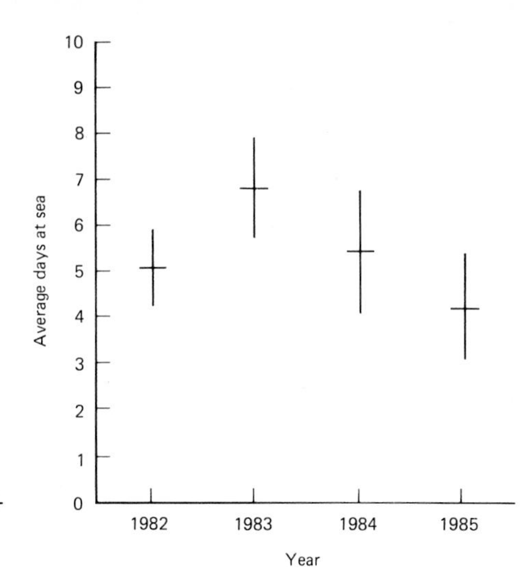

Fig. 3. Mean length of first two at-sea feeding cycles for 1982–1985 (*vertical bars* are 95% C.I., i.e., ± approx. 2 SD errors)

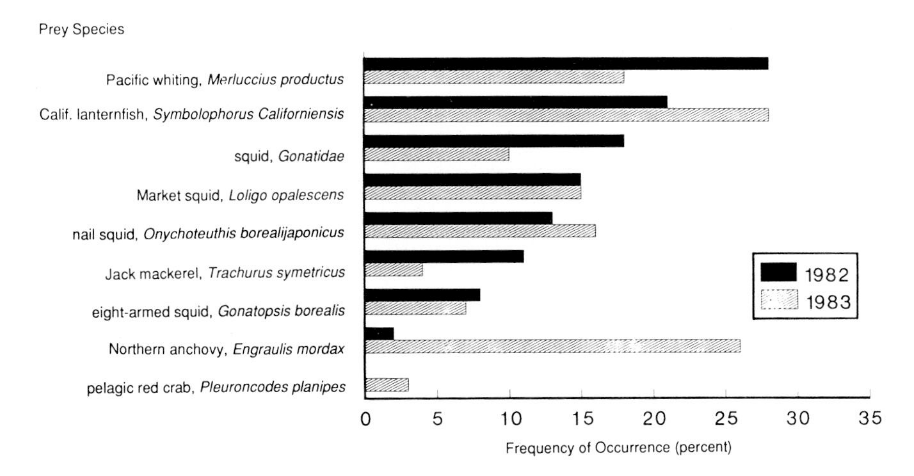

Fig. 4. Frequency of occurrence of prey of female northern fur seals, Adams Cove, San Miguel Island, 1982 and 1983

most dramatic increase in occurrence and pelagic red crab occurred as prey of the northern fur seal for the first time on record.

A comparison of the 1983 fur seal population structure at Adams Cove with that of previous years suggests that there was a substantial loss of adults and subadults prior to June 1983. Adult female and juvenile male maximum counts from 1983 were low compared to those of 1982, and remained low into 1987, suggesting substantial loss of females through emigration or death (Table 1). Adult male counts, however, did not decrease from 1982 to 1983. Tags were resighted on ani-

Table 2. San Miguel Island tagged northern fur seal pups from 1976 through 1984 cohorts found dead or observed alive after weaning through 5 years of age

Year	Tagged	Found dead	Observed alive	Survival index (%)
1976	400	5	15	4
1977	200	3	18	9
1978	200	0	14	7
1979	300	4	29	10
1980	530	2	48	9
1981	500	6	47	9
1982	300	13	12	4
1983	300	7	1	0.3
1984	100	0	6	6[a]

[a]Tag resightings only until animals were up to 4 years old.

mals from all cohorts from 1976 to 1984 (Table 2). The 1983 cohort experienced significantly lower survival than all other years (Z-test of binomial proportions, $P < 0.05$). Survival of pups born in 1983 was also lower than for pups born in 1976 or 1982 ($Z = 2.37$, $P < 0.05$). Survival of the 1982 and 1976 cohorts were significantly lower ($P < 0.10$) than survival in all but the 1984 cohort. Survival of none of the non-El Niño cohorts was different from any other.

Based on tag recoveries along the Pacific coast (Table 2), there was a significant increase in juvenile (0–4 years of age) mortality of pups from the 1982 ($P < 0.03$) and 1983 ($P = 0.07$) cohorts (logistic regression analysis of the probability of mortality, Baker and Nelder 1978). Tag recovery data suggest that a northward shift may have occurred in the distribution of fur seals from San Miguel Island during the El Niño. Seven of 20 (35%) tagged animals were found ashore between October 1982 and December 1984 north of California, whereas only 4 of 25 (16%) of all other animals tagged between 1976 and 1985 were found north of California (Table 3); however, the difference in the proportion of animals recovered north of Califor-

Table 3. Distribution of dead fur seals which had been tagged as pups on San Miguel Island between 1976 and 1985. Animals found during the period influenced by El Niño are compared with all animals found during the remaining time period

Location	Oct. 1982–Dec. 1984	Oct. 1976–Sept. 1982 and Jan. Dec. 1985
	(n = 20)	(n = 25)
Southern California	25%	24%
Central California	25%	44%
Northern California	15%	16%
Oregon	20%	8%
Washington	0%	0%
British Columbia	15%	8%

nia is not significant (Chi-square, $P > 0.1$). A tagged 3-year-old male was resighted on 31 July 1984 at St. Paul Island. This represents the first recorded movement of an animal from San Miguel Island to the Pribilof Islands (Antonelis and DeLong 1986).

7.4 Discussion

The greatest impact of the 1982–83 El Niño on northern fur seals at San Miguel Island was observed during the 1983 breeding season. In both the Adams Cove and Castle Rock colonies, pup production declined significantly largely because of a reduction in the numbers of adult females, births occurred later in the season, feeding trips of lactating females were longer than in other years, and weights of 3-month-old pups were 3 to 4 kg below the average weights of pups during years not influenced by an El Niño event. These observations suggest that San Miguel Island fur seal females were on a low nutritional plane during gestation and after birth. They foraged for significantly longer periods of time at sea and probably returned with less accumulated energy to be passed on as milk to the pups.

Although the survival indexes (Table 2) constructed from resightings of tagged animals are simplistic, they clearly indicate that cohorts born during the 1976 and 1983 El Niño events experienced lower survival than all other cohorts from 1976 through 1984. The pups of the 1983 cohort entered the pelagic phase of their annual cycle severely compromised by low weaning weights, the resighting of only one tagged pup from that cohort indicates that survival was near zero. The 1982 cohort also had a relatively low survival rate, an indication that even though the pups' weights were normal, food supplies were limited during their first winter at sea, when the initial stages of the El Niño event had begun. We presume that prey must have been widely dispersed or very deeply distributed to have so dramatically decreased the foraging efficiency of these fur seals. Lower quality (e.g., size of prey or caloric content) of the available prey could have also influenced fur seal foraging efficiency (Bailey and Incze 1985; Fiedler et al. 1986). The occurrence of dead pups farther north during 1982 and 1983 than in other years may be a reflection of general animal movement in pursuit of prey which had moved north (Pearcy et al. 1985; Arntz et al., this Vol.), or advection of pups farther north by an intensified Davidson Current during the winter months (McLain 1984; Huyer and Smith 1985).

In 1983, the prey species consumed by fur seals occurred at about the same frequencies as in 1982. The most conspicuous changes in the fur seal diet were the increased importance of northern anchovy and pelagic red crabs as food items in 1983. The species composition of the prey assemblage did not appear to be dramatically altered by El Niño conditions. However, judging from the foraging behavior of females and weights of pups, the availability and quality of prey must have been dramatically reduced in 1983. We presume that similar situations existed in 1984 and during the 1976 El Niño, but the degree of prey availability and quality was not as severely influenced as during 1983. Warm water conditions remained in much of the Southern California Bight well into 1984 (Norton et al. 1985; Cole and McLain 1989). Although fur seal pup production was not further depressed in 1984 and female feeding cycles were comparable with those of 1982 and 1985, the weights of 3-month-old pups were significantly lower than in most other years.

Pup production remained depressed in years following 1983. Adult females either died prior to the 1983 breeding season or emigrated to other populations and did not return. Northern fur seals are highly philopatric (Peterson 1965; Gentry et al. 1986a) and reproductive adult females would not be expected to leave the San Miguel populations. Yet, since most of the females in the San Miguel population actually immigrated from other populations (DeLong 1982), it is probable that when faced with foraging conditions which threatened their health and survival, they emigrated back to the northern populations in the Bering Sea. Although there is considerable observational effort at the rookeries of the Pribilof, Commander, and Robben islands, no females tagged at San Miguel Island have been observed. The movement of the juvenile male from San Miguel to St. Paul Island provides the only record of movement from San Miguel to northern populations. Although it is possible that adult females and juvenile males experienced much greater mortality due to low availability of food, it is also possible and perhaps more probable that when faced with poor foraging conditions animals from San Miguel simply emigrated.

In 1983, adult male abundance did not decline on San Miguel Island. A possible explanation of this discrepancy is that adult males typically depart the island earlier than the females, thereby having had the advantage of foraging for 2 or more months before prey abundance was altered by warm temperatures and deepening thermoclines (McGowan 1985; Tabata 1985). Also, due to their greater size (and corresponding physical and physiological capacities), the males were probably capable of diving to greater depths than either adult females or juvenile males (Gentry et al. 1986a). Female northern fur seals are capable of diving to about 200 m (Gentry et al. 1986c). Thermoclines were depressed to depths in excess of 200 m off California, Oregon, and Washington in 1982 and 1983. Adult males may have been able to forage on prey associated with thermoclines deeper (Cannon et al. 1985) than the diving ranges of either adult females or juvenile males. The 1975–1976 El Niño event did not cause a decline in numbers of fur seal pups born at San Miguel Island (Fig. 1) although their weights at about 3 months of age were significantly less in 1976 than in all years thereafter except 1983 (Fig. 2). The 1976 El Niño was less severe than the 1983–1984 event in the eastern North Pacific from California north to British Columbia, Canada (McLain et al. 1985). The observation that pup weights were down but pup births did not decrease in 1976 indicates that pup growth is very sensitive to prey availability for lactating adult females.

The rate of increase in pup production on San Miguel between 1983 and 1987 has been roughly twice that of the highest observed for nothern fur seals, 8%, during maximum population growth on the Pribilof Islands (T.D. Smith and Polachek 1981). If there were not a decline in females and depressed pregnancy rates (alone) accounted for low pup production and decreased female abundance during the El Niño, then pup births and maximum counts of females ashore at San Miguel should have increased dramatically in 1985 when foraging conditions were back to normal. Yet the female counts were about half the maximum levels of 1981 and 1982 through 1985. It appears, therefore, that reproductive females are immigrating back into the San Miguel population from northern populations.

The pattern of future northern fur seal population growth on San Miguel Island will probably continue to be influenced by the depressing effects of infrequent environmental perturbations such as the 1982–1983 El Niño and perhaps long-term changes in ocean climate (McLain 1984) countered by the immigration of mature females.

Acknowledgements. We thank the many field biologist who have contributed to the long-term studies of northern fur seals at San Miguel Island, but special thanks go to E. Jameyson, A.M. Johnson, and B. Stewart. H. Braham, M. Tillman, G. Harry and F. Wilke, all Directors of the National Marine Mammal Laboratory (and organizations which preceded it), provided support for long-term pinniped studies at San Miguel Island. Superintendent B. Ehorn, the rangers and staff of Channel Islands National Park provided logistic support and protection to the islands' pinniped populations. The paper was reviewed and improved by the comments of L. Jones, R. Gentry, R. Merrick, M. Perez, H. Braham, W. Pearcy, F. Trillmich, an anonymous reviewer, and the Publications Unit, Alaska Fisheries Center. A York provided statistical consultation and review.

8 El Niño Effects on Adult Northern Fur Seals at the Pribilof Islands

R.L. GENTRY

8.1 Introduction

El Niño-Southern Oscillation (EN) events apparently affect the Bering Sea through their effects on the intensity and position of the Aleutian low pressure system during the northern winter (Niebauer 1988). During the 1982–83 EN event, this low was displaced farther eastward than usual which allowed cool northerly winds to prevail across the Bering Sea instead of the warm, southerly winds that lesser displacements of the low pressure system foster. It was because of these northerly winds that the strong 1982–83 EN event had little apparent effect on the Bering Sea (Niebauer 1988). The sea surface temperature (SST) anomaly in the Bering Sea peaked at +2 °C from January to April 1983, declined to +1.5 °C through March 1984, and thereafter declined to +0.5 °C (U.S. Dept. Commerce 1983).

The question that this chapter addresses is whether the 1982–83 event caused measurable changes in the behavior, reproduction, or survival of northern fur seals, *Callorhinus ursinus*, breeding on the Pribilof Islands. The question is reasonable because even though oceanographic changes in the Bering Sea were acknowledged to be slight, and evidence for changes in primary productivity is unavailable (Sambroto 1985), a possibility exists that the EN caused widespread mortality of adult seabirds in Alaska in 1983 (Hatch 1987).

Because EN usually affects pinniped populations through the food web, changes in local fishery landings are often observed during EN years. If the 1982–83 event affected Bering Sea fishery landings (summarized by Bakkala 1988), such effects were hard to identify because of yearly changes in fishery effort, location, and quota (Arntz et al., this Vol.).

This chapter considers only adult fur seals; for effects on juveniles, sea York (this Vol.). Data are presented on the same parameters in which clear EN effects were found in pinniped populations from California to Peru. For female fur seals, the parameters examined are the onshore/offshore movement patterns (i.e., attendance behavior), proportion of each year's population that suckled a pup (natality rate), and diving behavior. For males, the parameters examined are territorial tenure and the proportion of each year's marked population that failed to return the following year (interannual loss rate). Data are compared for the 1982 through 1984 seasons. Because the peak SST anomaly began after the 1982 fur seal reproductive season ended (in November), and faded before the 1984 season began (in May), I consider data from the 1982, 1983, and 1984 seasons to represent pre-EN, EN, and post-EN conditions, respectively.

Female fur seals at St. George Island, Alaska forage in both the neritic zone shoreward of the Bering Sea shelf break (Fig. 1), and the epipelagic zone seaward of the shelf break. The zones in which animals forage can be deduced from their diving pattern. So-called deep divers dive throughout day and night to depths of 125–200 m (Gentry et al. 1986b). Tracking studies have shown that such animals are diving (presumably foraging) in the neritic zone (Goebel et al., 1991). "Shallow" divers forage only between dusk and dawn at depths of less than 100 m, and

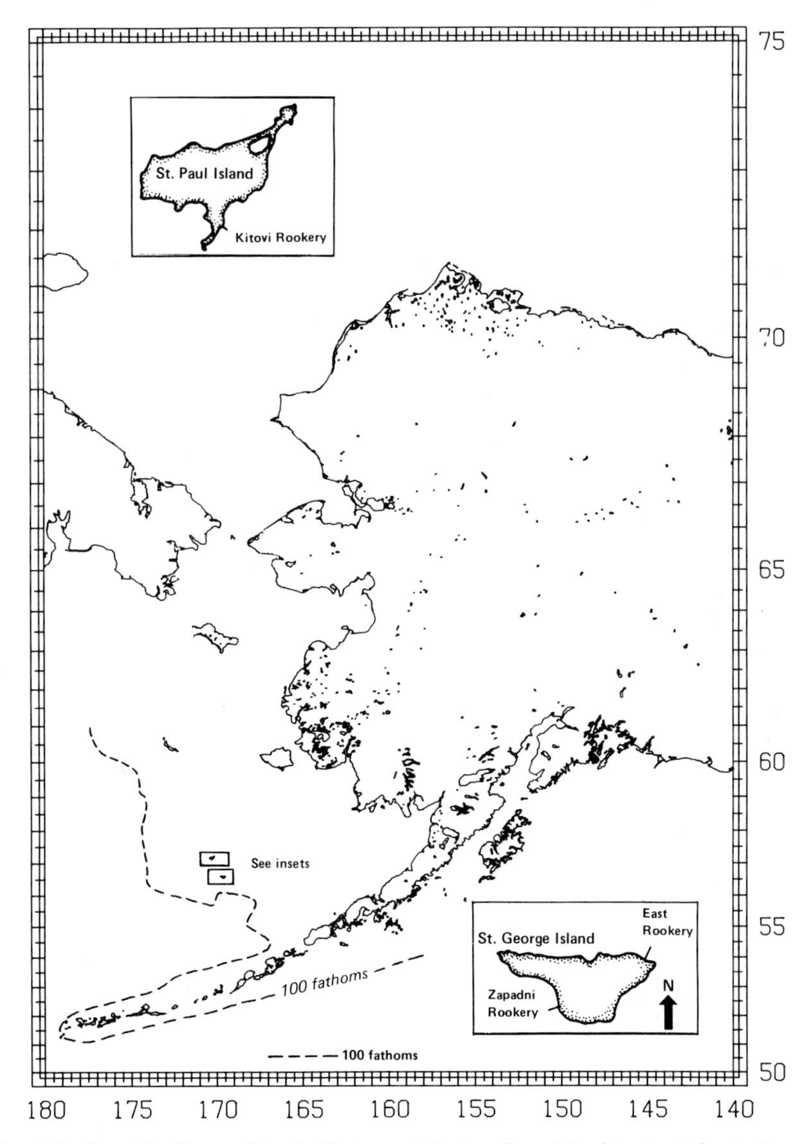

Fig. 1. The study site at St. George Island, Alaska, and the location of the island relative to the continental shelf break

change dive depth by hour, similar to the way the Deep Scattering Layer moves. Tracking studies have located shallow divers in the epipelagic zone at various distances beyond the shelf break (Goebel et al., 1991.). Females with a "mixed" diving pattern perform the deep pattern on some days and the shallow pattern on others within the same trip; by extension, they alternate between the neritic and epipelagic zones. In past sampling, deep, shallow, and mixed divers comprised 30, 30, and 40%, respectively, of the St. George Island population (Gentry et al., 1986a). Loughlin et al. (1987) suggest that individuals may return to the same foraging area on successive trips to sea. The extent to which animals might shift between neritic and epipelagic zones in response to EN is not known.

Deep and shallow divers usually behave differently, and this difference must be considered when testing for EN effects. At St. George Island, deep divers have shorter transit times and trips to sea, perform only one-third as many dives per trip, and have longer-lasting and deeper dives (average >125 m compared to <75 m, Gentry et al. 1986b) than shallow divers. Deep divers are also more efficient than shallow divers in that they expend less energy per unit of food energy gained (Costa and Gentry 1986). Females at St. Paul Island appear to differ from those at St. George Island in some of these traits (Goebel et al., 1991).

8.2 Methods

The study area, East Reef Rookery, is a narrow, rocky beach on the north shore of St. George Island (Fig. 1). Data were collected from a 4-m-high blind in the center, and a 6-m-high tower at the west end of the study area. The study area was marked by a grid painted on the rocks. Sightings were facilitated by a high degree of female site fidelity (they never moved their pups to another beach), and highly predictable attendance behavior (they rarely landed at other rookeries).

Only 18 of 261 males in this study were identifiable by flipper tags; these were applied before 1982. The remaining males were identified by natural marks and scars. All females in this study were given numbered flipper tags prior to 1982. Tag numbers were read using a 60 power telescope.

Attendance and reproductive records were kept on history cards for individual females. An observer (M.E. Goebel in all 3 years) searched for tagged females about 8 h/day beginning in late June and ending in August. Annual observer effort for females was 311, 275, and 270 h in 1982, 1983, and 1984, respectively.

Because the number of years females returned may have influenced attendance behavior and natality rate, differences among years were tested by analyzing female records in the following groups: Group 1, those present for only a single year; Group 2, those present in 2 years, with subgroups for 1982/83, 1983/84, and 1982/84; and Group 3, those present in all 3 years. The analysis of attendance behavior was restricted to females with pups because those lacking pups have erratic attendance (Gentry and Holt 1986). Analysis was further restricted to the most easily identified and observable females, namely, those bearing easily read plastic tags, and those within sight of the study grid. Only the first three trips to sea and the first four visits to shore were analyzed for each female so that all record lengths would conform to the short 1984 season (data collection ended August 1). Comparing identical numbers of trips for all years was necessary because the duration of trips to sea usually increases within a season (Gentry and Holt 1986). The duration of the first visit to shore (the one on which parturition and copulation occur), the mean of the first three trips to sea, and the mean of the three subsequent visits to shore were calculated for each female. Yearly differences in each of these three parameters were tested in the three groups of females using a Friedman's two-way analysis of variance (ANOVA).

To compare natality rate, individual females were scored as suckling only if suckling was observed, and as nonsuckling if they were seen on fewer than 6 days total or on more than 6 days without suckling (the criteria developed by Vladimirov 1987). Differences in natality rate were tested by first calculating the average number of pups born per female for all years. This value was then used to estimate the expected numbers of females with various natality histories, assuming that natality was independent over years and number of times the female was seen. Natality history refers to the number of possible combinations of suckling and nonsuckling over years. For example, Group 2 females had four possible histories: suckling in both years, nonsuckling in both, suckling in the first and nonsuckling in the second, and nonsuckling in the first and suckling in the second. The expected value for each group was estimated by multiplying the number of females in the group, times the number of years the group was present, times the probability of each natality history.

Male territorial tenure, recorded on history cards, was calculated as the number of consecutive days each male spent on territory between 21 May and 31 July, the maximum period of data collection that the years 1982–84 had in common. The tenure of a few males exceeded this 72-day period; these extra days were disregarded in data analysis. To focus on the records of adult, stationary, breeding males, records were considered only for males which were seen:

1. Inside the study grid on more than 50% of the days they were observed:
2. For at least 5 days in July (the month of peak copulations); and
3. For a total of at least 10 days.

Because tenure lengths may have depended on the number of years males returned to territory, yearly differences in average tenure were examined in the same three groups as for females: Group 1, those with only a single year of tenure; Group 2, those with 2 years of tenure, with subgroups for 1982/83 and 1983/84, but not for 1982/84 (it was not represented); and Group 3, males seen in all 3 years. The hypothesis that tenure lengths were similar among years was tested in Group 1 using a Kruskal-Wallis one-way ANOVA test, and in Groups 2 and 3 using Friedman's two-way ANOVA test. Similarities within Group 3 males were further defined using the paired comparison test for Friedman's rank sums (Hollander and Wolfe 1973).

A true survival rate for males could not be calculated due to possible emigration. Interannual loss rate was used instead, because it could be calculated without knowing the fate of lost animals. Trends in interannual loss rate were tested by finding the proportion of the known male populations from the 1981, 1982, and 1983 seasons that failed to return in the following year (hence, loss rates are reported for 1982, 1983, and 1984), and comparing years using a Chi-square test.

Diving behavior was measured by instrumenting foraging females with photomechanical time-depth-recorders (Gentry and Kooyman 1986b). All females to be instrumented were captured at the open west end of the rookery, beyond the study grid. Because of the distances involved, capture disturbances did not affect the attendance behavior of noninstrumented females.

Dive records were analyzed for the number of dives made per trip to sea, and mean and maximum depths attained. Previous analysis of one northern fur seal record (Gentry et al. 1986b) and two South African fur seal, *Arctocephalus pusillus*, records (Kooyman and Gentry 1986) showed that dive durations were closely correlated with dive depths ($r^2 = 0.71 - 0.81$). Therefore, the present records were not analyzed for dive duration on the assumption that any trend by year would be reflected in the analysis for dive depth by year.

Statistical significance is accepted at the alpha = 0.05 level on all statistical tests.

8.3 Results

8.3.1 Female Attendance Behavior and Natality Rate

The number of easily observed females in this study was 204 (83, 68, and 53 respectively in 1982, 1983, and 1984). For the analysis of attendance behavior, the sample sizes of Group 1 females in 1983 (n = 1) and 1984 (n = 5), and Group 2 females in 1983/84 (n = 3) were too small to make comparisons within groups. Therefore, tests of the effects of EN on female attendance behavior were limited to Group 2 in 1982/83 (n = 10) and Group 3 females (n = 11) only. The requirement that females be suckling in all years had the strongest influence on group sample size.

The attendance behavior of Group 2 and 3 females was not different before, during, or after the EN event (Table 1). Specifically, no significant differences were found for either Group 2 or Group 3 females in duration of the first visit, mean duration of the first three trips to sea, or mean duration of the subsequent three shore visits (Friedman's two-way ANOVA by ranks, all $p \geq 0.13$). Therefore, these groups were merged to create Table 1.

The natality rate among the 204 females for which suckling was assessed did not change before, during, or after the EN (Chi-square = 10.6; $p = 0.44$). The estimated average number of pups born per female per year ($\hat{m}$) was 0.744.

8.3.2 Male Tenure and Interannual Loss Rate

All 255 males that met the criteria for tenure calculation were grouped by year, irrespective of the number of years each was on territory. The mean territorial tenure for 1982 was 43.7 days (n = 87, SD = 19, range 10–72 days), for 1983 it was 39.8 days (n = 91, SD = 14.3, range 13–72 days), and for 1984 it was 46 days (n = 77,

Table 1. Attendance behavior of female northern fur seals at St. George Island, Alaska in years before, during, and after the 1983 EN event. All units are days except N, which is number of females. Means are the means of individual means. Table 1 combines Group 2 and Group 3 records (see text)

	1982			1983			1984		
	Mean	SD	N	Mean	SD	N	Mean	SD	N
First shore visit[a]	7.1	2.2	21	7.3	1.9	21	6.7	2.6	11
Trips to sea[b]	4.7	1.4	21	4.5	1.4	21	4.7	1.7	11
Visits to shore[c]	1.9	0.4	21	1.8	0.4	21	1.5	0.4	11

[a]The first shore visit includes parturition and copulation.
[b]Mean duration of the first three trips to sea for the number of females indicated.
[c]Mean duration of three visits to shore (subsequent to the first visit) for the number of females indicated.

SD = 15.7, range 13.72 days). The mean male tenures for 1982 and 1983 were not different from each other, and both differed from tenure in 1984. Not all groups showed this trend. Specifically, Group 1 males showed no significant differences in any years (Kruskal-Wallis test, H = 0.4796, 2 df, p = 0.7868), nor did Group 2 males differ from 1982/83 (Friedman's test, Chi-square = 0.3333, n = 12, 1 df, p = 0.5637). Group 2 males for 1983/84 showed a significant difference (Friedman's test, Chi-square = 4.6538, n = 26, 1 df, p = 0.0310), as did Group 3 males, analyzed for all 3 years (Friedman's test, Chi-square = 6.1071, n = 14, 2 df, p = 0.0472). The difference in Group 3 males was attributable to the differences between 1983/84 and 1982/84 only (paired comparison for Friedman's test; differences in the rank sums for 1982/83, 1983/84 were 10.5, 134.96, and 145.46, respectively, test statistic = 12.40; Hollander and Wolfe 1973). In summary, tenure in 1983 was shorter than in 1984 because of differences in Group 2 and Group 3 males that were on territory in both of those years. Tenure in 1982 was shorter than in 1984 because of differences only in Group 3 males.

There was no significant difference in interannual loss rate before, during, or after the EN event (Table 2; total N = 261 males, Chi-square = 4.33, p = 0.8853. Similar numbers of males failed to meet the criteria for analysis of tenure calcula-

Table 2. Loss rate of known adult male fur seals from the population at East Reef Rookery, St. George Island, Alaska, before, during, and after the 1983 EN event

Season interval	1981–82	1982–83	1983–84
No. males (year 1/year 2)	83/87	87/91	91/77
No. year 1 males returning in year 2 of interval	29	37	46
Percent of year 1 males not returning in year 2 of interval	65.1	57.5	49.5

Chi-square test on number not returning vs number returning; Chi-square = 4.33; not significant at 0.05 level (p = 0.8853). Total N for all years = 261 different males.

tion in the sample years (38, 31, and 33 males were rejected from the 1982, 1983, and 1984 data, respectively).

8.3.3 Diving Behavior

Dive records were obtained for 11 tips to sea for 9 individuals in the 3 years. No individual was instrumented in all 3 years, but one was instrumented in both 1982 and 1983 and another in 1983 and 1984. Statistical tests included values for both females for both years because discarding data to avoid possible repeated measures problems would have made the sample too small to compare by year.

The number of dives and the mean and maximum dive depths did not change during the EN year compared to the years before and after (Table 3, Kruskal-Wallis tests; the differences were not significant at the 0.05 level, test statistic comparable to Chi-square with 2 df). However, there was a weak tendency (significant at the 0.10 level) for the number of dives to increase from 1982 to 1984 (p = 0.09). One factor contributing to this apparent trend was that the proportion of shallow divers, which have more dives per trip to sea than deep divers, progressively increased throughout the three sample years. (This change in proportion also accounted for the suggested but nonsignificant change in mean depth over years.) Another factor was that one female (592) had an unusually long trip to sea in 1984 (17 days), compared to a normal trip (5 days) in 1983. (The change in number of dives for female 1789 also resulted from different trip lengths; 8 days in 1982, 6 days in 1983). These females each had slightly greater mean dive depths in their respective second years of

Table 3. Eleven diving records for nine female northern fur seals and yearly averages at East Reef Rookery, St. George Island, Alaska, before, during, and after the 1983 EN[a]

Year	Female	No. of dives	SD	Max. depth	SD	Mean depth	SD
1982	579	175		193		126	40
1982	542	324		172		46	46
1982	1789	365		99		45	14
1982	540	122		192		150	22
	Average	246	116.4	164	44.4	92	30
1983	1789	148		147		69	23
1983	592	154		101		34	10
1983	P9	175		206		128	47
	Average	159	14.2	151	52.6	77	27
1984	2775	250		101		50	17
1984	207	421		132		46	13
1984	592	477		133		48	15
1984	J8	200		168		75	37
	Average	337	132.9	134	27.4	55	20

[a]Kruskal-Wallis tests: number of dives, H = 4.513, n = 11, p = 0.8964; mean depths, H = 0.1023, n = 11, p = 0.0501; maximum depths, H = 1.119, n = 11, p = 0.4285. Females with mean depths >125 m are "deep" divers; others are "shallow" divers.

being instrumented, but these increases did not both occur in the EN year. Records for the proportion of time at sea spent resting could not be analyzed for yearly trends because of small sample size.

8.4 Discussion

Compared to the years before and after the 1983 EN, no changes were found for females at St. George Island in the

1. Duration of first visits to shore;
2. Mean duration of trips to sea;
3. Mean duration of subsequent visits to shore;
4. Natality rate;
5. Number of dives; or
6. Mean or maximum diving depths.

No changes were found for males in interannual loss rate.

The only change in any parameter was that territorial tenure of some adult males (those present in all 3 years and those present in 1983/84 only) was 14% longer in the year after the EN than during or before it. This could be interpreted as an EN-induced increase in food available to some males before they began fasting for the 1984 breeding season. Increased food availability was suggested by York (this Vol.) to account for the correlation between survival of juvenile fur seals and SST. Whether increased tenure of adult males and increased survival of juveniles are attributable to the same enhancement effect on EN is problematic. Adult males winter farther north (they remain in the Bering Sea and northern Gulf of Alaska, Kajimura 1980) than juveniles do, and would have been exposed to less extreme oceanographic changes in the 1982–83 event.

It is also possible that territorial tenure of adult males was the same every year, but that in some years territorial occupation began earlier, or ended later, than the sampling cutoff dates of May 21 to July 31. Such changes would cause a false calculation of tenure.

A suggested increase was found in the number of dives during trips to sea from 1982 through 1984 (significant at the 0.10 level only). The changing proportions of shallow and deep divers in the sample, magnified by small sample size, probably explains this finding. Because shallow divers at St. George Island average three times more dives per trip to sea than deep divers, and because deep divers progressively disappeared from the sample from 1982 to 1984, the number of dives appeared to increase.

Does the change in the proportion of shallow and deep divers in the sample signify a shift to increased feeding in the epipelagic zone, perhaps from an increase in productivity there? Unfortunately, the sample size for dive records from these and subsequent years is too small to conclude that such a shift has occurred. Oceanographic sampling was inadequate to show whether primary productivity in these zones was affected during the EN (Sambrotto 1985).

From these results I conclude that the 1982–83 EN had such reduced effects in the Bering Sea, and its timing was such, that neither positive (York, this Vol.) nor negative effects on adult female fur seals were detectable in the parameters measured. The major surface warming had ceased by at least 2 months before female fur seals arrived in the Bering Sea in 1983, and the SST anomaly did not exceed +1.5 °C through the rest of that season. By July 1984, when the majority of females returned to the Bering Sea, the SST anomaly had declined to only +0.5 °C, a rise that occurs in some non-EN years. An increase in adult male territorial tenure in 1984 was consistent with a possible preseason food enhancement, but other explanations of increased tenure cannot be ruled out.

The question of whether female attendance behaviour in the Bering Sea can reflect environmental change (such as fishing pressure) was first raised by Chapman (1961). Gentry et al. (1986c) refuted this suggestion on the general grounds that female foraging behavior is flexible enough to mask all but disastrous food fluctuations, and that these are not as likely to occur in subpolar latitudes as in more equatorial ones. Loughlin et al. (1987) upheld Chapman's view, citing the EN response of fur seals in California (DeLong and Antonelis, this Vol.) as evidence. The present results confirm our earlier view with specific data. The strongest EN of the century had no measurable effect on attendance or diving behavior of female northern fur seals in the Bering Sea. For fishing pressure to be measurable through seal behavior, its effects would have to exceed those of the 1983 EN event.

The flexibility in fur seal foraging behavior, referred to by Gentry et al. (1986c), may reside in at-sea metabolic expenditure. Costa et al. (1989) show that when food is scarce, female Antarctic fur seals, *Arctocephalus gazella*, increase feeding trip length without altering their field metabolic rate (FMR), whereas northern fur seals hold trip length constant and increase FMR. Maximum FMR in the two species is quite similar. Costa et al. (1989) conclude that the difference between northern and Antarctic fur seals lies in their respective abilities to modify at-sea time budgets. Northern fur seals, which normally spend about 17% of a foraging trip resting, can compensate for reduced food by resting less and diving more: Antarctic fur seals, which normally spend only 5% of sea time resting, are nearer their "metabolic ceiling" and cannot work harder. By this reasoning, northern fur seals should be less likely to alter trip length (and hence should be less likely to show EN effects in attendance behavior) than the five other otariids for which time spent resting has been reported *(Arctocephalus gazella, A. australis, A. galapagoensis, Zalophus californianus wollebaeki, Gentry et al. 1986c; Z. c. californianus, Feldkamp et al.* 1989). Unfortunately, sea time spent resting could not be compared in the present study.

Northern fur seals also derive flexibility in foraging behavior by being feeding generalists. Their diet in the Bering Sea is diverse, comprising 23 species of fish and three species of squid (Kajimura 1984). Various prey species predominate, depending in part on where fur seals feed and time of season. Feeding specialists, such as the Antarctic fur seal (Doidge and Croxall 1985), have less potential than feeding generalists for switching prey species when environmental change reduces their major prey species.

In any otariid species, attendance behavior should not be used as an index of food availability without measuring diving behavior, because food supply is only one determinant of attendance behavior. Other determinants may be the size or age of females sampled; young female northern fur seals have longer, more variable initial trips to sea than older females (Goebel 1988). Another determinant, at least for northern fur seals, is feeding location; deep divers in the neritic zone have shorter trips than shallow divers in the epipelagic zone. Because of these variables, especially with small sample sizes, attendance behavior may appear to change irrespective of trends in food supply. On the other hand, attendance behavior can fail to reflect a change in food supply if the seal can alter its foraging effort (metabolic expenditure), prey species, or feeding location without adding time to the trip to sea.

Caution should be used in interpreting any single change in attendance or diving behavior as an indication of changed food supply. For example, the present study found a weak trend toward increased numbers of dives per trip to sea. Without knowing that the number of dives on a trip to sea depends on whether the female is a deep or a shallow diver, and that the proportion of these types in the sample changed over years, this trend might have been attributed to a change in food supply.

Acknowledgments. I thank Michael E. Goebel for his patient dedication and consistency in collecting attendance data of high quality. My thanks also to Wendy E. Roberts for preliminary data analysis, and to Camille A. Goebel-Diaz for the data analysis that appears here. Anne York gave valuable advice on statistics and assistance in data analysis. The data were checked and edited by Deborah Horn and Vivian Casanas. All phases of the project were supported by the National Marine Mammal Laboratory of NOAA's Northwest and Alaska Fisheries Center.

9 Sea Surface Temperatures and Their Relationship to the Survival of Juvenile Male Northern Fur Seals from the Pribilof Islands

A.E. York

9.1 Introduction

Several chapters in this volume have compared attendance patterns of female fur seals and the growth and survival of fur seal pups during normal years and during the El Niño event of 1982–83. DeLong and Antonelis have shown that northern fur seals (*Callorhinus ursinus*) breeding on San Miguel island in California were adversely affected by the 1982–83 El Niño: females foraged for longer periods than normal, and weights of pups at age 3 months and survival of pups during their first 3 months were significantly less than other years. On the other hand, Gentry found no such effects for northern fur seals breeding on St. George Island (part of the Pribilof Islands) in Alaska; this is consistent with Niebauer's (1988) analysis that the 1982–83 El Niño had only a limited effect in the Bering Sea. Both the Gentry and the DeLong and Antonelis studies were conducted during the May-September breeding seasons.

The present study examines data on a broader temporal and spatial scale and shows that there is a significant, detectable relationship between changes in the environment, measured by sea surface temperatures (SSTs) and the survival rates of juvenile male fur seals from the Pribilof Islands. In effect, it is a long-term study of animals distributed over large areas of the North Pacific Ocean, outside the time limits of the breeding season; it is an analysis of survival from the time of weaning to age 2 years of northern fur seals from the 1950-1979 cohorts. A broad-scale effect on survival of juvenile northern fur seals from the Alaskan population would seem likely because of their wide distribution in the North Pacific Ocean (Fig. 1) and their opportunistic feeding behavior (Kajimura 1984; Perez and Bigg 1986).

The initial motivation for this work was the observation that the two cohorts which exhibited the highest and the lowest juvenile survival estimates were also born in years in which the average annual SSTs in the Bering Sea and the Gulf of Alaska were extreme. The 1958 cohort had the highest survival and the 1956 cohort the lowest; 1958 was a strong El Niño year with very warm SSTs extending into the North Pacific Ocean, and 1956 was an abnormally cold year. This suggests that the effect of an El Niño might be positive for the Alaskan population of northern fur seals – unlike the San Miguel Island population, which was affected very negatively by the 1982–83 El Niño event. A correlation between SST and juvenile survival and the possibility that SST has a long-term effect on juvenile survival has been discussed in York (1985a).

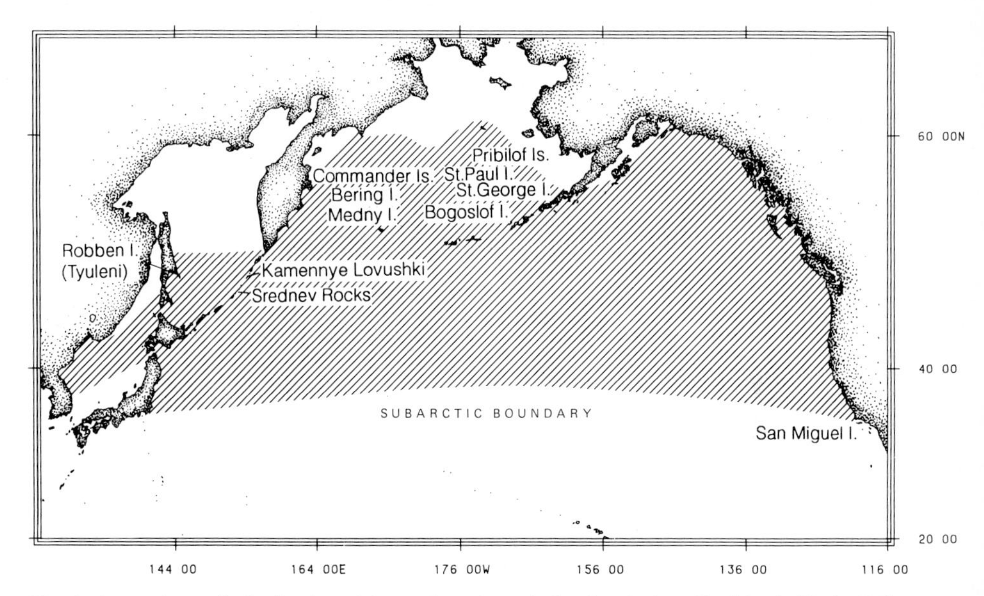

Fig. 1. General oceanic distribution of the northern fur seal, showing the breeding islands (York 1987)

I analyze a long-running series of SSTs from Pine Island, British Columbia (Queen Charlotte Strait, 50°56′N 128°44′W), examine the structure of the correlation of the SSTs with the estimated survival rates of cohorts of subadult male fur seals born on St. Paul Island, Alaska, and develop a simple model relating early survivorship and SSTs. These results are discussed in terms of the known feeding ecology of the fur seal and responses of fish stocks in Washington, British Columbia, and Alaska to changing temperatures; in addition, some suggestions for interpreting their interrelationship are given.

9.2 Basic Natural History

The basic biology and natural history of the northern fur seal is summarized in York (1987). The northern fur seal is pelagic during most of the year. The breeding islands and general oceanic distribution are shown in Fig. 1. Fur seals are not uniformly distributed throughout their range, but tend to accumulate in biologically productive areas, such as regions of upwelling along the coast of North America.

Pups are weaned during October and November and then begin their pelagic life. Mortality among juvenile fur seals is variable and often high. At least 50%, and sometimes more than 90%, die before the age of 2 years (Lander 1979; York 1985b). Mortality during the first 2 years occurs in one of two stages: (1) 0–4 months, when the animals are on land, or (2) age 4 months–2 years, when they are at sea. Typically, 5–20% of pups die during their first 4 months, while 50–80% of the remainder die during the following 20 months (Lander 1979; York 1985b). Mortality on land during the first 4 months is correlated with population size and is considered density-dependent (York 1985b). The time at sea between age 4 months and 2 years is a period of high mortality.

From 1958 through 1974, scientists from the United States and Canada conducted research cruises during which they collected northern fur seals. The locations of collection of all northern fur seals younger than 2 years are shown in Fig. 2A and those older than 2 years in Fig. 2B. Kajimura (1979) suggested that the distribution and migration routes of young fur seals were strongly affected by ocean currents and prevailing wind regimes. The diagram in Fig. 3 shows the major currents and possible routes of pups journeying from the Pribilof Islands to the areas where they have been found, mostly in coastal areas of Alaska, British Columbia, and Washington (Fig. 2B).

Some work has been done on environmentally induced mortality on northern fur seals. Scheffer (1950a) reported large numbers of dead pups washed up on the Oregon and Washington coasts after a very stormy period in January and February 1950. Trites (1984) suggested that storm regimes and cold weather in the mid-1970s in the North Pacific Ocean adversely influenced the survival of young fur seal pups. York (1985a) showed a positive correlation between early survival for the 1950–79 cohorts of fur seals and SST at Pine Island, British Columbia, averaged over the 5-year period from 4 years before the birth year to the birth year.

9.3 Methods

9.3.1 Available Population Data and Estimates of Early Survival

A commercial harvest of subadult male fur seals (mostly ages 2 to 5 years) was conducted on the Pribilof Islands from 1918 to 1984. Biological samples from the harvests have been collected for the determination of life history parameters since 1950. Scheffer (1950b) discovered that the age of fur seals could be determined from counts of numbers of growth rings in dentine layers of teeth, and a 20% subsample of animals from the harvest was routinely aged after 1950. Estimates of numbers of pups born (Fig. 4) and estimates of mortality on land are available for most years since 1950 (York and Kozloff 1987). I calculated age-specific harvest rates of 2-, 3-, 4-, and 5-year-old male fur seals (Fig. 5) for the 1950–79 cohorts (except 1971) by dividing the number of males killed in each age class by the estimated number of male pups alive in September of their birth year. The total harvest rate (Fig. 5) of a cohort is calculated by summing harvest rates for ages 2 to 5 years (i.e., it is the total kill from a cohort divided by the number of male pups alive in September of the birth year). I also smoothed those time series using the method of Velleman (1980). Applying Lander's (1975) method of using harvest rates to estimate survival from age 4 months to 2 years, I estimated the survival rates of juvenile males from most cohorts born during 1950 through 1979 (Fig. 6).

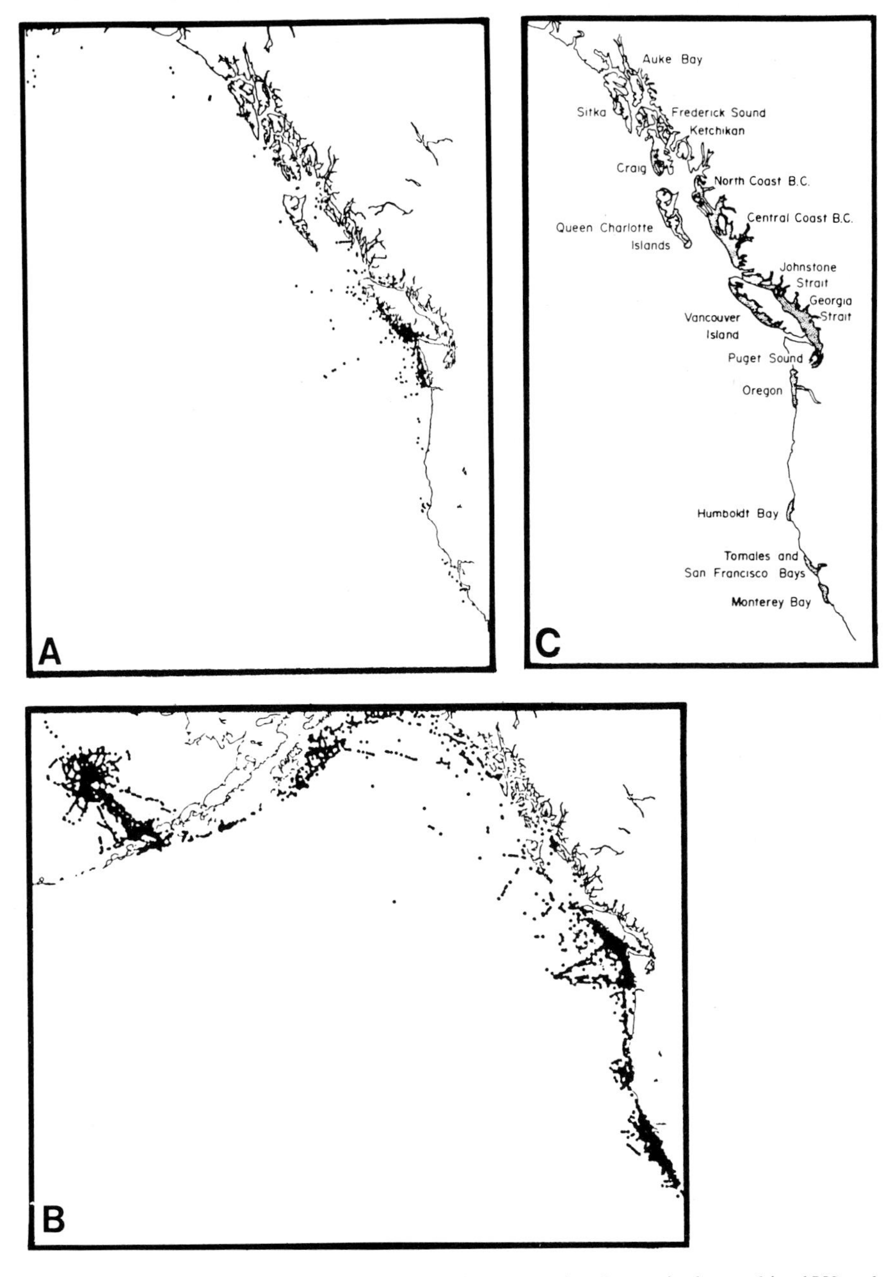

Fig. 2A Locations of collection of northern furs seals younger than 2 years in the combined US and Canadian pelagic collections (1958–74). **B** Locations of collection of northern furs seals, 2 years and older in the combined US and Canadian pelagic collections (1958–74). **C** Locations of spawning grounds of Pacific herring from Auke Bay to Monterey Bay (Haegele and Schweigert 1985)

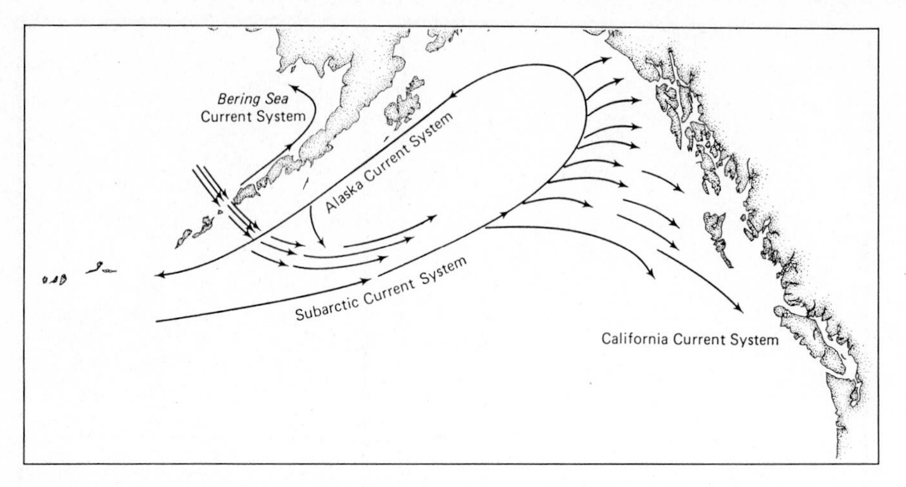

Fig. 3. A schematic diagram showing the major ocean currents (after Reed and Schumacher 1985) with possible migration routes of fur seal pups (Kajimura 1979)

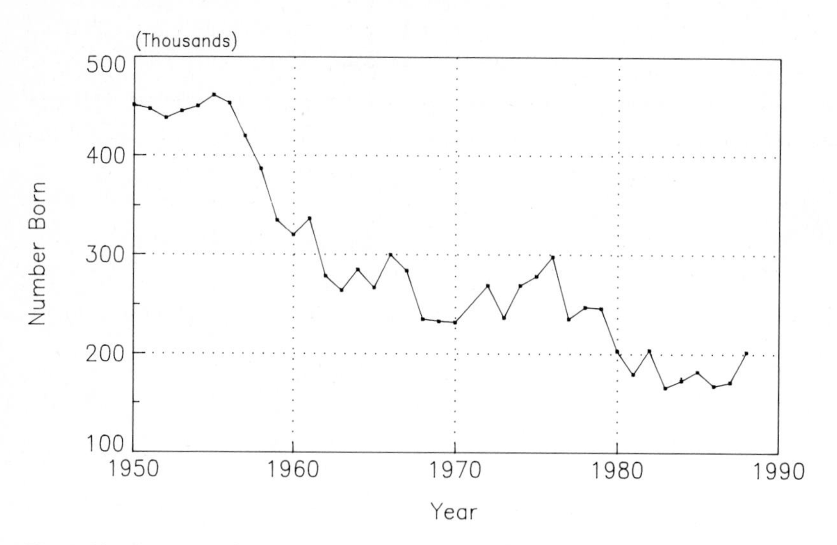

Fig. 4. Numbers of northern fur seal pups born on St. Paul Island, Alaska, 1950–88

9.3.2 Sea-Surface Temperature Data

Monthly anomalies and monthly mean SSTs for Pine Island, British Columbia for 1939 through 1982, based on data from Tabata (1985), are shown in Fig. 7. This time series was chosen for several reasons: Pine Island is central to the known distribution of young animals (Fig. 2A); the time series is sufficiently long to cover the period for which the fur seal data exist; the series has few missing values over that

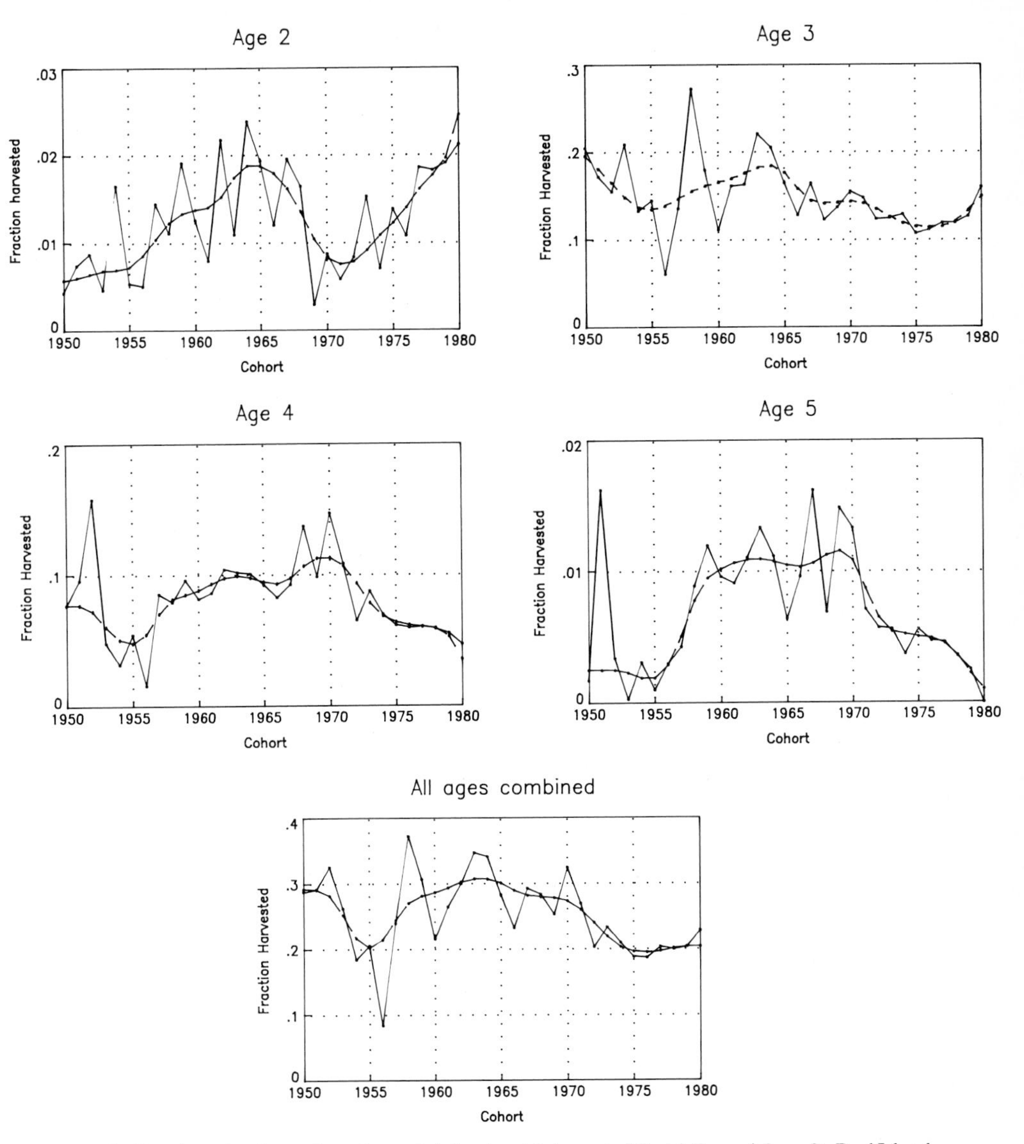

Fig. 5. Fraction of male northern fur seals (alive in mid-August of the birth year) from St. Paul Island, Alaska, appearing in the commercial harvest at ages 2, 3, 4, 5 years, and all ages combined for the 1950–79 (except 1971) cohorts. A robust smooth of the time series is also shown

period; and the area near Pine Island is considered representative of the near-shore SSTs in British Columbia (S. Tabata, pers. comm.). I also examined the relationship of two other time series of temperatures (the Bering Sea near the Pribilof Islands, and Canadian Station P, 50°N 145°W, in the Gulf of Alaska) to the early survival of fur seals.

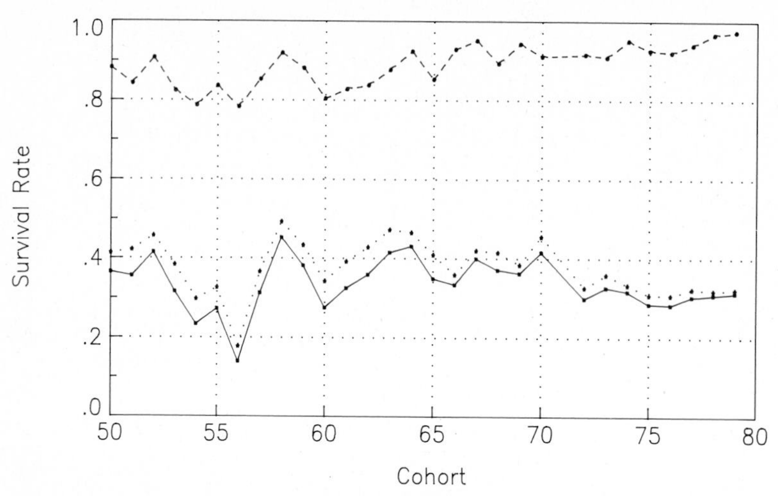

Fig. 6. Survival on land, birth to age 4 months (*dashed line*); survival at sea, age 4 months to 2 years (*dotted line*); and survival from birth to age 2 years (*solid line*) for the 1950–79 year classes (except 1971) of male northern fur seals, St. Paul Island, Alaska

The relationship between the time series of early survival of cohorts of northern fur seals and the monthly SSTs from Pine Island was studied by constructing the lagged cross-correlations (Chatfield 19750 of the two series. The cross-correlation between juvenile survival and SST (lagged m months from January of the cohort year, say r_m) is the simple correlation between the time series of survival rates and the sea-surface temperatures m months after January of the cohort year. Thus, r_0 is the correlation between juvenile survival and the SSTs for January of the cohort year; r_{-1} is the correlation between juvenile survival and the SSTs for December of the year before the cohort year; r_{48} is the correlation between juvenile survival and the SSTs for January 4 years after the cohort is born. The cross-correlations were calculated for m in a range of –60 to 60 months from January in the cohort year. There were no significant ($P \leq 0.95$) autocorrelations (up to lag 10) in the time series of early oceanic survivals, and therefore, no correction for integral time scale (cf. Kundu and Allen 1976) was necessary. The break point between statistical significance and nonsignificance ($p = 0.95$) was estimated using Fisher's Z-transformation (Kendall and Stuart 1977); the level of correlation, which is significantly different from 0, changes after 36 months because SSTs were not available after 1982, and the sample size was reduced. A regression equation modeling juvenile survival as a function of two temperature series was fit using the 2 months which gave the largest combined value of r^2, the fraction of variability explained by the regression.

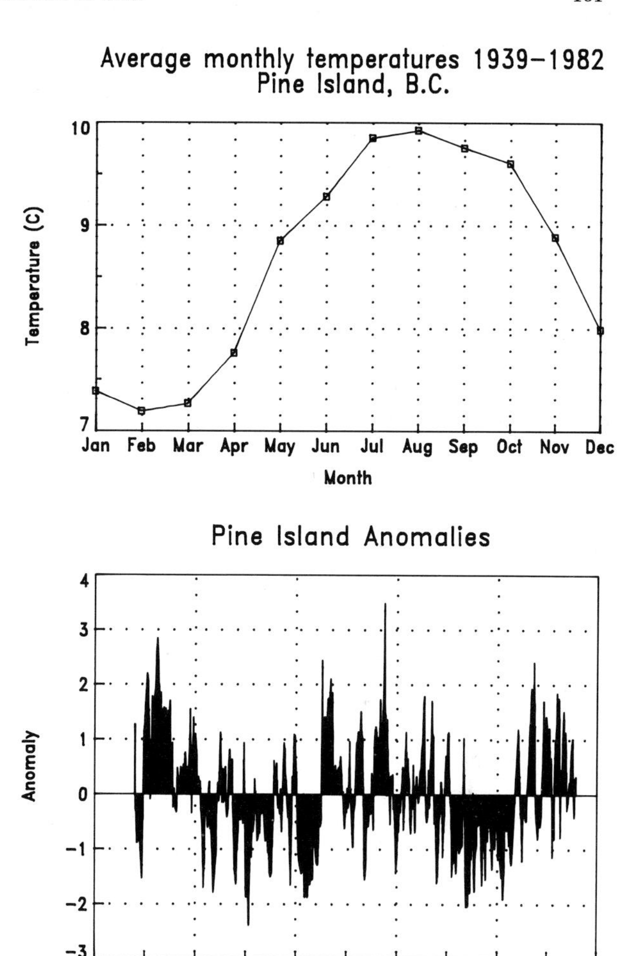

Fig. 7. *Above* Monthly mean SSTs, Pine Island, British Columbia, 1939–83. *Below* Monthly anomalies of SST, Pine Island, British Columbia, 1939–83 (after data from Tabata 1985)

9.4 Results

Lander's estimates of mortality at sea (Fig. 6) are highly correlated (r = 0.98; $P < 0.01$) with estimates of total harvest rates of cohorts of subadult male fur seals calculated by summing the age-specific harvest rates (Fig. 5). Estimates of survival, based on Lander's method, were produced for cohorts that were at least age 5 years before the commercial harvest ended in 1984. These estimates appear to provide a rough index of the relative survival of the various cohorts and have been useful for modeling the population dynamics of the St. Paul Island and Roben Island populations (e.g. York and Hartley 1981; Frisman et al. 1982; Trites 1984). The increase in pup production during 1970–75 and the decline in pup production during 1975–80 (Fig. 4) are consistent with the pattern of changing harvest rates during the mid-1960s to the early 1970s (Fig. 5) insofar as the changing harvest rates imply a similarly changing level of the survival of juvenile females as well as males. Following the pattern of numbers of pups born (Fig. 4), the numbers of males harvested during 1956 through the late 1960s declined. The total numbers harvested from the post-1975 cohorts did not follow the pattern of the number of pups born, but the average age of harvested males did decrease significantly during 1975–84 (based on a linear regression of the average age of the harvested animals on year, $P < 0.05$). The pattern of harvest rates of the various age classes has varied over time. The harvest rate of 2-year-old males increased in the late 1970s and was the highest among all cohorts since 1950. The total harvest rate reached a high of about 30% during the early 1960s, although it was higher for some cohorts; the low point occurs for the 1975 to 1979 cohorts (Fig. 5).

The lagged cross-correlations of early ocean survival and SST at Pine Island in Fig. 8 show significant positive correlations for most months in a broad window extending from July of the year before birth through August of the birth year (corresponding to $m = -5$ to $m = 8$ in Fig. 8); several significant correlations are also apparent in a narrower window from January to June, 4 years before the cohort is born (corresponding to $m = -48$ to $m = -42$ in Fig. 8). There are two other significant correlations but they are only significant within a 1-month window. Two other time series of SSTs (on the Pribilof Islands and at the Canadian Oceanic Station P) were also investigated and their lagged cross-correlations with the survival data were very similar to the Pine Island data. No detailed analysis is included here because those series were not as long-running and had missing values.

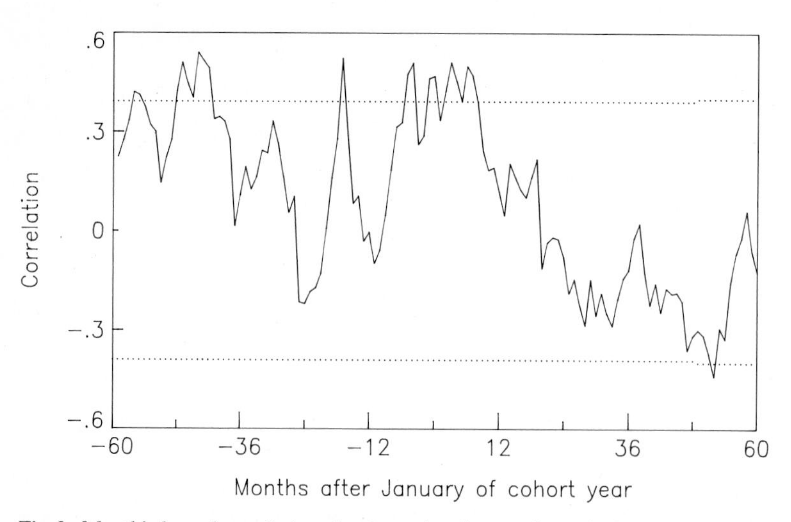

Fig. 8. Monthly lagged correlation of estimated early oceanic survival of male northern fur seals (St. Paul Island, Alaska) and SSTs Pine Island, B.C. Limits of the 95% confidence interval are indicated by the *dotted lines*

The two best predictors of survival (in terms of r^2) were the SSTs from April of year –4 and March of the cohort year (months –44 and 2 in Fig. 8). A regression of early ocean survival on those temperatures is given in Table 1. Both series were significantly correlated with survival but were not significantly correlated with each other. The regression model explains 50% of the variability in survival and predicts that the raw survival rate increases 5.5% for each degree increase in SST in April of year –4 and 6.9% for each degree increase in SST in March of the cohort year. Figure 9 shows the relationship between a weighted average of temperatures in April of

Table 1. Estimates, standard errors, and analysis of variance of the regression of early oceanic survival of northern fur seals (s–0-2) on the Pine Island sea surface temperatures in March of the birth year (Mar_0) and April 4 years before the birth year (Apr_{-4})

The regression equation is:

$s_{0\text{-}2} = -0.535 + 0.0689\ Apr_{-4} + 0.0549\ Mar_0$

Parameter	Estimate	Standard deviation
Constant	–0.5353	0.1799
Apr_{-4}	0.0689	0.0196
Mar_0	0.0549	0.0167

Analysis of variance

Source	DF	Sum of squares	Mean square	F	P
Total	28	0.1309			
Apr_{-4}	1	0.0382	0.0382	15.28	<0.01
Mar_0	1	0.0272	0.0272	10.88	<0.01
Error	26	0.0655	0.0025		

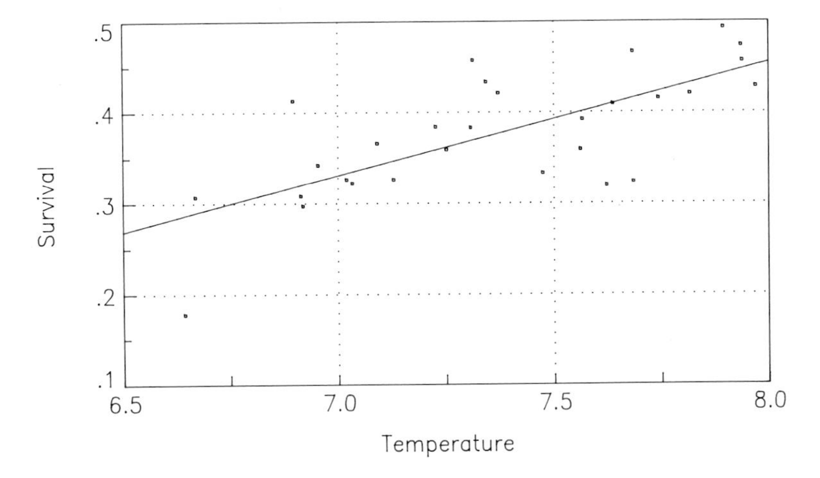

Fig. 9. Survival from age 4 months to 2 years versus the Pine Island combined SST for April 4 years before the birth of the cohort and March of the cohort year; the combined SST is computed using the regression equation in Table 1

4 years before the birth year and of March in the birth year; the abscissa contains the regressors divided by the sum of the coefficients: (0.069 Apr_{-4} + 0.0549 Mar_0)/0.1239 = 0.556 Apr_{-4} + 0.443 Mar_0. The weights are proportional to the regression coefficients in Table 1.

9.5 Discussion

Estimates of early survival of subadult male fur seals were not measured directly. Lander (1975) developed methods of using harvest rates to estimate survival, but there are many difficulties with interpreting a harvest rate as an index of natural survival. These difficulties are related to the accuracy and consistency of the basic data: the numbers of pups born were estimated in different ways and the harvesting of subadult males was conducted under somewhat different regimes since 1950. Escapement from the harvest has no doubt varied from year to year, but Lander's method of estimating the survival rate of juveniles from the composition of the harvest does not take this variability into account. When the harvest strategy remains constant and the estimates of numbers of pups born are unbiased, harvest rates may be nearly proportional to natural survival rates. Since limits on the size of the harvest were imposed only by the availability of animals during the harvest season (late June-early July), the harvest rate is a rough but consistent index of early survival.

Most of the statistically significant correlations between SST and early oceanic survival (Fig. 8) occur in two windows: a wide one extending from the July before birth through July of the year of birth (corresponding to $m = -6$ to $m = 7$ in Fig. 8), and a narrower one extending from January through June, 4 years before the birth of the cohort (corresponding to $m = -48$ to $m = -43$ in Fig. 8). Many significant correlations occur during important months in the life of cohort: included are the months of conception, implantation, the fetal period, and the birth of most pups.

One hypothesis explaining the biological relationship of SST to early oceanic survival of fur seals is through the thermal effects of a change in SST on young pups. If the metabolic rate were higher in colder water, then exposure to colder water over the pup's first winter at sea could result in significantly higher total food requirements over the year; this, in turn, could lead to lower survival rates during the colder years if prey were not sufficiently abundant. Some data in Kooyman et al. (1976) suggest that the metabolic rate of northern fur seals increases in colder water; recent experiments (Worthy, pers. comm.) show that the metabolic rate of California sea lions increases in colder water. However, the pattern in Fig. 8 argues against this explanation because there is no significant positive correlation between SST and oceanic survival during the period that the animals are at sea during their first 4 years ($m = 11$ to $m = 60$ in Fig. 8).

Several studies in this volume show that food availability directly affects survival of young seals; it is clear that the lack of food diminishes survival. High availability of preferred prey should then enhance survival. If the pattern in Fig. 9 is related to food resources for young fur seals, we must identify responses in the food chain caused by events 5 years and 1 year before the animals are weaned. An important step in explaining the timing of the bands of statistically significant correlations between SST and survival would be an analysis of the life history of prey of the fur seal in order to generate hypotheses concerning the likely effects of a change in SST on food availability for young fur seals. However, given the fur seal's wide-ranging diet and the lack of data on the abundance of its prey, especially of fish which are not commercially important, this is not possible at present. Fisheries are complicated, and although one environmental variable cannot be used to predict the year-class strengths of stocks of fish, environmental variables are significant factors and affect the size of a stock (Arntz et al., this Vol.; Favorite and McLain 1973; Sutcliffe et al. 1977). In the following section, I speculate on the possible role Pacific herring (*Clupea harengus pallasi*) could play in the survival of northern fur seals, relating some details on herring life history and their relationship to temperature change and how the correlations in Fig. 8 and the model for Fig. 9 might be related to the availability of herring or related fish.

Herring is an important prey species for fur seals (Kajimura 1984; Perez and Bigg 1986); it ranks first or second in importance (based on percentage of volume in stomachs) for most months in British Columbia, the Gulf of Alaska, and Washington. Herring is used to feed captive fur seals and may be close to ideal for young animals: it is small, has high fat content, swims in large schools, and it would be readily available to young fur seals – especially during spawning. In Washington and British Columbia, herring spawn in March and April. Female herring reproduce for the first time at age 3 years in the waters of British Columbia and Washington and at age 4 years in Alaskan waters.

The distribution of herring spawning grounds from Auke Bay to Monterey Bay (Fig. 2C from Haegele and Schweigert 1985) is similar to the distribution of young northern fur seals (those less than 2 years old) collected by scientists from the United States and Canada during 1958 through 1974 (Fig. 2A). In contrast to the distribution of all other fur seals taken during the same research cruises (Fig. 2B), no animals younger than 2 years were found near the Aleutian chain or in the Bering Sea. In general, the younger animals were found closer to shore than the older animals; many were found near the mouths of bays, although this may be partially a result of sampling effort. Fur seals, with the exception of young animals or groups of animals following a spawning fish population, do not usually move into restricted bays (Kajimura 1984). Kajimura (1984) noted that the area off the Strait of Juan de Fuca and Hecate Strait are principal feeding grounds for juvenile and immature herring. Most young fur seals were found in areas near spawning or feeding grounds of herring.

Herring is also an important commercial species, and there are data on year-class strengths for some stocks. Macy et al. (1978), Schoener and Fluharty (1985), and Bailey and Incze (1985) have reported that year classes of most stocks of herring in Washington, British Columbia, and Alaska have been quite strong in years following El Niños, while those in California have been very weak. Favorite and McLain (1973) showed that strong cohorts of herring near the Queen Charlotte Islands followed years with anomalously high SST. Increases in SST during the summer are associated with higher growth rates for juvenile Pacific herring (Haist and Stocker 1985). Hourston et al. (1981) showed that larger herring are more fecund, Alderdice and Velsen (1971) that water temperature has an important effect on the rate of development of eggs and larvae of herring, and Hay (1985) that changes in SST affect the timing of spawning. An increase in SST the summer before the birth of a cohort could therefore increase the availability of herring for young fur seals during the following winter and spring. Furthermore, a strong year class of herring could give rise to another strong year class n years following, when a large portion of the cohort reproduces at age n – a "baby-boom" effect. Herring are multiple spawners and are sexually mature in Alaska and British Columbia between ages of 3 and 5 years. It is commonly believed among herring fishermen in Alaska that large runs occur about every fifth year (Macy et al. 1978).

Other prey of fur seals also seem to respond to occurrences of El Niños. Schoener and Fluharty (1985) reported that the abundance of Pacific squid (*Loligo opalesens*) off Washington increased sharply following El Niño events in 1940–41, 1957–58, and 1982–83, and that the Pacific mackerel (*Scomber japonicus*) increased in abundance during the 1982–83 El Niño.

Thus, there is evidence that the availability of herring and, perhaps, squid and mackerel as food for fur seals is affected by changes in SST.

The effect of an increase in SST could be even more general. Frost (1983) showed that levels of zooplankton productivity in the Gulf of Alaska increased during the 1958–59 El Niño, whereas the abundance of zooplankton off California was reduced (Bernal and McGowan 1981). Bailey and Incze (1985) suggested that the warm water and increased productivity in the north were beneficial to species of fish in the north. The implication of these studies for fur seals is that the effect of an El Niño could be positive for the northern populations of fur seals but deleterious for the southern populations. It is not known if these apparent differences in the effect of El Niños cause shifts in the distribution of northern fur seals.

This work is speculative and shows a correlative relationship, not a causative one, between SST and early oceanic survival of fur seals. It is not known how changes in SST affect the assemblages of prey available to the fur seal over its extensive range in the North Pacific Ocean. If aged samples of prey items were available over several years from a large sample of fur seals over its range, it would be, at least conceptually, possible to do a multispecies virtual population analysis of the prey. It may then be possible to understand how changes in environmental parameters, such as SSTs, and other important variables, such as fishing effort, affect the availability of the prey species and how that is related to vital parameters of fur seals. Our knowledge of the origins and the age and sex distribution of fur seals feeding in the transition zone between the Alaska current system and the sub-Arctic current system (see Fig. 3) outside the United States 200-mile Exclusive Economic Zone is confined to a few samples from Japanese research cruises. This region could be very important for young fur seals, and research on fur seals there would fill an important gap in our knowledge of fur seal distribution and feeding ecology.

Acknowledgments. I wish to thank S. Tabata for sharing the Pine Island, British Columbia and Canadian Station P SST time series; R. DeLong for suggesting that I consider the relationship between herring availability and environmental conditions; H. Kajimura for sharing his knowledge of the fish of the North Pacific Ocean; R. Merrick, M. Goebel, and H. Huber for fruitful discussions on possible effects of environmental changes on northern fur seals; and H. Friedman for the continual prodding without which this manuscript would never have been completed. Helpful reviews of the manuscript were provided by P. Bedeno, C. Fowler, L. Jones, T. Loughlin, M. Perez, J. Schumacher, and E. Sinclair.

Part III
California Sea Lion

10 Introductory Remarks and the Natural History of the California Sea Lion

K.A. Ono

The following section is an in-depth analysis of the effect of the El Niño on one species, the California sea lion (*Zalophus californianus*). The organization of this section differs from the others in that the papers are based upon topics rather than by studies of one or two species at a particular location. The large number and diversity of studies on the California sea lion which were being conducted simultaneously during the El Niño year enables us to view the phenomenon from many different angles. In order to accomplish this the authors involved have combined data sets from several studies, sites, or islands wherever possible. Since the effects of El Niño were more subtle at higher latitudes, concurrence between several studies strengthens the conclusion that observed effects were indeed due to the El Niño. The papers may be considered, then, as pieces of a larger puzzle which collectively allow us to deduce the mechanisms which shape the immediate behavior, as well as evolution, of this species. The topics range from studies of population dynamics (Chaps. 11, 12, 13, 17) to female attendance (Chap 14), foraging behavior (Chap 15), energetics (Chap. 16), and diet (Chap. 17). Pup development with respect to growth (Chaps. 17, 18), milk intake (Chap. 19), and behavior (Chap 20), as well as yearling behavior (Chap. 21) are also considered. Figure 1 is a map of the study sites included in this section.

To minimize repetition in introductory sections a brief summary of the natural history of this species will be given here. The California sea lion is a member of the family Otariidae, and the genus *Zalophus* which contains only one species and two extant subspecies (Odell 1981). *Zalophus californianus californianus* (Lesson 1828) is distributed north from British Columbia south to the tip of Baja California and into the Gulf of California (also called the Sea of Cortez). *Z. c. wollebaeki* (Sivertsen 1953) occurs on several of the Galapagos Islands (see Chap. 6). The following focuses on the subspecies *Z. c. californianus*.

The average weight and length of adult male *Z.c.c.* is 375.1 kg and 213.8 cm, respectively (n = 36, from Lluch-Belda 1969). Adult males are typically dark brown, almost black, in color, but interindividual variation extends all the way to the tawny brown color of females. Males also have a prominent sagittal crest that is covered with lighter pelage. Juveniles and subadults of both sexes have the tawny brown color of adult females. The average weight and length of adult females is 93.6 kg and 164.1 cm, respectively (n = 84; Lluch-Belda 1969). This gives an adult male to female sexual weight dimorphism ratio of about 4:1. Pups are dark brown to black at birth, and molt to the juvenile/female color at 4–6 months. Newborn male and female pups (<7 days of age) have an average weight of 9.09 kg (n = 59, SD =

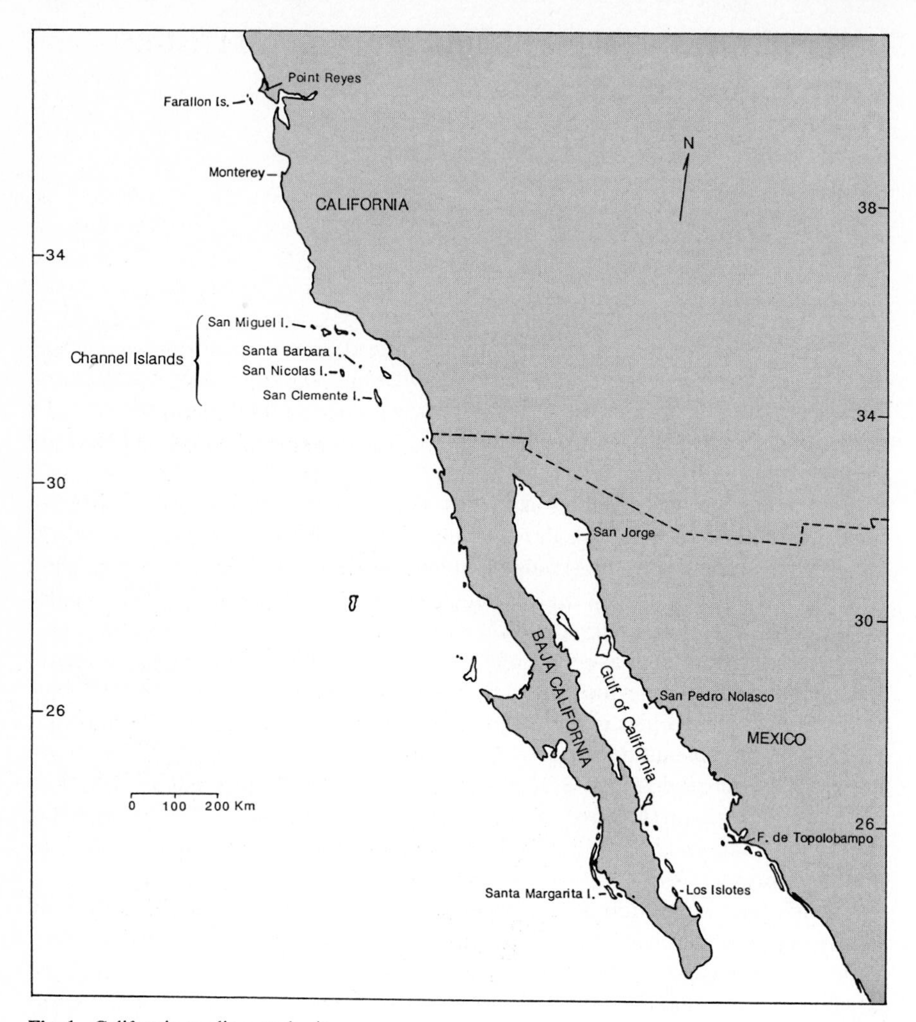

Fig. 1. California sea lion study sites

0.95) and 7.73 kg, respectively (n = 78, SD = 0.86, weighted $\bar{x}$ and SD calculated from Table 4, Ono et al. 1987).

Although a few pups are born on more northerly islands (Farallon Islands, Pierotti et al. 1977; Año Nuevo Island, Keith et al. 1984), the majority of breeding occurs on the Channel Islands and extends to the southern tip of the range, including the Gulf of California (Odell 1981). During the nonbreeding season, males tend to disperse north of the breeding areas (Bigg 1973; Mate 1973), while females and pups tend to either remain on the island of parturition, or disperse to other islands within the breeding range. These patterns are not absolute, with some females migrating north, and with some males (mostly nonbreeding males) remaining on the breeding islands, or migrating south (see Chap. 11). The population size has been increasing steadily since the end of harvesting in the late 19th century (Bonnot et al. 1938). DeMaster et al. (1982) have estimated the increase of the California population prior to the El Niño at about 5% per year.

Breeding begins in mid- to late May with adult males and females hauling out onto traditional breeding areas simultaneously (Peterson and Bartholomew 1967; Odell 1981). Females remain on land for several days before giving birth to a single pup. The mother and pup remain together continuously for about 1 week postpartum, at which point the female departs on her first feeding trip. While at sea a female's time is partitioned into time spent traveling to the foraging site, diving and actively foraging, and resting on the surface (Chap. 15). Females are absent for 1–3 days and then return to suckle their pups for 1–2 days. The cycle of female presence and absence continues throughout the mother-young relationship. Females are able to recognize their own pups soon after birth via auditory and olfactory cues and, except in rare cases of adoption, normally suckle only their own pup. Postparturient females typically copulate once approximately 3 weeks postpartum. Not all adult females breed each year. Some maintain a suckling relationship with their pups from the previous season (yearlings) and a small number of females continue to suckle 2-year-old juveniles. Females without dependent young also haul out on the breeding areas during the breeding season although their feeding cycles differ from those of females with pups (Chap. 14).

Adult males obtain and maintain territories via fighting and ritualized displays. Territorial males remain in the breeding area for their entire tenure, i.e., they do not feed during this time, relying on stored body fat accumulated prior to the breeding season. Peterson and Bartholomew (1967) as well as Odell (1972, 1981) observed relatively short periods of tenure for males on San Nicolas Island $\bar{x}$ = 27 days, Odell; $\bar{x}$ = 9 days, Peterson and Bartholomew). Male tenure appears to vary on different parts of the island, since elsewhere on San Nicolas male tenure was found to be much longer $\bar{x}$ = 45 days, Boness and Ono, unpubl.). Males may maintain territories for up to six or more breeding seasons. Territorial males are not disturbed by adjacent territorial males when copulating within their territories. In the breeding areas, copulations by nonterritorial males are rare, although females may possibly copulate at sites which are not traditional pupping and breeding areas. Subadult males are not tolerated in breeding areas, being chased away by resident adult males; they tend to gather in large aggregation in nonbreeding areas.

Pups have considerable social contact while their mothers are away on feeding trips. Beginning with the first or second absence of their mothers, pups gather in pods, usually away from adult females or males. In these groups, pups sleep, play (including mock fighting), and begin to swim. When their mothers are present in the breeding area, young pups spend most of their time next to her, suckling and sleeping. Older pups will leave their mother's side to swim or play with other pups, then return to their mother to suckle. Most pups probably begin foraging on their own as a supplement to milk intake at about 7 months of age (Chap. 18). Weaning appears to occur shortly prior to the birth of the next pup (10–12 months, Boness et al. unpubl. data).

11 Effects of the El Niño 1982–83 on California Sea Lions in Mexico

D. AURIOLES and B.J. LE BOEUF

11.1 Introduction

Before, during, and after El Niño (EN) 1982–83, several aspects of California sea lion biology were being studied in the Gulf of California (Sea of Cortéz) and the Pacific coast of Baja California, Mexico, as part of a long-term program initiated by representatives of Centro de Investigaciones Biologicas in 1978 (Aurioles et al. 1983; Le Boeuf et al. 1983; Aurioles and Sinsel 1988). The program involved monitoring the population distribution and abundance, estimating annual pup mortality rates, and estimating winter immigration rates of subadult males at major rookeries and resting sites. During winter, a large number of subadult males move into rookeries and resting sites on both coasts of the southern Baja California peninsula (Aurioles et al. 1983). These movements may be analogous to the northward movements of sea lions inhabiting southern California waters (Fry 1939; Mate 1975) which are associated with fluctuations in the distribution of Pacific hake (*Merluccius productus*; Ainley et al. 1982). Subadult males from the central Gulf may follow the southward fall-winter migration of the Monterey sardine (*Sardinops sagax*), the threadfin herrings (*Ophistonema libertate, O. bulleri,* and *O. medirastre*) and the mackerel (*Scomber japonicus*) which are the basis for sardine fishery in the Gulf (Aurioles, in press).

An interesting comparison in EN effects exists between the relatively isolated Gulf of California sea lion population and the population residing on the Pacific Ocean side of the Baja California peninsula. Based on geography, we would expect the animals inhabiting the Gulf to be more buffered against effects associated with EN than the animals living in the Pacific Ocean. Primary productivity in the northern and central Gulf may be independent of major ecological perturbations on the Pacific coast where the ecosystem is dependent upon coastal upwelling (Lara-Lara et al. 1984; Mee et al. 1985). We present data suggesting that EN 1982–83 had no deleterious effect on the California sea lion population in the Gulf of California.

11.2 Methods

A total of 171 censuses were conducted between June 1978 and July 1985 at key rookeries and hauling out sites of Baja California (Table 1; Fig. 1 of Chap. 10, this Vol.). See Aurioles et al. (1983) and Le Boeuf et al. (1983) for description of age/sex categories. Censuses were conducted throughout the summer breeding season (late May to late July); we emphasize those taken at the peak of the pupping period (Table 1). Winter censuses were conducted from September through March when subadult males are most abundant. Subadult males start to appear at the southern part of Baja California at the end of September, and their numbers increase with the advance of the winter season (Aurioles et al. 1983). The mean number of subadult males each winter was estimated for two sites, Santa Margarita and Los Islotes (Table 2).

In order to track the mortality of pups during the first year of life, a total of 190 sea lion pups born on Los Islotes were hot-branded during the second week of July, near the end of the breeding season. Twenty-five pups were branded in 1980, 35 in 1981, 37 in 1982, 50 in 1983, and 43 in 1984. The brands were located on the back and could be identified with binoculars from a distance of 50–70 m. We dedicated 2–4 h on each 2 days/month to record marked pups. Observations were conducted monthly during the first year of the study, bi-monthly during the second and third, and at least four times a year in the last 2 years. Censuses were made simultaneously on land and in the water around the island, using binoculars (20 × 50) and scuba equipment. We assumed that pups not seen again had died. Estimates of annual pup mortality were based on the following assumptions:

1. All marks were permanent;
2. Branding did not decrease survival;
3. Pups remained in the area during the first year of life.

The first assumption was supported by resightings of branded individuals during subsequent years. The second assumption was supported by comparing the mortality rates from years when branding was conducted to years when branding was not conducted. The mortality rates were no different (Aurioles and Sinsel 1988). The last assumption was supported by the observation that no pups categorized as dead were ever resighted. Pups classed as survivors were in two categories; some were seen throughout the first year of life and others disappeared for short periods during the first year of life, but were resighted subsequently during the next 6 months of the second year of life. In contrast, pups considered dead disappeared in the first year of life and were not resighted during 15 additional surveys of branded pups.

11.3 Results

The first sea lion births occur in Baja California during the last week in May. The number of pups increases through June and reaches a peak during the first week in July (Fig. 1). The number of pups censused after the second week of July remains stable through August when pups are 1–2 months old. Pup mortality in the Gulf of California during the first 6 months of life is about 20% (Aurioles and Sinsel 1988). The total annual pup mortality is similar from year to year and is estimated to be in the range of 50–60%, with most of the mortality occurring during the second half of the year (Aurioles and Sinsel 1988). Thus, the number of pups counted after the first week of July through the end of August can be used as a measure of annual pup production.

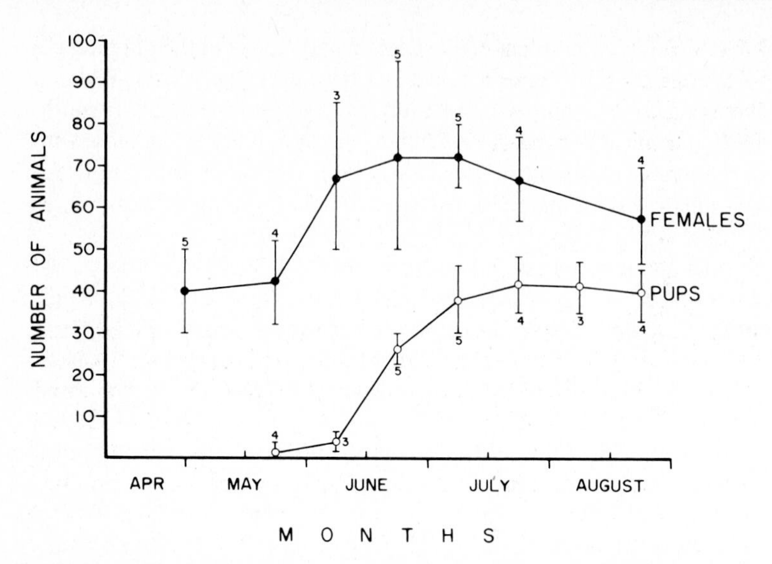

Fig. 1. Variation in the mean numbers of adult females and pups of California sea lions during the months around the breeding season at Los Islotes, B.C.S. Mexico. Average numbers of each month were arbitrarily placed at mid-month. *Numbers* over the vertical lines represent censuses. *Vertical lines* are standard deviations

11.3.1 Pacific Coast

The only sea lion rookery monitored on the Pacific coast was Isla Santa Margarita. Pup production in 1983, estimated from censuses, was 45% lower than the mean value calculated from preceding and succeeding years (Table 1). This difference is statistically significant ($t = 7.23$, $df = 2$, $P < 0.05$). Similar decreases in other animal categories in 1983 are difficult to interpret due to the late date of census.

We conclude that the low census in 1983 was due to reduced pup production rather than a high mortality of newborn pups. This is based on two observations: First, we did not record unusual numbers of carcasses on the beaches during our visit in August. A possible explanation is that the bodies were washed out to sea. However, a mortality rate around 50% in such a short time would be evident from carcasses on the beach above the high tide mark. Secondly, the number of females present during the 1983 breeding season was small (Table 1), suggesting that fewer females hauled out to give birth. Part of the decrease in the number of females was due to the late date of the census. An approximately 10% reduction is expected (Fig. 1). However, the number of females decreased 59% from 1982 to 1983. It is possible that increased duration of foraging at sea could have accounted for the decrease in females observed in 1983. If the low pup production in 1983 was caused by a reduced number of females on the island, then a complementary number of females were unable to finish the energetic investment in utero or these females went to other areas to give birth. The last possibility seems unlikely given that females of this species show high site fidelity (Peterson and Bartholomew 1967; Heath and J. Francis, pers. comm.).

Table 1. Population censuses of California sea lions on the Pacific coast of Baja California, and the Gulf of California

Location	Date	AM[a]	SA	AF	J	P	M	Total
Santa Margarita	7/14/79	265		2676	114	1202	–	4257
	8/20/83	3	37	465	142	409	20	1157
	7/27/84	23	80	1908	267	1080	116	3474
	7/03/85	33	38	1108	166	925	–	2270
Los Islotes	7/08/78	4	19	78	21	40	3	165
	7/10/79	8	4	71	27	39	8	157
	8/23/80	6	35	54	26	38	–	159
	7/07/81	10	13	63	18	46	–	150
	7/10/82	12	4	68	29	54	10	177
	7/16/83	11	10	53	19	50	–	143
	8/19/84	3	41	42	13	49	6	154
	7/19/85	5	10	79	22	52	8	176
Topo-lobampo	6/29/82	26	22	322	73	248	22	702
	7/01/83/	23	6	226	40	222	34	551
San Pedro	7/24/83	36	56	362	147	150	81	832
Nolasco	7/04/84	48	19	325	116	127	122	757
San Jorge	7/06/79	121	46	1398	632	1030	26	3253
	7/06/83	136	129	1785	670	907	36	3663

[a]AM, Adult males; SA, subadult males; AF, adult females; J, juveniles; P, pups; M, miscellaneous.

11.3.2 Gulf of California Rookeries

Pup production at Los Islotes, the rookery near the mouth of the Gulf closest to the Pacific, and at a similar latitude as Isla Santa Margarita, ranged from 38 to 54 before and after EN 1982–83 (Table 1). The number of pups born in 1983 was 10% higher than the mean of the other 7 years (50 vs 45.4 ± 6.5, respectively). Clearly, there was no significative difference or direct immediate deleterious effect of EN 1982–83 on breeding at Los Islotes in 1983. However, long-term survival may have been affected since an exceptionally low number of juveniles was counted the following year.

The data from other rookeries higher up in the Gulf are scantly and inconclusive. Farallon de Topolobampo is also located at the mouth of the Gulf but on the exposed east side. The 1982 census at this rookery was taken early in the breeding season when the number of pups was still increasing. The 1983 census taken 2 days later in the season should have produced slightly more pups, but instead the count decreased by 10.5%. We cannot conclude that this was due to EN 1982–83 since this difference is within the margin of error associated with the method of censusing. The decreases in numbers of females and subadult males in 1983 were greater and may be of significance.

The census data from San Pedro Nolasco and Isla San Jorge were taken at different times in the breeding season and, in the case of the latter, the two censuses were separated by 5 years. This makes the data difficult to interpret. It is not clear whether EN 1982–83 had a deleterious or a salubrious effect on either colony.

Pup mortality during the first year of life for pups born on Los Islotes was estimated for each of five breeding seasons, 1980 to 1984. On average, 10% of the annual mortality at Los Islotes occurred from mid-July to December; the majority of the annual mortality, about 39%, occurred from January to June. Fluctuation in annual mortality at Los Islotes is shown for these years in Fig. 2. Since pups are born in summer and mortality is estimated over the period to the next summer, we were interested in determining whether the start of EN 1982–83 could have affected the survival of the 1982 cohort near the end of their first year of life. The pup mortality rate in 1982 was the lowest of the 5 years recorded; the 1983 pup mortality rate was the highest recorded. However, the 1983 pup mortality rate is not significantly different from other years (Chi square = 2.46, df = 4, $P > 0.05$).

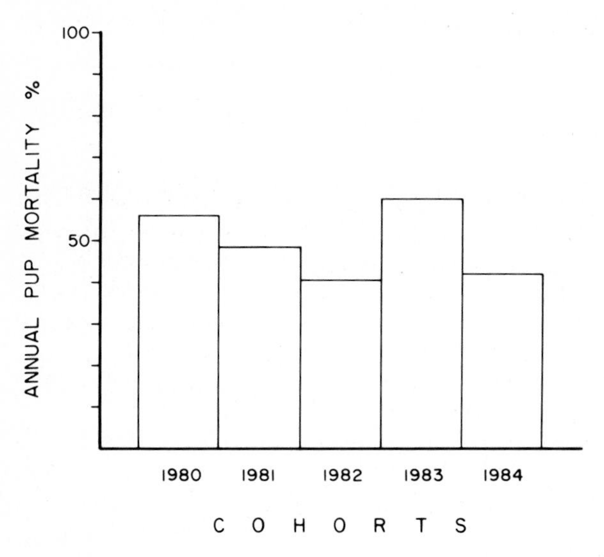

Fig. 2. Mortality at the end of the first year of life of California sea lions during five continuous years at Los Islotes, inside the Gulf of California

11.3.3 Movements and Abundance of Subadult Males

If we assume that the appearance of subadult males around the southern tip of Baja California in winter is associated with feeding in the area, then the number of males present may be interpreted as a reflection of feeding conditions. If food is scarce locally, we would expect the males to avoid the area or spend little time there.

Table 2 shows the maximum number of subadult males present in two rookeries in south Baja California over the period of EN 1982–83. If EN 1982–83 had a deleterious effect on foraging, the critical winters would have been 1982–83 and 1983–84. The first, because of a possible EN effect starting at the end of 1982, and the second, because of a potential delayed negative effect during 1983 that extended into winter 1984. The number of males at Los Islotes during the winter of 1982–83

Table 2. Mean number of subadult male California sea lions during winter from 1978 to 1985 at two locations of Baja California, Mexico (N = number of censuses)

	Los Islotes		
Winter	$\bar{X}$	SD	N
1978–79	44.2	14.1	7
1979–80	72.6	25.3	3
1980–81	51.0	22.0	7
1981–82	99.6	57.4	3
1982–83	68.3	14.8	3
1983–84	85.3	18.7	3
1984–85	123.0	37.1	3
	Santa Margarita		
1982–83	240.3	65.4	3
1983–84	132.1	43.2	3
1984–85	217.6	30.3	3

was not significantly different from other years ($t = 0.577$, $df = 7$, $P > 0.05$). The number of males recorded in the next winter was even greater. In contrast, the Santa Margarita rookery had a high number of males present during the winter of 1982–83 but a low number present the following winter. The latter figure represents a 45% reduction in the number of males over the previous year and 39% lower than the following year, 1984–85. A comparison of the EN year to the pooled data from the previous and following year (six censuses versus three censuses) shows a statisti- cally significant difference ($t = 2.442$, $df = 7$, $P < 0.05$).

11.4 Discussion

California sea lions at Isla Santa Margarita, located on the Pacific coast of Baja California, were negatively impacted by EN 1982–83. The effect was a 50% reduction in pup production, a reduction of parturient females during the breeding season, and a reduced number of subadult males present in the rookery the following winter.

In contrast, representative sea lion colonies in the Gulf of California were affected less, if at all, by the event. This is supported by the following points:

1. Pup production at Los Islotes did not change significantly over the period, 1978 to 1985.
2. The number of females during the breeding season was about the same during the period of study.
3. Pup mortality during the first year of life did not change significantly during the years 1980 to 1984.
4. The number of subadult males migrating into the southern Gulf of California area, presumably to feed, did not change significantly over the years that included EN 1982–83. Fragmentary data from rookeries high up in the Gulf do

not confirm significant decreases in population number or increases in pup mortality.

The lack of an adverse effect of EN 1982–83 on sea lions in the Gulf of California is consistent with similar observations on other species. The mean number of brown pelican (*Pelecanus occidentalis*) chicks fledged in Bahia de La Paz over a 7-year period was no different in 1983 as in non-EN years (Jiménez and Guzmán 1986). The frequency and abundance of cetacean species in the vicinity of Bahia de los Angeles in the north-central part of the Gulf was equal to or higher than in non-EN years (B. Tershy, pers. comm.). Studies on primary productivity in the Gulf of California during the years following the EN 1982–83 seem to confirm the hypothesis that the productivity in the Gulf not only remained unchanged but in some areas may have had higher values during the EN than in previous and subsequent years (Baumgartner et al. 1987; Lara-Osorio and Lara-Lara 1987; but see Lara-Lara et al. (1984) for disagreement regarding increase). According to Mee et al. (1985), the tidal mixing process (Alvarez-Borrego 1983), which discharges fertile waters to adjacent areas, would occur despite EN events and should protect the Gulf against the catastrophic effects recorded in other areas.

11.5 Conclusions

Pup production, pup mortality and number of breeding females in the summer, and number of subadult males in winter were no different during the El Niño year (1983) than during previous and subsequent years inside the Gulf of California. In contrast, pup production and the number of females and subadult males decreased significantly in 1983 at a rookery off the Pacific coast of Baja California.

California sea lions in the Gulf of California seem to be protected from El Niño events (considering the last as the strongest of the century). Fertilization of the Gulf of California is due to a strong mixing process which appears to be independent of El Niño effects and is responsible for the benign conditions recorded in this area. In contrast, upwelling fertilization in the Pacific coast of Baja California was affected by the warm water carried by El Niño, as recorded in other areas of the Eastern Pacific coast.

Acknowledgments. We thank Dr. Félix Cordoba Alva for help with permits and logistics; George Shor for his help with ship support; Esteban Alvarado, Francisco Sinsel, and Eduardo Muñoz for assistance in the field; and Drs. Fritz Trillmich and Kathy Ono for comments on the paper. This research was supported in part by grants from Consejo Nacional de Ciencia y Tecnologia, UC Mexus, and the National Science Foundation (DEB 77–17063 AO1).

12 Population Abundance, Pup Mortality, and Copulation Frequency in the California Sea Lion in Relation to the 1983 El Niño on San Nicolas Island

J.M. FRANCIS and C.B. HEATH

12.1 Introduction

Large-scale oceanographic events such as EN can be expected to affect pinnipeds in various ways, one of which is the number that return to their island breeding areas during a year of environmental perturbation. Altered food supplies may lead to changes in the distribution of animals throughout their range, in the amount of time spent at sea on feeding forays from the breeding areas, in the fecundity of females and of males, and in the mortality of animals at sea and on land. Each of these variations can lead to changes in the population size and composition observed on the rookeries. By examining in detail the population of one island (San Nicolas Island) we will attempt to clarify not only the net changes in sea lion numbers that occurred during EN, but by what mechanisms (mortality, migration, copulation frequency, etc.) these changes are brought about.

12.2 Methods

12.2.1 Censuses

The numbers of adult males, females, subadult males, juveniles, and pups on San Nicolas Island were determined by censuses conducted from land. Censuses of all sea lions on the island (total island censuses) were performed every 2 weeks during the breeding season (May-August) 1982–1984, and at less frequent intervals during other periods from 1981–1984. During the breeding season, censuses were also conducted one to three times daily at our two study sites, areas 232 and 233 at 0800, 1400, and/or 1800 h. Area 232 is a broad, sandy beach, while area 233 is an uplifted rocky reef area. The two sites typify the island topography. Age classes of animals were assigned as in Heath and Francis (1983).

12.2.2 Pup Mortality

Descriptions and locations of all dead pups sighted on study areas 232 and 233 were recorded daily from 17 May through 26 July, 1982–1984. In addition, during each total island census, the number of pups that had died since the previous census (determined by degree of decomposition) were recorded. Also, in early August at both study sites, two observers walked on the beach and independently counted dead pups for comparison with data collected daily from the observation blinds.

12.2.3 Male Tenure and Copulation Frequency

Individual males were identified by natural scars, and records were kept on arrival and departure dates and copulations during the 1982–1984 breeding seasons. Tenure was calculated for those males holding territories for more than 4 days, regardless of copulatory status. Copulation rates were calculated using the sum of all copulations observed from 20 June through 25 July of each year, divided by the number of observation hours during that interval. Since the 1981 observations began after many males were on territory, tenure was not calculated for that season.

12.3 Results

12.3.1 Censuses

The number of adult males at San Nicolas Island during the breeding season changed relatively little from 1981 through 1984 as measured both over the total island and at the study sites. Since total island census dates varied slightly between years, and because the number of males varied over each breeding season (see Fig. 1; Heath and Francis 1983) exact comparisons between some years are not possible. However, the 1982 and 1983 June censuses were only 1 day apart (20 and 19 June) and differed by only one male. Further censuses on 24 June in both 1981 and 1984 show 1984 as 11% higher. The peak censuses in 1983 and 1984 are also comparable (5 July and 6 July) and show the male population to be 21% higher in 1984 than in 1983. By mid-July, however, the censuses for 1983 and 1984 were nearly identical (511 and 526, respectively). Censuses at the study areas, since they were conducted daily, provide a better basis for comparisons of male abundance between years. There were no significant differences in the number of territorial males at areas 232 or 233 during the peak breeding season (28 June-11 July) from 1981–1984 (Table 1).

Subadult male (SAM) censuses for the entire island are also difficult to compare between years due to slight changes in census dates. Two comparable censuses, in July of 1983 and 1984, show almost identical numbers of SAMs ashore (Fig. 1). The mid-June censuses, however, show more of a change: the number of SAMs ashore on 24 June 1984 was 27% higher than on the same date in 1981. The 1982 and 1983 counts fall in between, suggesting a steady increase over the 4 years. The daily censuses at areas 232 and 233 are quite variable and show no significant differences during the peak breeding season (28 June-11 July) from 1981 through 1984 (Table 1).

The number of females on San Nicolas Island was lower in the EN year as compared to the years preceding and following (Fig. 1). There was a 16% increase from 1981 to 1982, a 40% decline in the (1983) EN year, and a partial recovery in 1984 to 74% of the 1982 value. Peak female counts at the study areas changed sig-

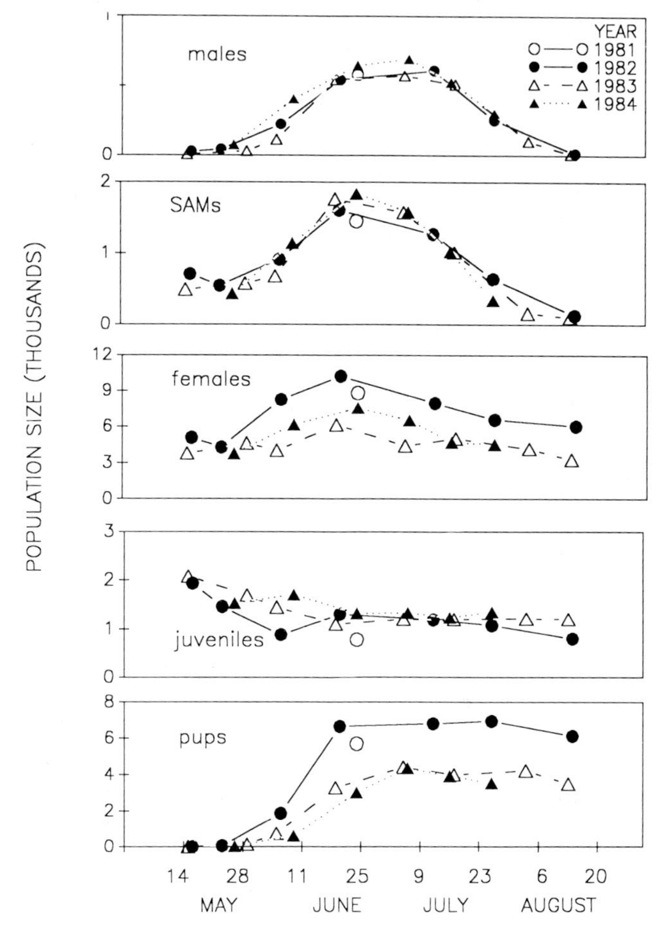

Fig. 1. Numbers of adult and subadult males (SAM), females, juveniles, and pups on San Nicolas Island, 1981–1984

nificantly and followed the same trend with a 38% decline in 1983 and a 1984 increase to 71% of 1982 values (Table 2).

The number of juveniles counted on San Nicolas increased slightly over the 1982 to 1984 breeding seasons. Mean values during the months of June and July (during which juvenile numbers varied little; Fig. 1) showed an increase of 10% per year. Maximum juvenile counts, occurring just prior to the breeding season, increased by 8% from 1982 to 1983. Yearlings, judging by size, comprised the vast majority of juveniles on San Nicolas. Juvenile counts also indicate an increase from 1982 through 1984 at one of the two study areas, although the difference was not significant (Table 2).

The number of pups on San Nicolas Island was substantially lower in the EN year and the year following as compared to the 2 years preceding (Fig. 1). Counts of

Table 1. Number of California sea lion males and subadult males at study areas 232 and 233, San Nicolas Island during peak breeding season (28 June to 11 July) 1981–1984. Means are calculated from all midday censuses during the interval. Interannual comparisons are made using these yearly means with a one-observation-per-treatment two-way ANOVA (Neter et al. 1985). Values presented are mean, standard deviation, and (n)

Area	1981	1982	1983	1984
Males[a]				
232	10.0 1.6 (6)	12.8 1.8 (11)	11.6 1.2 (11)	11.7 1.5 (11)
233	23.8 2.0 (9)	24.5 1.6 (10)	23.1 2.5 (10)	22.6 1.8 (10)
Subadult males[b]				
232	8.4 6.7 (5)	10.9 4.2 (11)	14.2 11.3 (11)	12.8 9.1 (11)
233	3.7 4.6 (6)	4.3 1.5 (10)	3.7 2.2 (10)	1.2 1.2 (10)

[a]Year effect not significant: $F_{3,3} = 1.57, p > 0.10$.
[b]Year effect not significant: $F_{3,3} = 0.56, p > 0.50$.

Table 2. Number of California sea lion females, juveniles, and pups at study areas 232 and 233, San Nicolas Island. Means are calculated from all midday censuses during the week ending July 4 for juveniles and pups, and the week ending June 20 for females. Interannual comparisons are made using these yearly means with a one-observation-per-treatment two-way ANOVA (Neter et al. 1985). Values presented are mean, standard deviation, and (n)

Area	1981	1982	1983	1984
Females[a]				
232	334.0 8.5 (2)	480.7 64.4 (7)	270.8 36.5 (6)	316.3 60.1 (6)
233	499.0 — (1)	566.8 95.8 (5)	365.6 62.8 (5)	396.0 18.0 (5)
Juveniles[b]				
232	89.5 53.0 (2)	55.0 15.0 (5)	61.6 11.9 (5)	80.5 14.8 (4)
233	69.0 46.7 (2)	20.3 5.0 (4)	18.7 6.4 (7)	17.8 7.9 (4)
Pups[c]				
232	437.5 29.0 (2)	567.6 14.9 (5)	371.6 11.2 (5)	383.7 23.3 (4)
233	593.0 52.2 (2)	741.8 37.0 (4)	423.3 45.5 (7)	383.0 41.3 (4)

[a]Year effect significant: $F_{3,3} = 20.6, p > 0.025$.
[b]Year effect not significant: $F_{3,3} = 4.7, p > 0.100$.
[c]Year effect not significant: $F_{3,3} = 9.1\ p = 0.053$.

pups vary considerably between days depending on weather and consequent location and visibility of pups. Consequently, for this age/sex class, it is most useful to compare the highest pup count obtained following the cessation of pupping (early July). The peak pup count for 1983 was 37% lower than our highest 1982 count. Pup numbers were essentially unchanged in 1984 (Fig. 1). Counts at the study areas show the same trend as for the total island, and the differences approached significance (Table 2).

12.3.2 Male Tenure and Copulation Frequency

The average male stayed on land at least as long or longer and copulated less frequently during the EN year and the year following as compared to the 2 years preceding. For all males at the study areas with tenure greater than 4 days, the mean duration of tenure was significantly longer in 1983 than in 1982, with no significant difference between 1983 and 1984 (Table 3). In contrast, the number of copulations observed per hour was significantly lower during the 1983 EN breeding season and the year following as compared to the 1981 and 1982 breeding seasons (Table 4). The copulation rate decreased at both areas by about 46% from 1982 to 1983, and was still 34% lower than the 1982 rate in 1984. In addition, at area 233 only, the copulation rate was significantly higher during 1981 than in the other years (Table 4). Since the number of territorial males remained relatively constant at each study area over the years, the number of copulations per hour per male also decreased significantly from 1981 to 1982 to 1983 and 1984 (Table 4).

We were unable to take any direct measurements of male weight or health; visual inspection showed no discernible differences among years.

Table 3. Mean tenure of territorial adult males at areas 232 and 233, San Nicolas Island, 1982–1984. Comparisons between years were made for males with tenure > 4 days (with or without copulations) using each male's tenure in a two-way ANOVA corrected for unequal sample size. Adjustments for multiple comparisons were made using the Bonferroni method[a]

Study area		1982	1983	1984
232	$\bar{x}$	22.5	30.2	24.3
	SE	2.4	3.2	3.3
	n	29	16	22
233	$\bar{x}$	20.8	29.6	27.1
	SE	1.9	2.6	2.5
	n	53	28	40

[a]ANOVA for year effect: $F_{2,182} = 4.54$, $p = 0.0119$, 1983 > 1982; $p < 0.05$.

Table 4. Copulation rates at areas 232 and 233, San Nicolas Island, 1981–1984. Number of males are means of all midday censuses during the study period

Area	Year	No. of males	Copulations/h[a]	Copulations/h/male[b]
232	1981	10.0	0.44	0.044
	1982	12.8	0.43	0.034
	1983	11.6	0.24	0.021
	1984	11.7	0.28	0.024
233	1981	23.8	1.15	0.048
	1982	24.5	0.79	0.032
	1983	23.1	0.43	0.019
	1984	22.6	0.54	0.024

[a]Chi-square – area 232: 1981, 1982 > 1983, 1984. $p < 0.001$; area 233: 1981 > 1982 > 1983, 1984. $p < 0.001$.
[b]Chi-square – area 232: 1981, 1982 > 1983, 1984. $p < 0.001$; area 233: 1981 > 1982 > 1983, 1984. $p < 0.001$.

12.3.3 Pup Mortality

Pup mortality was higher during the 1983 and 1984 breeding seasons as compared to the 1982 breeding season (Table 5). This is true for the daily dead pup counts at the study sites but is more apparent in the estimates from biweekly total island dead pup counts. These estimates include correction factors for dead pups that were either overlooked or had disappeared. Because carcasses decomposed, were buried, eaten, or otherwise rendered indistinguishable after an average of 8 days (n = 384 carcasses), biweekly counts of dead pups overlooked a proportion of the pups that died. Daily records at the study sites showed that on average 57% of pups dying in a 2-week period were visible at the end of that period, and that only 59% of those present were counted during census scans (Heath and Francis 1983, 1984, unpubl. data). Biweekly counts of dead pups thus only accounted for, depending on the year, 25 to 41% of pups that died over the breeding season, and total dead pups counted were corrected by this factor. It is interesting to note that single counts of

Table 5. Pup mortality[a] during the first 2 months of life in the California sea lion for study sites 232 and 233 and for the total population on San Nicolas Island (number of dead pups in parentheses)

	1982		1983		1984	
Area 232	12%	(92)	17%	(79)	13%	(62)
Area 233	10%	(78)	12%	(66)	13%	(66)
Total island[b]	10%	(813)	18%	(983)	17%	(883)

[a]Calculated as No. of dead pups/(peak pup count + No. of dead pups).
[b]Total pup numbers corrected for disappearance of carcasses and scanning error (see text for explanation).

dead pups at the end of the breeding season underestimated pup mortality by 30–70% depending on the year and, alone, provide inaccurate estimates of pup mortality in this species (Heath and Francis 1983, 1984).

As with pup mortality, the frequency of abortions was higher in 1983 than 1982. Beginning in January 1983, aborted fetuses were counted during all of our censuses. Seven were found on 30 January, 1 on 27 February, 52 on 16 April, and 109 on 15 May. We do not have census data for the 1982 winter, however, during our 16 May 1982 census we counted 31 fetuses as compared to 109 in 1983. Similar data on the frequency of abortions are unfortunately not available for the 1984 breeding season.

12.3.4 Female Fecundity

Female fecundity appeared to decline in the year following EN as estimated from female and pup counts. Assuming that peak female counts are equally representative of the total island population in a given year (discussed below), relative reproductive rates can be calculated as the ratio of peak pup to peak female counts. In 1981–84 these values were 0.76 (6704/8829), 0.68 (6952/10224), 0.72 (4405/6128), and 0.58 (4360/7555) pups produced per female. However, even at peak population, a proportion of the breeding female population is out feeding and any changes in female attendance patterns between years will differentially affect this proportion. Our data on attendance patterns indicate that an average of 60% (59, 62, and 61% for 1982–84, respectively) of a female's time is spent at sea (Heath et al., this Vol.). When these values (or the average, in the case of 1981) are used to modify peak female counts, fecundity is estimated as 0.30, 0.28, 0.27, and 0.23 pups per female for 1981–84, respectively.

12.4 Discussion

12.4.1 Censuses

It is apparent from these data that adult males and females at San Nicolas Island were affected differently by EN. Except for a possible increase in the first half of the 1984 season, adult males were present in roughly equal numbers in all years, while the number of females decreased by 40% in 1983 and recovered to only 74% of the 1982 values in 1984.

Subadult numbers varied greatly on a daily basis at the study sites and no clear interannual trend is apparent in comparisons of either the study site or total island censuses. Trends in population size for SAMs may be obfuscated by their movement patterns. Very little is known about the degree of site fidelity or duration of time ashore for individual SAMs. Their ties to the breeding islands appear to be much weaker (pers. obs.) than those of adult, territorial males, or of females with dependent offspring. More information is necessary on SAM behavior if we are to draw any conclusions from census data.

Juvenile numbers on San Nicolas Island appeared to increase in 1984, a trend unsubstantiated with tag resight data (Francis and Heath, Chap. 21, this Vol.). It is unlikely that this reflects higher juvenile survival, since concurrent resights of yearlings at nonbreeding sites decreased in 1984. Instead, these changes may have been a product of a higher reliance of yearlings on maternal milk which contributed to higher numbers remaining near their birth site in the year following EN (Francis and Heath, Chap. 21, this Vol.).

Pup numbers declined precipitously from 1982 to 1983 and remained low in 1984. The drop in 1983 was likely a result of the high abortion rate observed in addition to higher postnatal mortality. Female mammals reduced to a low nutritional plane are more likely to abort (Brambell 1948; Stein and Susser 1975) and there is ample evidence that female pinnipeds of the eastern tropical Pacific were so limited during this year (see Chap. 26, this Vol.).

The absence of an increase in pup numbers in 1984 appears to have been a latent effect of the 1983 environmental conditions. Reduced female numbers at San Nicolas in 1983 resulted in low copulation frequencies. This may have created a higher than normal proportion of nonparous females in the population which remained at sea in 1984, not having to give birth. However, while this explains low numbers of pups and females in the 1984 breeding season, it does not explain why in 1985 and 1986 on San Nicolas there was no rebound in pup production to pre-EN numbers as there was on the other Channel Islands (DeLong et al., this Vol.).

It is possible that female numbers and consequent copulation frequencies on the other Channel Islands did not drop as they did on San Nicolas. Thus, the lagging decline in pup production may not have occurred to the same extent as on San Nicolas. It is also possible that the other Channel Islands are closer to the best foraging areas and the pressures of EN caused individual females to relocate in response to foraging difficulties. Further, it has been suggested that human disturbance caused emigration of San Nicolas females and a drop in pup numbers (DeLong et al., this Vol.). We need to know more about the plasticity of female site fidelity and about female foraging behavior in order to test these hypotheses.

12.4.2 Copulation Frequency and Tenure

Even though the average male stayed on land as long or longer in 1983 and 1984 than in the years preceding, the benefits of this increased tenure, in terms of copulation frequencies, declined. However, the benefits of establishing or maintaining prior residence, as has been documented for the Steller sea lion (Gisiner 1985) must also be considered. If a male's continued presence during 1983 increased his chance of obtaining a territory and copulations in the future, this, plus the copulations achieved in 1983, could have offset the costs of maintaining a territory in a year of low productivity.

It is not clear which factors, such as physical condition or expected reproductive benefit, affect duration of tenure in otariids. Further, whether males were actually in the same physical condition in all years cannot be determined. It is possible that their food supplies to the north were unchanged, decreased, or even improved as a result of EN (see York, this Vol.). Assuming that tenure reflects physical condition, that male tenures and numbers did not decrease during the same time period when female numbers decreased and female feeding trips to sea increased indicates that males were not as adversely affected by EN as were females.

12.4.3 Pup Mortality

Pup mortality was higher in the 1983 EN year and the year following than in 1982. While no data were collected on the causes of pup deaths, there were no obvious differences between years in the occurrence of high waves (e.g. the northern elephant seal, Le Boeuf and Reiter, this Vol.) which could account for the higher mortalities observed. Higher temperatures were observed on San Nicolas Island in 1984 as compared to 1983 (unpubl. data) as was reported for San Miguel Island (DeLong et al., this Vol.). The extent to which pup mortality was caused by heat exhaustion on San Nicolas is unknown. Lower pup weights in 1983 and 1984 (Francis and Heath, Chap. 21, this Vol.) indicate poorer nutrition of pups in general and suggest a higher probability of death related to emaciation. This is the predominant cause of breeding season pup mortality in northern fur seals (Keyes 1965), the otariid for which we have the best data on causation of pup mortality. Densities of adult males and females, if anything, decreased on San Nicolas Island (unpubl. data) during EN and the year following, so that increases in mortality due to crowding and related trauma (e.g. in the Antarctic fur seal, Doidge et al. 1984) is unlikely.

12.4.4 Female Fecundity

Though female fecundity appeared to decline following EN, as measured through pup/female ratios, these results must be considered tentative. Even with modifications to peak female counts based on attendance patterns of parturient females, the accuracy of these fecundity estimates is limited by the unknown and likely variable component of the female population that remains at sea during the breeding season. Problems arise in interyear comparisons, for example, if more females abort in a particular year, and these females only visit the rookeries briefly in the summer to mate, if at all. In fact, the low numbers of females on the rookeries in 1983 and 1984, if indicative of a large nonparous population at sea, suggest that the actual fecundity was much lower than estimated for these 2 years.

Despite the potential inaccuracy introduced by females remaining at sea, the pup/female ratios provide believable estimates of female fecundity in this species, given the other data available. Data on the reproductive status of a sample of naturally marked females indicate a fecundity of 52% for 1982 females on San Nicolas Island (Heath and Francis 1983), higher than the 28% female/pup ratio estimate for the same year. The estimate derived from marked females, however, is subject to bias since marks take time to accumulate and may be more common on older females. Female fecundity in the northern fur seal has been shown to increase with age over the first 10 years of life (York and Hartley 1981) which may explain the higher estimates for the sample of naturally marked sea lion females.

The 30 to 50% female fecundity of the California sea lion is much lower than in the northern fur seal, where pregnancy rates range from 70 to 90% among adult females (York and Hartley 1981). This reduced fecundity, associated with high abortion rates, low copulation frequencies, and females caring for their young into a second year (Francis and Heath, Chap. 21, this Vol.) appears to be a product of variations in environmental quality associated with EN.

Acknowledgments. We thank Lt. Cdr. Eugene Giffin and Mr. Ron Dow of the U.S. Navy Pacific Missile Test Center for logistical support on San Nicolas Island. Wyatt Decker, Charles Deutsch, Jane King, and Mark Lowry contributed to data collection in the field. Dr. Kathy Ono and one anonymous reviewer commented on the manuscript and Drs. Bill Rice and Kathy Ono provided statistical advice. Funding was provided by the Southwest Fisheries Center, Coastal Marine Mammals Program under the direction of Dr. Doug DeMaster and by the Biology Board of the University of California, Santa Cruz.

13 Changes in the Distribution of California Sea Lions North of the Breeding Rookeries During the 1982–83 El Niño

H.R. HUBER

13.1 Introduction

At southern California rookeries and at nearby nonbreeding areas, the number of California sea lions peaks during the summer breeding season (Bonnell et al. 1983). At central California haulouts north of the breeding rookeries, numbers of adult and subadult males (Bartholomew 1967) peak during the spring and fall migration (Orr and Poulter 1965; Mate 1975; Bigg 1985; Brown 1988; Huber et al. in press) and decline during the summer when adults are at the breeding rookeries.

In the 1960s, most California sea lions north of the rookeries in the nonbreeding season were adult and subadult males; it was hypothesized that females and immatures remained near the rookeries or moved south after the breeding season (Bartholomew 1967). As the population increased in the 1970s, the distribution of immature animals in central California changed. The difference was first noted in 1977 when the number of immature California sea lions hauling out during fall at South Farallon Islands increased (Ainley et al. 1982). A change was also documented at Año Nuevo Island where an increase in the number of immatures hauled out during the summer began in 1981 (Bonnell et al. 1983). Few immature animals were reported north of central California, although Mate (1975) found 4 to 10% of the California sea lions in Oregon were under age 5 (in 1968–70) and Everitt et al. (1980) reported at least 11% immature California sea lions at one site in Puget Sound, Washington (in 1979). In British Columbia, only adult and subadult males have been reported (Bigg 1985).

During the summers of 1983 and 1984, the number of California sea lions in central California increased dramatically over previous summer counts. In this chapter, I compare the results of censuses at two sites on the Point Reyes Peninsula, California, between March 1982 and March 1984, with weekly censuses from 1973 to 1986 conducted at the Farallon Islands (Huber et al. in press) and censuses from 1982 to 1985 at the Monterey Breakwater (Nicholson 1986).

13.2 Methods and Study Area

The Farallon Islands are three groups of granitic islands and rocks 43 km west of San Francisco: the North, Middle, and South Farallon Islands (Fig. 1). Farallon counts refer to sea lions on the largest of these groups, the South Farallon Islands (44 ha) where most wildlife is concentrated.

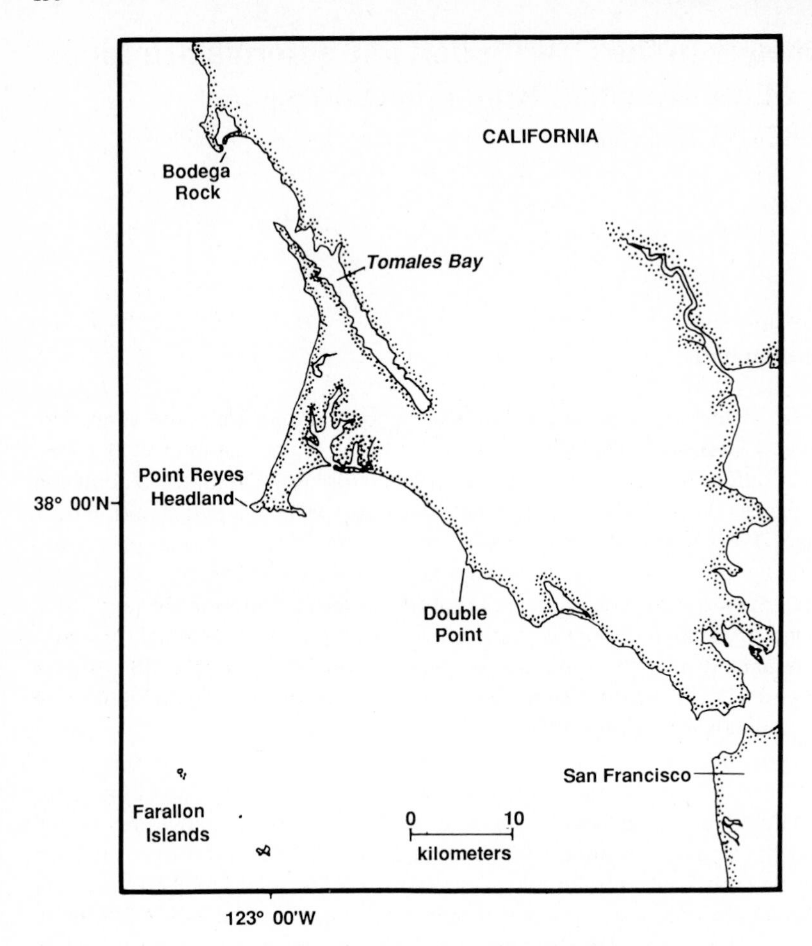

Fig. 1. Map of haulout sites of California sea lions on the Point Reyes Peninsula

Biologists from Point Reyes Bird Observatory began weekly censuses of California sea lions in late 1973. Throughout the year, sea lions were counted by telescope once a week from vantage points above the haulout sites. Sea lions were censused between 1300 and 1600 h when peak numbers of animals hauled out on the South Farallons (Ainley et al. 1982). The few animals in the water were included in counts. Because numbers of California sea lions fluctuated from week to week, monthly means were used in the analysis of population abundance.

Sea lions were also counted at their haulout site, a cobble beach on the southwest tip of the Point Reyes Headlands (Fig. 1). For the first year (March 1982-February 1983) animals were counted twice a month, except during the breeding season when censuses were conducted weekly to check for evidence of births. During the second year (March 1983-March 1984), they were counted each week. All counts on the Point Reyes Headlands were made from Sea Lion Overlook (35 m above the cobble beach) with a 25× spotting scope during midafternoon low tides when the maximum number of animals hauled out. In comparing censuses from Point Reyes and the Farallons, I compared only those censuses from the same time period.

California sea lions were also counted at Bodega Rock (Fig. 1), once a month during 1982 and twice a month during 1983. Bodega Rock is a 10-m-high rock, less than 0.5 km offshore at the mouth of Bodega Bay. In addition, sea lions were censused by boat in Tomales Bay (Fig. 1) during the Pacific herring (*Clupea harengus*) spawning period (December to March). Nine censuses were conducted during the first year and five during the second year.

To determine changes in abundance of different age and sex classes, sea lions were separated into three groups whenever possible: pups (young of the year, May through December), adult and subadult males (animals with at least the beginning of a sagittal crest), and others (immature animals of either sex that had no evidence of sagittal crest, and adult females). In the fall of 1976, a sample of 100 immature animals on the Farallons were sexed by inspection. All were males. In summer of 1983, a sample of 30 immature animals at the Farallons were sexed; a third were females. For analysis of changes in the proportion of age/sex classes present, I omitted censuses in which the age and sex of more than 10% of the animals were unidentified.

California sea lions tagged by researchers on Channel Island rookeries were resighted on the Farallons during migration. Analysis of tagging resights was based on the proportion of sea lions resighted on the Farallons compared to the number of animals tagged. Identification of island of tagging was based on tag color, which was unique for each island.

13.3 Results

During the summers of 1983 and 1984 marked changes occurred in the peak annual count, in seasonal abundance, and in the age and sex composition of California sea lions seen in central California compared to previous and subsequent years. Counts of sea lions were greater in 1983 than in 1982 at all three sites in central California: peak counts increased by 54% at the South Farallon Islands, by 125% at Point Reyes Headlands, and by 240% at Bodega Rock. At Double Point (10 km south of Point Reyes), up to 116 sea lions (95% immatures) were counted between May and August 1983 compared to the previous 7 years when no sea lions were observed (S.G. Allen, pers. comm.).

Seasonal abundance of sea lions at the South Farallon Islands and Point Reyes Headlands also changed after 1982. In 1982 and before, there were seasonal peaks at the Farallons corresponding to the spring and fall migrations (see Huber et al. in press). In 1983 and 1984, sea lion numbers began to increase in spring and continued to rise through the summer breeding season in contrast to all other years when there was a decline in numbers during the summer. From June through August the monthly mean counts of California sea lions at the Farallons were significantly greater in 1983 and 1984 than in 1982 (Mann-Whitney U-test, $P < 0.025$).

At Point Reyes there were three equal peaks in 1982: in spring at the beginning of the southward migration to the breeding rookeries, in fall during the northward migration, and in late December during the herring spawn (Fig. 2). In 1983, there was a large increase in animals which began in May and lasted through August, similar to the situation at the Farallons. From May through August 1983 the monthly mean number of California sea lions at Point Reyes was significantly greater than in 1982 (Mann-Whitney U-test, $P < 0.05$). A smaller increase, coinciding with the herring run in late December, also occurred in 1983 (Fig. 2). No pups or evidence of births were observed in either 1982 or 1983.

In contrast, at Bodega Rock the seasonal patterns for 1982 and 1983 were similar: in both years the number of sea lions peaked during the spring and fall migration and were low during the breeding season (Fig. 3). The greatest abundance of sea lions occurred in the fall, in September 1982 and in October 1983 (Fig. 3). During December 1983 there was a third peak that coincided with the Tomales Bay herring spawn. A winter peak was not recorded in 1982; probably because a series of severe storms limited observations. Almost all animals seen at Bodega Rock were adult and subadult males.

In Tomales Bay, peak counts occurred during the herring spawn in winter 1982–83 (37 animals) and in winter 1983–84 (32 animals). Twenty percent of all animals seen in Tomales Bay in 1982–83 were adult males compared with 5% in 1983–84.

Between 1982 and 1983, significant changes occurred in the age composition of California sea lions at the Farallons. The mean number of California sea lions present during the summer nearly doubled from 2034 in 1982 to 3990 in 1983. Thus, even though the mean number of adult and subadult males present in both years was similar (435 in 1982 and 475 in 1983), the proportion of adult and subadult males present during the summer decreased significantly from 21% in 1982 to 12% in 1983 ($X^2 = 78.08$, df = 1, $P = 0.0$). As a corollary, the mean number and

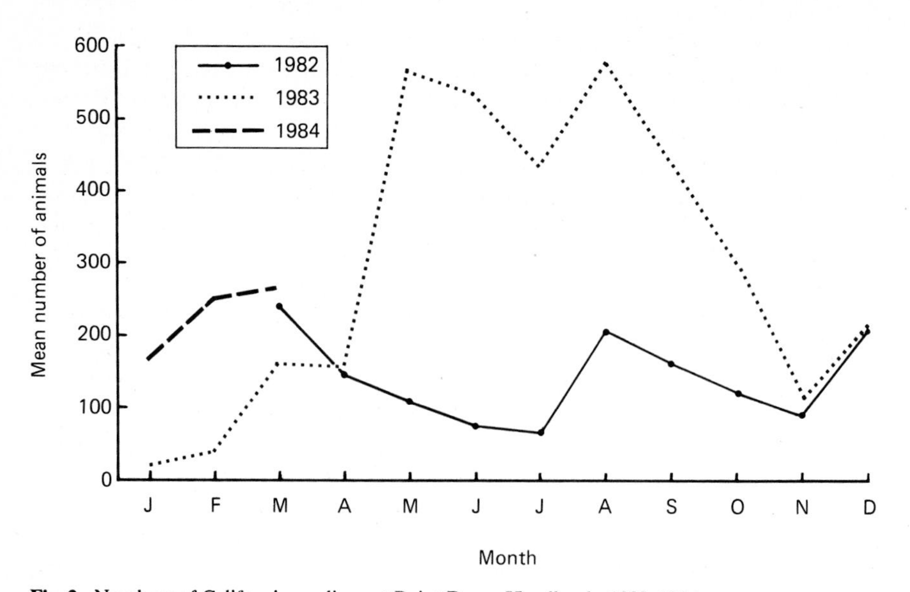

Fig. 2. Numbers of California sea lions at Point Reyes Headlands, 1982–1984

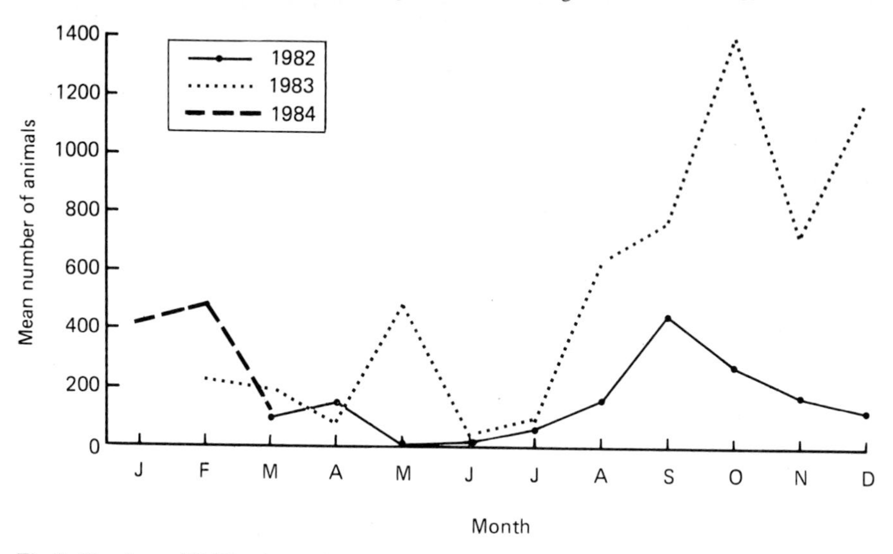

Fig. 3. Numbers of California sea lions at Bodega Rock, 1982–1984

proportion of immature animals present in the summer also changed significantly, increasing from 1603 in 1982 (78%) to 3305 (88%) in 1983. At least ten adult females were also present. At Point Reyes, the number of adult and subadult males present during the summer increased fivefold, from 13 in 1982 to 70 in 1983. However, because the total number of California sea lions present also increased fivefold, the proportion of adult and subadult males present was not significantly different between the 2 years ($X^2 = 0.33$, df = 1, $P = 0.56$).

Of the 114 tagged California sea lions resighted at the South Farallon Islands from 1983 to 1985, 71 of the animals (19 females, 52 males) were tagged as pups at San Nicolas Island and 43 animals (11 females, 32 males) were tagged as pups at San Miguel Island (Table 1). In analyzing resighting data for differences between the two islands where pups were tagged, I found that yearlings in the 1981 cohort from San Miguel Island were less likely to be seen on the Farallons than animals from the 1981 cohort tagged at San Nicolas Island ($X^2 = 4.0$, df = 1, $P = 0.046$; Table 1), although the difference may have been caused by the type of tag used on the 1981 cohort from San Miguel, a monel tag which is more difficult to read from a distance. Otherwise, there were no interisland differences in resighting on the Farallons of animals tagged on San Nicolas Island or San Miguel Island. From both islands, the proportion of tagged males observed was significantly greater than the proportion of tagged females ($X^2 = 29.03$, df = 1, $P = 0.001$; Table 1). Based on animals tagged only on San Nicolas Island (because of the interisland differences between yearlings in the 1981 cohort), there was no difference in the proportion of females seen as yearlings between the 1981 and 1982 cohorts. However, a significantly lower proportion of tagged males was resighted as yearlings from the 1982 cohort than from the 1981 cohort ($X^2 = 5.04$, df = 1, $P = 0.025$). The ratio of males

Table 1. Number of California sea lions tagged as pups on the Channel Islands resighted as yearlings or 2-year-olds on the Farallon Islands (SNI = San Nicolas Island, SMI = San Miguel Island)

		Resight period			June 82–May 83			June 83–May 84			June 84–May 85		
		M	F	Total	M	F	Total	M	F	Total	M	F	Total
		(No. tagged as pups)			(Yearlings)			(2-year-olds)					
Cohort													
1981	SMI	47	53	100	4	2	6	9	2	11			
	SNI	107	82	189	20	6	26	21	1	22			
								(Yearlings)			(2-year-olds)		
1982	SMI	66	83	149				12	5	17	7	2	9
	SNI	187	162	349				14	9	23	8	6	14
											(Yearlings)		
1983	SMI	134	150	284							6	2	8
	SNI	195	201	396							5	4	9

to females, therefore, decreased by half, from 3.3:1 (resightings from June 1982 to May 1983) to 1.6:1 (resightings from June 1983 to May 1984). In addition, a significantly higher proportion of tagged yearlings and 2-year-olds were resighted between June 1983 to May 1984 compared to the proportion of tagged animals resighted between June 1984 to May 1985 (Table 1; $X^2 = 18.42$, df = 1, $P = 0.001$ for yearlings; $X^2 = 9.26$, df = 1, $P = 0.002$ for 2-year-olds).

Ten immature California sea lions tagged on San Miguel Island and 12 tagged on San Nicolas Island were observed during censuses at Point Reyes. Tag numbers were unreadable because of distance. Although pups were also tagged on Santa Barbara Island, none of these were observed at either the Farallons or Point Reyes Headlands.

Between March 1983 and March 1984, at least 30 immature animals washed up dead or died on shore on the South Farallon Islands (none with tags). The animals were generally emaciated and, of 19 stomachs examined, all were empty; 13 were males and 6 were females. In previous years one to two dead animals washed ashore.

13.4 Discussion

A summary of available peak annual counts at California sea lion haulout areas north of the breeding rookeries before, during, and after the 1982–83 EN is given in Table 2. It appears that in the 1960s, California sea lions north of the rookeries were concentrated at Año Nuevo Island. As the population increased rapidly in the 1970s and early 1980s (DeMaster et al. 1982), animals dispersed northward; numbers at Año Nuevo dropped and numbers further north increased. In 1983 and 1984 significant changes in sea lion distribution occurred. Numbers of adult females declined at southern rookeries during the summer breeding seasons of 1983 and 1984 (DeLong et al., this Vol.) at the same time that numbers in central California increased. It is possible that females that normally pupped on the Channel Islands suppressed re-

Table 2. Timing of annual peaks of California sea lions north of the breeding rookeries

Area	Peak numbers	Season	Census period	Source
Vancouver Is., B.C.	473	Winter	1972–78	Bigg (1985)
	4500	Winter	1982–84	Bigg (1985)
	3000	Winter	1985–87	Bigg (pers. comm.)
Puget Sound, WA	108, 296	Spring, winter	1976–79	Everitt et al. (1980)
	961	Spring	1984	Gearin et al. (1986a)
So. Jetty, OR	181	Spring	1980–82	Jefferies (1984)
	277	Spring	1984–85	Brown (1988)
Bodega Rock, CA	1077, 1096	Fall, winter	1980–82	Bonnell et al. (1983)
	1560, 1295	Fall, winter	1982–84	This study
Point Reyes, CA	314	Spring	1980–82	Bonnell et al. (1983)
	338, 270	Spring, winter	1982	This study
	763, 304	Summer, winter	1983	This study
So. Farallon Is., CA	1403	Spring	1974–76	Huber et al. (in press)
	3609	Fall	1978–82	Huber et al. (in press)
	6783	Summer	1983–84	Huber et al. (in press)
	2478	Spring	1985–86	Huber et al. (in press)
Año Nuevo Is., CA	13000	Fall	1961–74	Le Boeuf and Bonnell (1980)
	7319	Fall	1980–83	Bonnell et al. (1983)
Monterey Breakwater, CA	2000	Spring	1980–81	Miller et al. (1983)
	1176, 868	Spring, summer	1983	Nicholson (1986)
	666	Spring	1984–85	Nicholson (1986)

production in those years and summered in central California. Numbers of adult females increased again at the Channel Islands during the breeding seasons in 1985 and 1986 at the same time that numbers returned to lower levels in central California. Although numbers of immature California sea lions also increased in central California during 1983 and 1984, it does not appear that large numbers of adult female or immature California sea lions moved as far north as Oregon or Washington. However, few censuses were conducted there in 1983 and 1984 (Table 2). Censuses in British Columbia indicate that numbers of adult and subadult male California sea lions increased until 1984, then decreased and, from 1985 to 1987, stabilized at about 3000 (Bigg, pers. comm.; Table 2).

In central California, prior to the effects of the 1982–83 EN, few adult females were seen. The greatest monthly mean count and the greatest proportion of adult and subadult males occurred during the same months – May and October – indicating that peak numbers were the result of males stopping briefly at haulout sites at the Farallons and Point Reyes during migration. In 1983, however, greatest monthly means were in June, July, and August, when adult and subadult males made up only 12 to 14% of the population, indicating that peak numbers in those months were due to higher numbers of immature animals and adult females.

Censuses were conducted at the Monterey Breakwater (see Fig. 1, Chap. 10, this Vol.) by Miller et al. (1983) prior to and by Nicholson (1986) during and after the 1982–83 EN. Similar male migrational population peaks and an increase in the number of immatures during 1983 occurred, as well as a decrease in immatures in

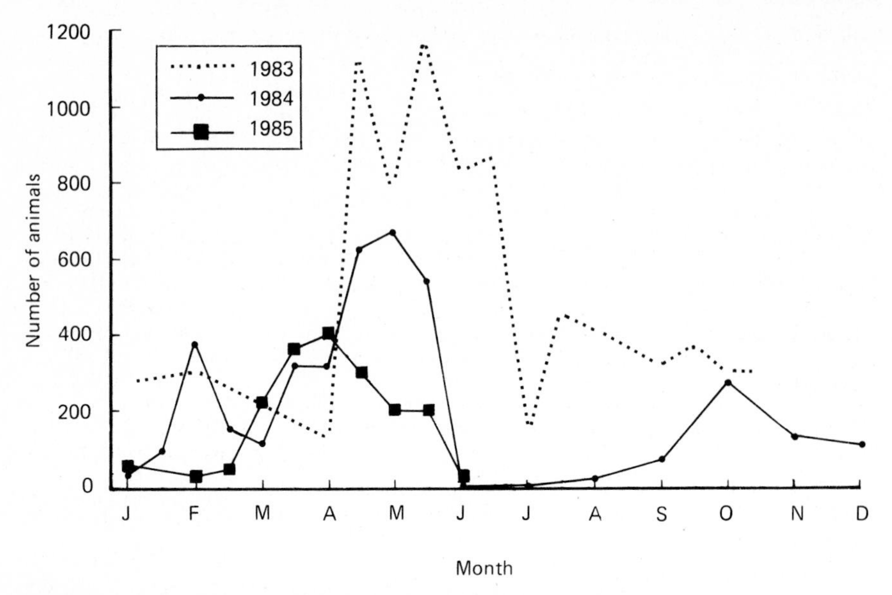

Fig. 4. Numbers of California sea lions at the Monterey Breakwater 1983–1985 (data from Nicholson 1986)

1984 (Fig. 4). Thus, at the Monterey Breakwater, the Farallons, and Point Reyes, the increased abundance of animals present in 1983, particularly in the summer, over the number of animals present in 1982 was the result of an influx of immature sea lions (age 1–4 years) and some adult females. At Bodega Rock, where only adult and subadult males hauled out, numbers did not increase in the summer of 1983.

In 1984 numbers of immatures and adult females remained high at the Farallons, particularly in the summer. At the Point Reyes Headlands, censuses were not conducted past March of 1984 and it is not known whether numbers of sea lions there increased as they did at the Farallons or decreased as they did at the Monterey Breakwater. Increased abundance at the Farallons in 1984 was not a result of yearlings moving northward from the breeding rookeries because, at least on San Nicolas, the 1983 cohort apparently remained in the Channel Islands as yearlings rather than migrating north (Chap. 21, this Vol.). It is possible that the immature animals that moved to the Monterey Breakwater in the summer of 1983 left that area in 1984 and moved north to the Farallons where, in 1984, numbers of immature animals increased over numbers present in 1983. Numbers of sea lions at the Farallons did not begin to decrease until 1985.

Most California sea lions seen in central California in the summer of 1983 were immature animals younger than 5 years of age and more adult females were also present than in other years. Abundance was greatest during the summer months, June to September, but more immatures than usual were present throughout the winter months as well. This influx was a temporary phenomenon that coincided with the warmer waters of the 1982–83 EN, which lasted until 1984 in central California (Huber et al., this Vol.). As a result of these warmer waters, various prey species moved northward (Pearcy and Schoener 1987) and it is presumed that immature California sea lions and some adult females, which normally feed in southern California waters, followed prey northward and concentrated in central California in 1983.

Acknowledgments. Studies on the Farallons were supported by the National Marine Fisheries Service (National Marine Mammal Laboratory and Southwest Fisheries Center), the Marine Mammal Commission, and the Point Reyes Bird Observatory (PRBO). Studies at Point Reyes Headlands were funded by the Point Reyes/Farallon Islands Marine Sanctuary. Permission to work on the Farallons was given by U.S. Fish and Wildlife Service, San Francisco Bay Refuge, and the U.S. Coast Guard, Twelfth District. Support during data analyses and write-up was provided by the National Marine Mammal Laboratory.

Tagging data for San Nicolas Island were supplied by John Francis and Carolyn Heath (University of California at Santa Cruz) and Daryl Boness (National Zoological Park), and from San Miguel Island by Robert DeLong and George Antonelis (National Marine Mammal Laboratory). R. Allen, S. Allen, C. Connors, L. Fry, S. Peaslee, and A. Rovetta helped in data collection.

Comments from G. Antonelis, J. Francis, J. Harvey, T. Loughlin, R. Merrick, and M. Perez on earlier drafts improved the manuscript. Particular thanks to Robert DeLong for advice and encouragement.

This is PRBO contribution No. 444.

14 The Influence of El Niño on Female Attendance Patterns in the California Sea Lion

C.B. Heath, K.A. Ono, D.J. Boness, and J.M. Francis

14.1 Introduction

As is typical of otariids, California sea lion (*Zalophus californianus*) females alternate foraging trips to sea with nursing bouts on land (hereafter referred to as "attendance patterns") during the breeding season. Sea lion pups are entirely nutritionally dependent upon their mothers throughout the first several months of life (Ono et al. 1987). Also, a proportion of pups remain with their mothers into the next breeding season and continue to nurse as juveniles (Francis and Heath, Chap. 21, this Vol.). Alterations in attendance patterns may therefore affect the health and survival of both pups and older suckling juveniles.

Decreases in primary prey species, such as occurred during the 1982–1983 El Niño (EN) event, are likely to bring about modifications in attendance patterns. These decreases in prey began after the summer breeding season in 1982 and persisted into 1984 to some degree (McGowan 1985).

The response to a change in food availability appears to vary among otariid species and with the magnitude of the change (Gentry et al. 1986a, Trillmich et al., Chap. 26, this Vol.). The duration of trips to sea and visits on land by northern fur seal (*Callorhinus ursinus*) females did not change systematically over a 30-year period of increasing competition with commercial fisheries for the seals' primary prey species (Gentry and Holt 1986). However, the amount of energy expended while foraging differed between the 2 years in which it was measured (Costa and Gentry 1986). In contrast, foraging trips by female Antarctic fur seals (*Arctocephalus gazella*) were twice as long during a food-poor year as they were during a year of normal food availability, while foraging effort (metabolic rate) did not differ between the 2 years (Costa et al. 1989). Both trip duration and foraging effort of California sea lion females at San Miguel Island, CA, were significantly greater during the EN year than during 1982 (Costa et al., this Vol.). Finally, Galapagos fur seal females (*Arctocephalus galapagoensis*) had longer visits on land during years of unusually high prey availability than they did during normal years (Trillmich 1986a).

Here, we compare attendance patterns of female California sea lions at San Nicolas Island in years of normal food availability to those during the 1982–1983 EN. Specifically, we examine whether females spent more time at sea during the EN, as did Antarctic fur seals and California sea lions on San Miguel. Also, since Galapagos fur seal females spent more time on land during periods of high prey availability, we expected that California sea lion females might spend less time on land during a presumed decrease in prey availability. Similarly, due to decreased prey prior to parturition, we anticipated that females might spend less time with their pups immediately postpartum.

14.2 Methods

We recorded the presence or absence of individually recognized females from mid-May to early August at three study sites on San Nicolas Island, Channel Islands, CA. UC Santa Cruz (UCSC: CBH, JMF) kept records of females with pups or with juveniles at areas 232 and 233 from 1982–1984, while the National Zoological Park (NZP: DJB, KAO) kept records of females with pups at area 211 from 1982–1985 (area numbers are as in Bonnell et al. 1980). The study sites are separated by up to 1.5 km, and their populations differ in size (Heath and Francis 1983, 1984; K. Ono and D. Boness, unpubl. data).

Females at each site were identified by natural scars. In addition, some females without adequate natural scars were marked with either black dye (Lady Clairol: Clairol Inc., NY, NY; Nyanzol: Belmar Inc., N. Andover, MA) or paint (Nelson pellets: Nelson Paint Co., Iron Mountain, MI; Lenmar: Lenmar Inc., Baltimore, MD). Each year after most births had occurred at area 211 (late June), all pups were rounded up and individually marked with bleach. Pups at areas 232 and 233 were not handled or permanently marked until 2 months of age, but were often incidentally marked with the paint applied to their mothers. We will present data here only for females whose marks were highly visible (approximately 5% of the females at 232, 233, and 25% at 211) and whose pups or juveniles survived until early August.

A female seen during any part of a day was considered to have spent that day on land. If a female was not sighted on a given day, and the observers were present in the blind for at least 2 h during that day (the minimum time needed to locate all marked females that were ashore) then the female was assumed to be at sea.

Two potential sources of error exist in our measurements of attendance patterns. First, a female at the rookery may be overlooked. Since the likelihood of overlooking a female is approximately equal among years, this type of error should not weaken the interpretation of our analyses. Second, it is possible that some females considered to be at sea feeding were ashore at other beaches. The latter is probably a rare occurrence: despite our systematic searches during biweekly censuses of the entire island, females were almost never observed at any beach other than where their pups were located (JMF, CBH, unpubl. data).

Because postpartum trips (the first trip to sea following parturition) differed significantly from subsequent trips ("other trips") at areas 232 and 233, they are analyzed separately in interannual comparisons. At area 211 the first trips postpartum were not significantly different from subsequent trips, therefore, all trips were analyzed together in the interannual comparisons. See Heath and Francis (1983) and Ono et al. (1987) for other methods.

Statistical Methods. Since methods, topography, personnel, and other factors differed between (but not within) study areas, data for each area were analyzed separately, then the p-values were combined to obtain an overall p-value. Data on females with juveniles were combined for analysis due to small sample sizes. For those variables for which we had comparable data from all three study areas (211, 232, 233) statistical analysis was performed as follows:

1. When applicable, the mean of each variable was obtained for each female and then an overall mean was obtained for all females on a given study area in a given year. The sample sizes given, therefore, reflect the number of females used in the analysis;
2. A one-tailed Behrens-Fisher t-test (Behrens 1964) was performed for each study area;
3. A consensus p-value (Rice 1990) was obtained for all three study areas together;
4. A Bonferroni analysis was performed to correct for multiple tests (Neter et al. 1985). The exact p-value is reported except when $p > 0.5$; tests are considered significant when $p \leq 0.05$.

Attendance patterns of females with pups were compared against those of females with juveniles using an ANOVA corrected for unequal sample size. As we had no a priori expectation of how offspring age class would interact with EN and affect the duration of the females' trips to sea, visits on land, or the proportion of time at sea, two-tailed p-values are presented.

14.3 Results

Females with Pups. The number of days postpartum when the first trip to sea occurred, and the duration of the first trip showed no significant changes comparing EN to the other years (Table 1). The mean duration of visits on land showed little variation among years or study sites (1.7 to 1.9 days, Fig. 1a). The duration of trips to sea subsequent to the first, however, varied considerably among years (1.7 to 3.9 days). Mean trip length during the 1983 EN season was significantly longer than in 1982. Trip length had partially returned to 1982 values by the 1984 season, and was significantly shorter in 1985 than during EN (Fig. 1b). At area 211, the longest trip observed for a female with a surviving pup, 9 days, occurred during EN. Longer absences did occur at this area during the EN year ($\bar{x} = 11.3$, SD = 2.1, n = 3) but were associated with the deaths of the pups.

Females with pups spent about 5% more of their time at sea during EN compared to 1982 (Table 1). This is because during EN the duration of trips to sea increased relative to 1982 levels, while the duration of visits on land remained essentially the same. There was no difference in the proportion of time spent at sea between EN and 1984. In 1985 females spent significantly less time at sea compared to EN (about 9%).

Females with Juveniles. Females with juveniles showed no significant differences in mean days on land, mean days at sea, or percent time at sea comparing years 1982–1984 (Table 2). They spent a significantly greater proportion of time

Table 1. Attendance patterns at San Nicolas Island for females with pups. Values are mean number of days, standard deviation, and number of females. Comparisons between years were made with the Behrens-Fisher t-test, and a consensus p-value was calculated for all three study areas together. The Bonferroni method was used to adjust for multiple comparisons

	Area	1982 $\bar{x}$ SD (n)	1983 $\bar{x}$ SD (n)	1984 $\bar{x}$ SD (n)	1985 $\bar{x}$ SD (n)	Comparisons: years[a] p		
Days postpartum (of first trip)	211	9.0	6.2	8.6	8.6	83 < 82	83 < 84	83 < 85
		2.45 (5)	2.48 (6)	1.62 (7)	1.63 (16)	>0.5	0.071	0.083
	232	4.6	4.0	6.7				
		2.58 (11)	2.10 (6)	1.38 (7)				
	233	4.0	6.5	6.4				
		2.92 (5)	0.71 (2)	2.30 (5)				
Postpartum trip	211	1.3	1.8	1.1	1.1	83 > 82	83 > 84	83 > 85
		0.50 (4)	1.04 (8)	0.33 (9)	0.35 (15)	>0.5	0.340	0.073
	232	2.1	2.7	1.9				
		2.25 (14)	1.75 (6)	0.83 (8)				
	233	2.0	1.3	1.8				
		1.00 (5)	0.58 (3)	0.75 (6)				
Proportion of time at sea	211	0.53	0.58	0.57	0.49	83 > 82	83 > 84	83 > 85
		0.10 (13)	0.12 (15)	0.07 (20)	0.08 (20)	0.027	>0.5	0.011
	232	0.62	0.66	0.67				
		0.06 (20)	0.12 (11)	0.06 (11)				
	233	0.60	0.65	0.61				
		0.09 (15)	0.09 (8)	0.10 (9)				

[a]Comparisons between 1983 and 1985 are for area 211 only.

at sea than did females with pups (69 vs 63%) for the combined period of 1982 through 1984 (ANOVA: $F_{1,85} = 4.88$, $p = 0.03$). Mean trip length was not significantly different between the two groups ($F_{1,85} = 0.63$, $p = 0.43$); the greater proportion was due to shorter stays on land by females with juveniles (1.4 vs 1.7 days, $F_{1,89} = 5.37$, $p = 0.02$). There was no interaction between offspring age class and year for any of the attendance pattern variables (trips: $F_{2,86} = 0.01$, $p > 0.5$; visits: $F_{2,90} = 0.02$, $p > 0.5$; percent time at sea: $F_{2,86} = 0.13$, $p > 0.5$). Because of the small sample sizes a residual analysis could not be used with confidence to evaluate the ANOVA assumptions. Therefore, while the trends observed may be real, the exact p-values obtained should be interpreted with caution.

14.4 Discussion

The most pronounced change in attendance patterns during EN was the increase in time spent at sea by females with pups. Mean trip duration and percent time at sea were greater in 1983 than in 1982, and were still elevated in 1984. These changes were probably caused by the shifts in sea lion prey that began in late 1982 and persisted into 1984. The occurrence of market squid, jack mackerel, and north-

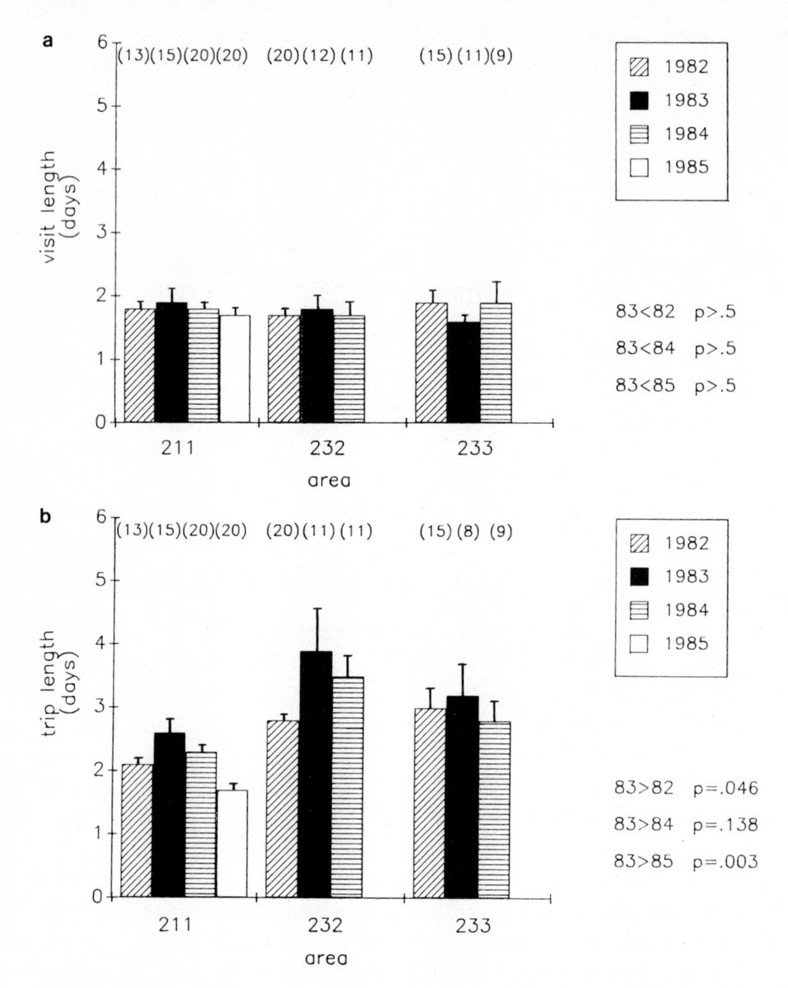

Fig. 1a,b. Mean (±SE) length of visits on land (**a**) and trips to sea (**b**) by females with pups at San Nicolas Island. For area 211, trips include postpartum and subsequent ("other") trips; at areas 232 and 233 postpartum trips were significantly different from other trips and are not included in this analysis. Sample sizes (n) are the number of females used in each analysis. Comparisons between years were made with the Behrens-Fisher t-test, and a consensus p-value was calculated for all three study areas. The Bonferroni method was used to adjust for multiple comparisons

ern anchovy in sea lion scats at San Nicolas Island decreased during the EN event (M. Lowry, unpubl. data) and commercial landings of squid and mackerel were very low (Wolf 1986; Worcester 1987). Northern anchovy females and young of the year were abnormally small in 1983 and 1984, and estimates of anchovy biomass for 1984 were low (Bindman 1986; Fiedler et al. 1986). Pelagic red crabs, which normally occur off Baja California, were abundant in the Southern California Bight during EN and simultaneously became common in sea lion scats (DeLong et al., this Vol.; M. Lowry, unpubl. data). Whether the increased time females spent at

Table 2. Attendance patterns of females with juveniles at San Nicolas Island, areas 232 and 233, combined. Values are mean number of days, standard deviation, and number of females. Comparisons between years were made with the Behrens-Fisher t-test, and the Bonferroni method was used to adjust for multiple comparisons

	1982 $\bar{x}$ SD (n)	1983 $\bar{x}$ SD (n)	1984 $\bar{x}$ SD (n)	Comparisons: years p	
Visits	1.4	1.4	1.4	83 < 82	83 < 84
	0.15 (5)	0.24 (5)	0.16 (3)	>0.5	>0.5
Trips	3.1	3.9	3.3	83 > 82	83 > 84
	0.81 (5)	2.07 (5)	1.44 (3)	>0.5	>0.5
Proportion of time at sea	0.68	0.70	0.69	83 > 82	83 > 84
	0.06 (5)	0.12 (5)	0.07 (3)	>0.5	>0.5

sea during EN was caused by a decrease in prey numbers or quality, or by altered prey distribution, the net result was the same: females had to spend more time at sea each trip to catch adequate prey.

Of what consequence was this increase in time spent at sea? Pup mortality rates paralleled the interannual changes in time at sea, increasing in 1983 over 1982 levels and remaining high in 1984 (Francis and Heath, Chap. 12, this Vol.). Two possible explanations for the increased pup mortality are (1) malnutrition and (2) increased exposure to tides, surf, and female aggression.

The lower pup weights in 1983 and 1984 (Boness et al., this Vol.; Francis and Heath, Chap. 21, this Vol.) suggest that emaciation due to poorer nutrition may have been an important factor in pup mortality during those years. Changes in female attendance patterns during 1983 and 1984 may have contributed to poorer pup nutrition, since increased trip durations required prolonged fasts by pups. The pups of three females at area 211 who made unusually long trips during EN did not survive the season. Galapagos fur seal pups are able to tolerate single, but probably not repeated prolonged fasts, and South American fur seal pups did not survive repeated prolonged fasts during EN (Trillmich and Dellinger, Chap. 6, this Vol.; Gentry et al. 1986a).

It does not appear that California sea lion females compensated for the longer trips to sea by providing more milk while ashore, since pups apparently obtained less milk during 1983 and 1984 than in 1982 (Iverson et al., this Vol.). It is possible that the females had elevated fat levels or otherwise altered milk composition which compensated for the prolonged trips. Across the otariids as a group, milk fat content correlates positively with the duration of maternal absences (Arnold and Trillmich 1985). Fat content has also been shown to change with age of pup in California sea lions (Oftedal et al. 1987b) and Galapagos fur seals (Trillmich and Lechner 1986). However, an increase in milk fat seems unlikely during a period of reduced prey productivity, and the decreased pup growth rates and increased pup mortality indicate that any changes in milk composition at most only partially compensated for the cost to the pup of the increased duration of maternal absences.

The prolonged female absences during 1983 and 1984 may also have contributed to pup mortality by increasing pup exposure to adverse conditions. While left alone on the rookeries pups are exposed to high tides and surf, and to aggression by other females. We observed lone pups being swept out to sea or incurring fatal injuries from adult females (Heath and Francis, unpubl. obs.). Our data are not detailed enough to determine exactly to what degree pup mortality from these factors was increased by EN.

We predicted that females would shorten their stays on land during a food-poor year in order to spend more time at sea foraging. However, the mean length of stays on land remained basically unchanged over the study period, indicating that this interval is less responsive to decreases in food supplies than is the duration of trips to sea.

Gentry et al. (1986a) suggested that the duration of trips to sea and visits on land by female otariids may be regulated by their adipose tissue stores or fat-lean ratios. The amount of reserves that a female has, and the rate at which her offspring depletes them will thus determine the length of her stay on land. In support of this idea, Gentry and Holt (1986) found that when returning fur seal females were withheld from making contact with their pups, their stay on land was equal to the normal 2-day stay plus the period of withholding. When pup demand was temporarily decreased by supplementing with milk from other females during their mothers' absence, the duration of mothers' visits to land also increased. Yearling Galapagos fur seals place greater energy demands on their mothers by requiring more milk than pups, and mothers of yearlings have shorter stays on land than do mothers of pups (Trillmich 1986a,b). We have no data for California sea lions on the relative energy demands placed on mothers by juveniles versus pups. While we would expect the juveniles' demands to be greater due to their larger size, we do not know to what degree suckling juveniles may also forage themselves, thus decreasing demand. If we assume that California sea lion juveniles do place greater energy demands on their mothers than do pups, then their mothers' shorter visits on land support the hypothesis that visit length is determined by the energy demands of offspring. Our finding that females' visits to land did not differ among years of differing food availability is also consistent with this hypothesis.

If trip duration is similarly controlled by fat reserves, females must increase foraging effort or trip duration to accumulate normal reserves in a food-poor year. Costa et al. (this Vol.) found that California sea lion females at San Miguel Island, CA, increased their foraging effort at sea during the EN year in addition to increasing the time spent foraging. We do not have comparable energetic data for San Nicolas Island, but the elevated pup mortality indicates that there is a limit to how much foraging effort can be increased. During food shortages of the extent that occurred during the 1982–1983 EN, increased time at sea was apparently necessary for female California sea lions to obtain adequate energy stores. Transit costs or some aspect of milk production may make returning to land before enough reserves are accrued energetically expensive.

While the effect of reduced food availability during EN is evident in the increase in mortality of pups, the effects on adult females and their reproductive success are less clear. Some of the females whose pups died in EN were able to continue investing in offspring by nursing their yearlings (Francis and Heath, Chap. 21, this Vol.). However, females must balance providing for existing offspring with the cost to their future reproductive efforts (Stearns 1976). This is especially important in times of food shortages, since a decline in female condition causes decreased fecundity in many mammals (Flowerdew 1987). Unlike the Galapagos, where EN was associated with a dramatic increase in adult female sea lion mortality (Trillmich and Dellinger, this Vol.), the severity of EN in the Southern California Bight apparently was not great enough to cause a large decline in female condition. Sea lion females at San Miguel Island were able to maintain their normal weight gains (and thus presumably condition) by increasing foraging effort and trip length during the EN year (Costa et al., this Vol.). While we have no similar quantitative data for San Nicolas Island, female emaciation or mortality did not visibly increase. Whether the females' future reproductive efforts were affected by small changes in body condition or by changes in attendance patterns during EN cannot be determined without long-term studies of individually marked females.

Acknowledgments. We would like to thank Lt. Cdr. E. Giffin and R. Dow of the Pacific Missile Test Center, Naval Air Station, for logistical support on San Nicolas Island. We are grateful to W. Decker, C. Deutsch, J. King, and M. Lowry for assistance in the field. Dr. W. Rice provided statistical advice, and Drs. G. Hall and J. Mulherin assisted with computer programs. M. Brown, Drs. J. Estes, R. Gentry, and F. Trillmich, and an anonymous reviewer made helpful comments on earlier versions of the manuscript. Funding was provided by the Biology Board of the University of California at Santa Cruz, the Friends of the National Zoo, and the Southwest Fisheries Center's Coastal Marine Mammals Program, under the direction of Dr. D. DeMaster.

15 Effects of El Niño 1983 on the Foraging Patterns of California Sea Lions (*Zalophus californianus*) near San Miguel Island, California

S.D. FELDKAMP, R.L. DELONG, and G.A. ANTONELIS

15.1 Introduction

During the 1983 El Niño event, California sea lions (*Zalophus californianus*) and Galapagos fur seals (*Arctocephalus galapagoensis*) responded to the reduction in food availability by extending their foraging trips to sea and delivering less milk to their pups (Trillmich and Limberger 1985; Ono et al. 1987). These observed changes likely reflected a change in foraging tactics as animals attempted to compensate for a reduced food supply. The manner in which these foraging patterns were altered, however, still remains unclear.

Foraging theory predicts that animals feeding on clumped or patchily distributed prey will leave a food patch when their rate of energy acquisition falls to a level equal to the average intake for the entire habitat (Charnov 1976; Krebs 1978). Furthermore, for an animal that forages from a central place, time spent in a food patch should increase with increasing travel time to that patch (Orians and Pearson 1979). Under EN conditions, when prey resources are depleted, these theories would predict that pinnipeds will forage longer in individual prey patches than during years when food is abundant. If animals spend greater amounts of time swimming out from shore or between prey patches, they would also be expected to stay longer in individual patches. Since pinnipeds must return to the surface (a central place) following every dive, this line of reasoning predicts that more dispersed prey will require longer and perhaps deeper dives. Minks (*Mustela vison*) respond to reduced prey encounter rates by lengthening the time spent in underwater searching and by increasing the number of dives per foraging bout (Dunstone and O'Connor 1979). Other breath-holding divers may show similar responses as food becomes more dispersed in space and time.

In 1983, during a severe EN, we examined the foraging behavior of California sea lions at San Miguel Island, California. Similar data were collected in 1982, a year with presumably normal food abundance. Changes in the 1983 diving patterns that might be attributed to reduced prey encounters were documented and compared to those predicted by optimal foraging theory. Although the small sample size obtained from 1983 precludes very firm conclusions, these preliminary data provide an indication of the behavioral responses sea lions make to compensate for lowered food availability.

15.2 Methods

Time-depth recorders (TDRs) were deployed on nine lactating female California sea lions during July and August of the 1982 summer breeding season on San Miguel Island, California, and on five lactating females during July 1983. Details of this instrument are given elsewhere (Kooyman et al. 1983).

Female sea lions that were observed nursing a pup were captured with hoop nets and physically restrained while the TDR and radio transmitter were attached using a nylon harness (Feldkamp et al. 1989). After instrumentation, animals were released to continue their feeding cycle, and were recaptured upon their return from sea.

Dive traces were digitized and then analyzed for depth, duration, and time of dive occurrence. Dive bouts, defined as groups of ten or more dives separated by surface intervals of 10 min or less, were determined using the methods outlined in Feldkamp et al. (1989). Interbout intervals were calculated as the time from the end of the last dive in a bout to the beginning of the first dive of the next bout. Transit time to the feeding area was calculated as the time spent between departure from shore and the beginning of the first dive bout. Transit time from the feeding area was determined as the time spent in returning to shore after the last dive bout.

In addition to providing dive depth information, the TDR acts as an activity recorder (Gentry and Kooyman 1986b; Feldkamp et al. 1989). Records with clearly demarcated traces were analyzed for percent time spent swimming, resting, and diving (Feldkamp et al. 1989).

15.3 Results

During July and August of 1982, under pre-El Niño (preEN) conditions, all nine animals returned to the study site. Because one TDR failed, records were obtained from eight individual animals (ZC-5 through ZC-12). Two animals (ZC-10 and ZC-12) made two trips to sea before recapture. ZC-10 also hauled out once at a location other than the study site, thereby making a total of three trips to sea. On average these eight animals were away from San Miguel (absence time) for a mean (± SD) of 4.5 ± 1.8 days (Table 1). A total of 11 feeding trips (i.e., time between stays ashore), with a mean duration of 3.2 ± 2.1 days, was recorded.

During the breeding season in July 1983, under warm water EN conditions, only two of the five instrumented females returned to San Miguel Island. Animal ZC-17 remained at sea for 16 days, and ZC-20 was away from San Miguel for 2.4 days (Table 1). It was evident from the TDR trace that ZC-17, like ZC-10 in the previous year, hauled out at locations other than San Miguel Island. If these stays ashore are assumed to be the end of trips to sea, then ZC-17 made five recorded feeding trips, averaging 2.1 ± 1.1 days. Only one trip was recorded from ZC-20. Therefore, we recorded six feeding trips lasting a mean of 2.1 ± 1.0 days from the two EN animals.

Activity patterns were determined for five of the preEN animals representing eight trips to sea, and for the six trips made by the two EN animals. On a per trip

Table 1. Absence times and foraging trip durations for California sea lions in this study

Animal	Days away from San Miguel	Recorded days at sea	Return to shore recorded?
Pre-El Niño (1982):			
ZC-5	5.5	1.2	No
ZC-6	4.75	3.5	No
ZC-7	7.75	7.5	No
ZC-8	5.5	3.1	No
ZC-9	3.75	2.8	No
ZC-10:			
Trip 1	3.0	0.7	Yes
Trip 2		1.6	Yes
Trip 3	2.0	1.7	Yes
ZC-11	6.5	6.5	Yes
ZC-12:			
Trip 1	3.2	3.2	Yes
Trip 2	3.5	3.5	Yes
Mean:	4.5	3.2	
± SD:	1.8	2.1	
El Niño (1983):			
ZC-17:	16		
Trip 1		3.5	Yes
Trip 2		1.8	Yes
Trip 3		1.3	Yes
Trip 4		0.9	Yes
Trip 5		2.9	No
ZC-20	2.4	2.4	Yes
Mean:	9.2	2.1	
± SD:	9.6	1.0	

basis, EN feeding trips consisted of significantly more time diving, and less time swimming than the preEN trips (Mann-Whitney U-test $p < 0.05$). The same trends were observed when individual animals were compared with one another, regardless of the number of feeding trips. Unfortunately, when compared in this way, the sample size was too small to test for significance. Nonetheless, the two EN animals spent 34% more time diving, and 22% less time swimming than did the five preEN animals (Table 2). Time spent resting was also slightly lower (1.6% compared to 4.1%) for the two EN animals (Table 2).

Because of the small sample size obtained in 1983, it was difficult to determine whether EN animals spent more time searching for feeding areas. ZC-20 had the longest transit time of any animal from San Miguel Island (11.5 h). ZC-17, however, spent only 3.5 h in transit to the first dive bout, a time intermediate in the range of the preEN animals (Table 3). In 1982, both ZC-10 and ZC-11 exhibited longer return transit times back to San Miguel Island than ZC-20 did in 1983.

PreEN sea lions had mean dive bout lengths that averaged 2.8 ± 0.9 h (n = 8). During EN, ZC-17 and ZC-20 had mean dive bout lengths averaging 4.5 ± 0.3 h

Table 2. Activity patterns of California sea lion females during trips to sea

	Time swimming (%)	Time diving (%)	Time resting (%)	Surface time between dive bout dives (%)
Pre-El Niño (1982):				
Mean	47.6	28.1	4.1	20.2
± SD	13.0	11.9	2.2	5.8
n	5	5	5	5
El Niño (1983):				
Mean	37.0	37.7	1.6	23.7
± SD	11.3	3.8	0.8	6.7
n	2	2	2	2

Table 3. Dive bout intervals and interbout intervals of female California sea lions

	Mean dive bout length (h) per trip	Mean dive bout length (h) per animal	Mean length between bouts (h) per animal
Pre-El Niño (1982):			
ZC-5	2.4	2.4	1.8
ZC-6	2.3	2.3	4.4
ZC-7	4.7	4.7	6.8
ZC-8	2.5	2.5	0.7
ZC-9	2.8	2.8	1.0
ZC-10:		2.3	6.2
Trip 1	2.8		
Trip 2	3.7		
Trip 3	1.6		
ZC-11	2.1	2.1	4.7
ZC-12:		3.4	2.5
Trip 1	3.9		
Trip 2	3.0		
Mean	2.9	2.8	3.5
± SD	0.9	0.9	2.3
El Niño (1983):			
ZC-17:		4.7	2.4
Trip 1	2.7		
Trip 2	3.0		
Trip 3	6.7		
Trip 4	9.6		
Trip 5	6.9		
ZC-20	4.3	4.3	1.6
Mean	5.5	4.5	2.0
± SD	2.7	0.3	0.6

(n = 2), significantly longer than in 1982 (T = 2.63; $p < 0.05$; Table 3). Furthermore, the percentage of time engaged in diving bouts during each foraging trip was significantly greater in 1983 (mean = 66.2 ± 21.2%; n = 6) than in 1982 (mean = 40.4 ± 12.7%; n = 11) (Mann-Whitney U-test: $p < 0.05$; Table 4). This was primarily because ZC-17 spent little time in transit during foraging trips initiated away from San Miguel Island. Interestingly, ZC-17 also increased the percentage of time it spent in diving bouts during each successive trip to sea, while in 1982 both ZC-10 and ZC-12 spent progressively less (Table 4).

Mean dive durations for sea lions in 1983 (ZC-17 and ZC-20) were within the range of those observed for the preEN animals, but they were in the upper end of the range (Table 5). The average mean duration for preEN animals was 1.9 ± 0.46 min while for the two EN animals it was 2.6 ± 0.28 min, suggesting that EN dive durations may have been increased. Frequency distributions of durations and dive depths show that EN animals had a greater percentage of deeper and longer dives than did preEN animals (Figs. 1, 2). While these distributions are significantly different (Kolmogorov-Smirnov test; $p < 0.05$), further studies with larger sample

Table 4. Transit times and dive bout activity of female California sea lions

	Transit time to first dive bout (h)	Number of dive bouts per trip	Feeding trips engaged in dive bouts (%)	Mean dives per bout	Total time at sea in dive bouts (%)
Pre-El Niño (1982):					
ZC-5	5.4	6	50	43	50
ZC-6	1.4	12	33	29	33
ZC-7	–	14[a]	–	51	–
ZC-8	–	15[a]	51	55	51
ZC-9	2.3	14	60	45	60
ZC-10:					27
Trip 1	0.5	2	36	57	
Trip 2	0.8	3	25	62	
Trip 3	3.8	6	25	29	
ZC-11	5.4	23	27	35	27
ZC-12:					48
Trip 1	5.4	10	51	67	
Trip 2	8.9	13	46	57	
Mean	3.8	10.7	40	48.2	42
± SD	2.8	6.1	13	13	13
El Niño (1983):					
ZC-17:					65
Trip 1	3.5	11	39	36	
Trip 2	1.2	9	64	39	
Trip 3	0.2	3	65	66	
Trip 4	0.6	2	91	98	
Trip 5	0.6	9	90	89	
ZC-20	11.5	7	48	64	48
Mean	2.9	6.8	66	65.3	56.5
± SD	4.4	3.6	21	25.2	12

[a]May have made an earlier, shallow dive bout.

Table 5. Dive depths and durations for female California sea lions (means ± SD are given where appropriate)

	Number of dives	Mean dive duration (min)	Maximum duration (min)	Mean dive depth (m)	Maximum depth (m)
Pre-El Niño (1982):					
ZC-5	287	1.7 ± 1.3	6.0	56.8 ± 43.8	207
ZC-6	489	1.5 ± 1.5	6.2	56.4 ± 53.0	207
ZC-7	838	2.8 ± 1.3	6.2	98.2 ± 43.3	196
ZC-8	931	1.5 ± 1.2	7.5	31.1 ± 35.1	207
ZC-9	720	2.4 ± 1.4	7.6	58.0 ± 49.3	274
ZC-10	543	1.9 ± 0.8	7.1	38.9 ± 18.8	207
ZC-11	1005	1.7 ± 1.6	9.9	76.3 ± 49.5	234
ZC-12	1478	2.0 ± 1.3	9.1	70.7 ± 41.3	235
El Niño (1983):					
ZC-17	2132	2.8 ± 0.8	9.3	77.2 ± 60.6	241
ZC-20	508	2.4 ± 1.5	8.2	54.0 ± 46.6	233

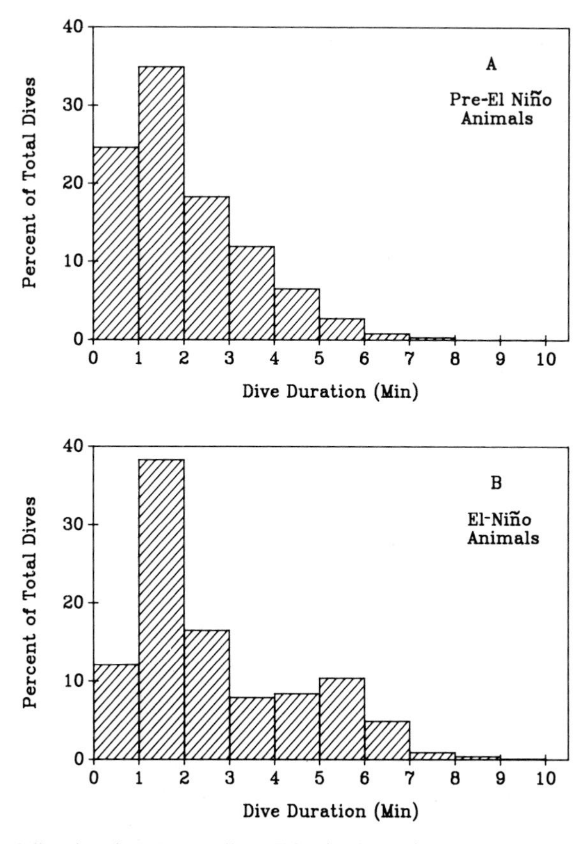

Fig. 1A,B. Frequency histogram of dive durations for: **A** all preEN animals combined, **B** both EN animals. Note the greater proportion of longer dives and relatively fewer number of dives 1 min or less made by the EN animals

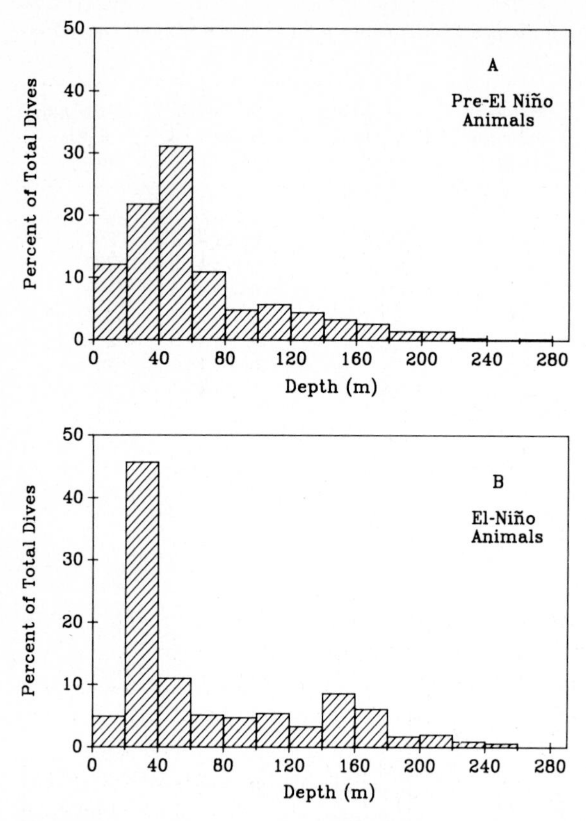

Fig. 2A,B. Frequency histogram of dive depths for: **A** all preEN animals combined, **B** both EN animals. Note how few dives were less than 20 m for the EN animals

sizes are needed to determine whether or not increased dive durations and deeper dives are a consistent behavioral response to reduced prey encounter rates.

The distribution of dive depths over a 24-h period also changed significantly between the 2 years (Kolmogorov-Smirnov test; $p < 0.05$. Mean dive depths for the two EN animals were deeper during the morning hours, from 0600 until 1200, than for the preEN animals (Fig. 3). During the afternoon and evening hours, sea lions dived to similar depths in both years. These observations also held true when animals were compared individually.

15.4 Discussion

It is evident from studies of attendance behavior and pup growth rates that the 1983 El Niño conditions adversely affected the ability of female sea lions to obtain food and to deliver milk to their pups (Ono et al. 1987; DeLong et al., this Vol.; Iverson

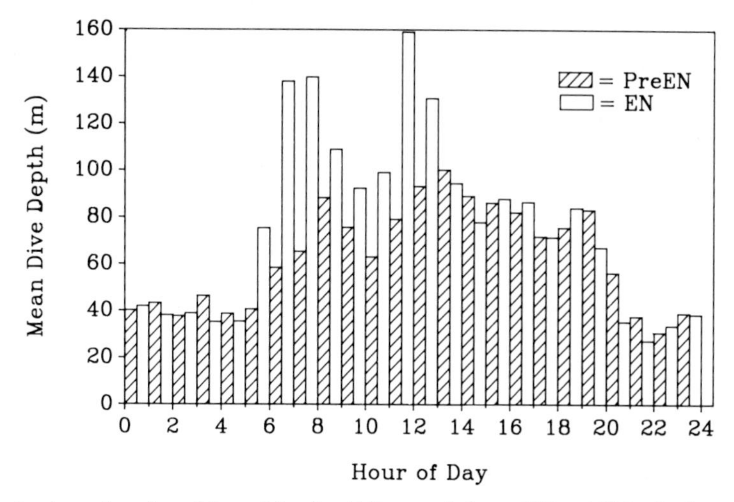

Fig. 3. Mean dive depth as a function of time of day for all dives made by preEN and EN animals

et al., this Vol.). Our observations also support this conclusion. Although all nine study animals returned to San Miguel Island in 1982, only 40% (two out of five) did so in 1983. Moreover, the two EN animals (ZC-17 and ZC-20) exhibited much greater variability in their time away from San Miguel Island than did preEN animals, being absent from their pups a mean of 9.2 days. Individual foraging trips were, however, on average shorter for the two EN animals. This may reflect the depauperate feeding conditions encountered during EN. The greater number of longer and deeper dives suggests that animals may have worked harder within dive bouts. They may have returned to shore sooner than animals in 1982 because they were exhausted. It is possible that ZC-17, which returned to shore at least four times before returning to San Miguel, may have avoided returning to its pup because her body reserves were not sufficiently replenished during these feeding trips.

Despite the small sample size obtained during 1983, certain generalities are apparent that may reflect behavioral consequences of foraging under conditions of reduced prey availability. Sea lions foraging under EN conditions spent a significantly greater percentage of time diving and less time swimming, had significantly longer diving bouts, and spent a greater percentage of their total trip time engaged in diving bouts. Furthermore, EN animals appeared to make a greater proportion of longer, and perhaps deeper, dives than did preEN animals.

These observations suggest that female sea lions were not swimming greater distances in search of prey, but instead were working harder to locate prey in the proximity of San Miguel Island. Data on transit times to the first diving bout and in returning from the last, however, complicate this generalization. Animal ZC-20 had the longest transit time of any animal (11.5 h; Table 4). Based on an estimated swim speed of 9 km/h (Feldkamp et al. 1989), ZC-20 traveled over 100 km before beginning to dive steadily. In contrast, ZC-17 traveled only 3.5 h (32 km) before beginning its first dive bout. On subsequent foraging sojourns, ZC-17 had outward transit times of 1.2 h or less (Table 4). The return transit time of ZC-20 to San Miguel, the only one recorded for an EN animal, was 8.9 h; shorter than those recorded for ZC-10 (11.3 h) and ZC-11 (15.2 h) in 1982.

Dive bouts, a characteristic feature of sea lion diving patterns, probably represent foraging effort concentrated on individual prey patches (Feldkamp et al. 1989). The significant increase in dive bout length during EN, coupled with a reduction in the proportion of time spent swimming, appears to indicate that prey patches still existed, but prey were probably more dispersed within these areas. Sea lions continued to feed on similar prey items during 1983 as they did in 1982, although in different relative amounts (DeLong et al., this Vol.). Evidently, preferred prey were present in the waters near San Miguel Island in both years, although their availability was undoubtedly reduced in 1983. It seems likely that the observed changes in sea lion foraging patterns may be accounted for, in part, by a reduction in fish density within individual schools.

A priori, one might also expect dive depths and durations to increase under these conditions because with lowered encounter rates, sea lions would need to spend greater amounts of time in underwater searching. This could be accomplished through two different foraging strategies; diving deeper in search of prey, or spending more time searching at a fixed depth. Because there is a finite physiological limit to the duration of any given dive, increasing the time spent searching for prey will necessarily decrease the amount of time available for pursuit, once prey are located. Calculations of oxygen reserves indicate that sea lions can sustain a 5.5-min dive aerobically (Feldkamp et al. 1989). Dives that exceed this estimated limit would result in a buildup of anaerobic metabolic by-products, necessitating extended surface times and reducing the proportion of time the animal can spend foraging underwater (Gentry et al. 1986a). Within this aerobic dive limit, however, there may be considerable behavioral flexibility in the ways that sea lions partition their time between search and pursuit. Under conditions of reduced prey availability, and because travel to and from depth is a cost associated with each dive, sea lions might be expected to extend their search time underwater, at the risk of losing prey because of finite pursuit times, in order to maximize the probability of a prey encounter (Dunstone and O'Connor 1979; Gentry et al. 1986a).

Dive durations were not exceptionally long for the two animals examined during EN, although they were in the upper part of the range for all sea lions studied. A frequency distribution of dive durations, however, shows proportionately more long dives by the EN animals than by preEN sea lions (Fig. 1). While a less pronounced trend was also evident for dive depths (Fig. 2), a plot of dives per hour reveals that EN animals did make deeper dives in the morning hours (Fig. 3). California sea lions normally dive deeper during the day as their prey move to deeper water (Feldkamp et al. 1989). The dives exhibited by EN animals during these hours appear to indicate an increased effort to search for vertically migrating prey. This seems to suggest that sea lions are compensating for a reduction in prey encounter rates by increasing search time, consistent with optimal foraging theory. However, more data are clearly needed to test this hypothesis.

In both years, sea lions spent very little time resting at sea ($\leq$4.1%), a fact that helps to explain the increased absence times observed for animals during EN.

Despite the fact the EN sea lions increased the amount of time diving, decreased the amount spent swimming, and took longer and deeper dives, there may be so little leeway in their activity budget that they could not increase their forage intake sufficiently by simply increasing effort. Unlike California sea lions, northern fur seals (*Callorhinus ursinus*) rest 17% of their time at sea (Gentry et al. 1986b). When prey resources are low, these animals can reduce time spent resting, elevate diving effort, and return to shore with adequate provisions within the normal range of absence times (Costa and Gentry 1986). Because sea lions do not appear to have this flexibility, they may be forced to extend their absence times as well as their foraging effort as a means to increased food intake.

It is possible that sea lions may have a greater ability to alter activity patterns under adverse conditions than our data show. In both years, the hydrodynamic effects of the TDR and harness may have influenced the observed activity partitioning. Drag arising from the instrument package and its contribution to the animal's overall resistance rises with increasing swimming speeds (Feldkamp et al. 1989). If prey capture was adversely affected because of the instrument package, the percentage of time spent resting at sea might be expected to decline as effort to obtain prey increased. Foraging California sea lions expended energy at a higher rate in 1983 than in 1984, indicating a moderate ability to increase daily foraging effort (Costa et al., this Vol.) However, observations by Ono et al. (1987) that documented increased absence times for sea lions during EN, again suggest that the capacity to increase foraging effort without increasing the time spent at sea is limited for sea lions.

Acknowledgments. This research was supported by the Marine Sanctuaries Program, the National Marine Fisheries Service, and by NIH grant USPHS HL17731 to Dr. G.L. Kooyman. We are grateful to Dr. G.L. Kooyman for his support in all aspects of this work. We also thank Dr. D. DeMaster for logistical support and field assistance, and Drs. D.P. Costa, R.L. Gentry, and F. Trillmich for important ideas and constructive criticism.

16 Effects of El Niño on the Foraging Energetics of the California Sea Lion

D.P. COSTA, G.A. ANTONELIS, and R.L. DELONG

16.1 Introduction

For a marine mammal foraging on widely dispersed prey items, the distribution and availability of prey and its species composition determine the rate of energy expenditure and acquisition. Models of central place foraging, optimal diet choice, and maximization of energy intake have attempted to predict how an animal should behave under a given set of circumstances (Charnov 1976; Dunstone and O'Connor 1979; Orians and Pearson 1979; Pyke 1984; Kramer 1988). However, few studies have documented how the rate of energy acquisition is modified when environmental circumstances change dramatically during a catastrophic event, such as the 1982/83 El Niño (EN).

Costa and Gentry (1986) and Costa et al. (1989) have proposed that lactating fur seals forage until a relative body set point of maternal mass is achieved. In northern fur seals (*Callorhinus ursinus*) foraging in the Bering Sea this body set point was maintained even though there were year to year variations in at-sea energy expenditure. Annual differences in the at-sea energy expenditure were thought to result from alterations in food availability, or the application of different foraging behaviors that were related to the pursuit of different prey (Costa 1988a). In this study changes in energy expenditure were not accompanied by modifications in the net mass gain or energy storage over a foraging trip. Apparently, differences in the intensity of foraging effort, as indicated by increases in field metabolic rate (FMR), allowed equivalent mass and energy gains without changes in trip duration (Costa and Gentry 1986). However, northern fur seals foraging in the Southern California Bight did increase foraging trip duration in an apparent attempt to compensate for alterations in food abundance that occurred during the 1982/83 EN (DeLong and Antonelis, this Vol.). In contrast, Antarctic fur seals (*Arctocephalus gazella*) foraging during a severe shortage of krill, achieved equivalent mass gains as fur seals foraging during a normal year but only after increasing trip duration with no change in at-sea FMR (Costa et al. 1989).

In the present study we quantified the energy budget of breeding female California sea lions (*Zalophus californianus*) at San Miguel Island, California, during the summers of 1983 and 1984. Here, we examine the possibility that, like fur seals, lactating California sea lions responded to changes in prey availability by increasing at-sea energy expenditure and trip duration.

16.2 Materials and Methods

Lactating California sea lions breeding on San Miguel Island, California were studied during July 1983 and 1984. Six were used in 1983 and eight in 1984. During each field season we recaptured all but one study animal, therefore, measurements of water influx using tritiated water (HTO) were made on five females in 1983 and seven in 1984. At-sea metabolism (CO_2 production) was measured on two females in 1983 and seven in 1984 using the doubly-labeled water method (Nagy 1980; Costa 1987). A randomly selected group of females previously marked with natural scars were used to determine which females had been suckling their pups and were about to feed at sea. At San Miguel Island females usually depart after 3 days of suckling their pup. Therefore, females that had been ashore suckling their pup for 3 days were captured using hoop nets. Animals were treated following the method in Costa et al. (1989), except that they received 10 ml of 95% oxygen-18 water and 0.6 mCi tritiated water in 3 ml of sterile saline. Body mass was measured with a hanging spring scale (±0.5 kg).

Radio transmitters which weighed 150 g (164–165 Mhz; Advanced Telemetry Systems, Bethel, Minnesota) were glued to the pelage on the females' middorsal area using a fast setting epoxy resin (Devcon "5 Minute Epoxy"). The arrival and departure of females were then monitored with a scanning telemetry receiver (Advanced Telemetry Systems, Bethel, Minnesota) interfaced to an Esterline Angus strip chart recorder. During 1984, the scanner malfunctioned and attendance patterns were determined by manually scanning the radio frequencies three times each day at morning (0600–0800 h), midday (1030–1230 h) and evening (1830–2030 h). Females were recaptured an average of 3 h (range 0.5 to 12 h) after returning from a foraging trip which lasted from 3 to 18 days. Upon recapture, body mass was recorded and blood samples taken.

Sample analysis and calculations follow the procedures detailed in Costa et al. (1989). Final TBW was calculated as the final mass times the initial TBW water/mass ratio. CO_2 production and water influx were calculated using Eq. (3) presented in Nagy (1980), and Eqs. (5) and (6) in Nagy and Costa (1980), assuming an exponentially changing body water pool. An energy yield of 23.9 kJ per l CO_2 produced for 1983 and 23.6 kJ per l CO_2 in 1984 was used to convert CO_2 production to energy consumption. This presumes that all of the fat and protein contained in the diet is oxidized (Table 1), assuming that fat metabolism yields 26.81 kJ/l CO_2 and protein metabolism 22.97 kJ/l CO_2 (Costa 1987).

The diet of southern California sea lions was estimated to be composed (by mass) of 45% market squid (*Loligo opalescens*), 22% jack mackerel (*Trachurus symeticus*), 19% Pacific whiting (*Merluccius productus*), and 14% northern anchovy (*Engraulis mordax*) in 1983, and 76% Pacific whiting, 22% juvenile rockfish (*Sebestes* spp.), 1.5% red octopus (*Octopus rubescens*), and 0.5% northern anchovy in 1984. These estimates were derived from the analysis of scat material collected in 1983 and 1984 from areas utilized by lactating sea lions (Antonelis and DeLong, unpubl. data). The proportion of each of the four most commonly occurring prey was multiplied by its average estimated size when preyed upon by California sea lions (Antonelis et al. 1984). We are aware of the biases associated with such esti-

Table 1. Proximate composition and the relative contribution to the diet of prey items consumed by California sea lions during 1983 and 1984. Except where noted otherwise data for proximate composition are from Sidwell et al. (1974)

Prey item	Proximate composition Water (%)	Fat (%)	Protein (%)	Energy (kJ/g)	Contribution to diet Mass (%) 1983	1984
Squid[a]	74.3	1.0	15.3	4.31[c]	45.0	0.0
Mackerel[a]	71.7	5.3	22.0	5.92[d]	22.0	0.0
Whiting	80.7	1.2	16.1	3.67	19.0	76.0
Anchovy[a]	74.1	2.4	20.2	5.08[e]	14.0	0.5
Rockfish[b]	78.9	1.8	18.9	4.06	0.0	22.0
Octopus[b]	82.2	0.8	15.3	3.05	0.0	1.5
Mean diet 1983	74.9	2.2	17.6	4.65	100.0	100.0
Mean diet 1984	80.3	1.3	16.7	3.75	100.0	100.0

[a]Water and energy content of prey items collected offshore San Miguel Island during July 1983 (Costa 1988b).
[b]Composition data from Watt and Merrill (1963).
[c,d,e]Metabolizable energy content (EC) determined by multiplying the energy content of the prey, determined by bomb calorimetry, times the metabolizable energy coefficient (MEC), determined from feeding trials (Costa 1988b). [c] Squid EC = 5.506 kJ/g wet weight, MEC = 0.783. [d] Mackerel EC = 6.481 kJ/g wet weight, MEC = 0.914. [e] Anchovy EC = 5.544 kJ/g, MEC = 0.916.

mates when derived from scat analysis (e.g., Prime and Hammond 1987; Dellinger and Trillmich 1988), but consider this the only available information for evaluating (at least in a relative sense) the dietary composition of female California sea lions during the 2 years of this study. These diets and their proximate compositions are summarized in Table 1.

FMR data collected over the entire measurement interval included variable amounts of onshore FMR. Data were normalized to estimate FMR while at sea, by correcting for the portion of time and, hence, FMR spent onshore. Time onshore was calculated as the portion of the measurement interval where the animal's radio signal was detected. Onshore FMR was assumed to be two times each animal's predicted BMR (Kleiber 1975). Doubly-labeled water measurements of onshore FMR of Australian sea lions (*Neophoca cinerea*), which are of equivalent mass and live in a similar thermal regime, were two times the predicted BMR (Costa, unpubl. data). We do not believe that there would be a marked difference in onshore metabolism between years, due to the similarity in thermal regime between seasons (Costa, pers. obser.). Furthermore, the error introduced in the calculation of at-sea FMR is minimal. The worst example is animal 284. If her onshore FMR was 1.1 or three times BMR, the at-sea FMR would be overestimated by 4.5% or underestimated by 7.5%. The mean deviation for all animals is between +2.5 to −4.1%. At-sea FMR was then calculated for each female by solving the following relationship for "at-sea FMR":

$$\text{Measured FMR} = (\text{onshore FMR}) \times (\%\ \text{time onshore}) + (\text{at-sea FMR}) \times (\%\ \text{time at sea})$$

Means are followed by ±1 standard error. Unless otherwise stated, differences between means were tested using the Mann-Whitney U-test and relations were tested by using least squares linear regression analysis.

16.3 Results

Female Body Mass. There was no difference in the initial mass of the females studied between years (1983; 76.8 kg ± 3, n = 6: 1984; 84.1 kg ± 3.2, n = 8, U = 29.5, $p > 0.05$). Similarly, no difference was observed in the TBW of females between years (Fig. 1). Therefore, the data for both years were pooled and yielded a significant regression equation of TBW with female body mass described by: TBW (L) = 0.659M (kg) – 0.166 (r = 0.922, $p < 0.01$, n = 13, Fig. 1).

Foraging Attendance Patterns. In 1983 foraging trips averaged 37% longer than trips in 1984 (1983 mean = 5.8 ± 0.8 days, n = 5; 1984 mean = 4.2 ± 0.4 days, n = 7, one-tailed, two sample t-test t = 1.82, $p < 0.05$; Tables 2, 3). Such increases in trip duration are consistent with more extensive observations of sea lion attendance patterns on southern California rookeries during 1983 (Heath et al., Feldkamp et al., this Vol.; Ono et al. 1987). There was no correlation between relative (kg/d) or absolute (kg/trip) mass change over the measurement interval and trip duration for both years combined (kg/trip; r = 0.143, n = 12, $p > 0.10$, kg/day; r = 0.125, n = 12, $p > 0.10$) or treated separately. During 1984 two of the females (584 and 684) made two trips to sea before being recaptured.

Foraging Energetics. During 1983, study animals expended energy at a faster rate and remained at sea longer than animals in 1984. At-sea mass-specific field meta-

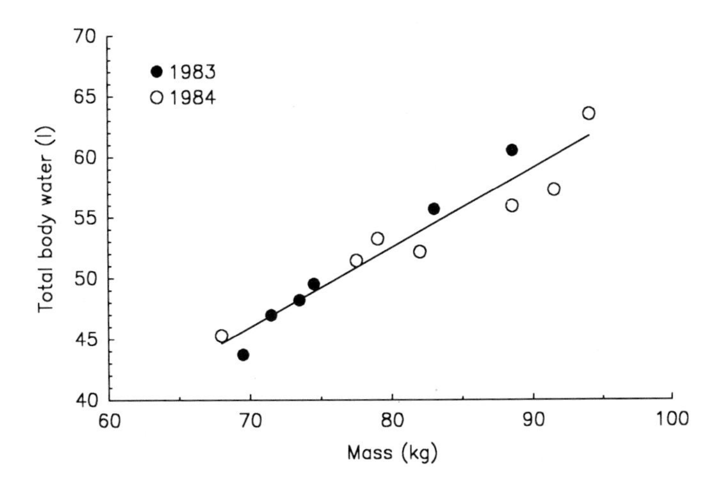

Fig. 1. Total body water content (l) in relation to body mass (kg) in female California sea lions followed a significant least squares linear regression (r = 0.960, $p < 0.01$, n = 13) and can be calculated as TBW (l) = 0.659 M (kg) – 0.166. There was no difference between years

Table 2. Measurement interval, time at sea, mass change, metabolic rate, and water influx of California sea lion females foraging during the 1983 El Niño conditions from San Miguel Island, CA

Overall Female No.	At sea Body initial (kg)	Mass final (kg)	Mean body mass (kg)	Measure-ment interval (days)	Time at[a] sea (days)	Water influx (ml kg^{-1} day^{-1})	CO_2 production (ml/kg-d)	Overall FMR (W)	At sea FMR[b] (W)
15	83.0	85.8	84.4	7.51	6.89	117	0.977	547	580
18	71.5	73.5	72.5	9.76	7.63	97	1.110	534	636
19	88.5	[c]		[c]					
23	74.5	70.0	72.3	4.85	3.51	71	[d]		
24	69.5	72.5	71.0	7.15	3.48	117	[d]		
25	73.5	78.5	75.9	5.60	4.25	109	[d]		
Mean =	76.8	76.1	75.2	6.97	5.75	102	1.044	541	608
s.e. =	3.0	2.8	2.4	0.85	0.80	9	0.221	6	28

[a]Mean trip duration was the same as time at sea, since over the measurement interval all animals made only one trip to sea.
[b]At-sea FMR was calculated from: at-sea FMR = {measured FMR – [(onshore FMR) × (% time onshore)]}/(% time at sea).
[c]Animal not recaptured.
[d]Animal not injected with 0–18 labeled water.

Table 3. Measurement interval, time at sea, mean trip duration, mass change, metabolic rate, water influx of California sea lion females foraging during 1984 from San Miguel Island, CA

Female No.	Body initial (kg)	Mass final (kg)	Mean body mass (kg)	Measure-ment interval (days)	Time at sea (days)	Mean trip[a] duration (days)	Water influx (ml kg^{-1} day^{-1})	CO_2 production (ml/kg-d)	Overall FMR (W)	At-sea[b] FMR (W)
184	79.0	80.0	79.5	4.78	4.71	4.71	93.3	0.911	475	479
284	68.0	65.0	66.5	7.75	5.96	5.96	87.8	0.928	405	478
384	92.0	[c]								
484	82.0	82.5	82.3	3.09	2.61	2.61	57.6	0.705	380	416
584	91.5	85.5	88.5	6.83	6.07	3.04	99.9	0.578	335	353
684	88.5	88.5	88.5	8.56	7.42	3.71	99.5	0.801	465	506
784	77.5	77.0	77.3	5.51	4.97	4.97	96.9	0.723	366	387
884	94.0	93.5	93.8	4.70	4.50	4.50	108.9	0.721	443	454
Mean =	84.1	81.7	82.3	5.89	5.18	4.21	92.0	0.767	410	439
SE =	3.2	3.5	3.4	0.73	0.57	0.44	6.2	0.047	20	21

[a]Animals 584 and 684 made two trips to sea over the measurement interval. The value given is the total time at sea divided by 2.
[b]At-sea FMR was calculated from: at-sea FMR = {Measured FMR – [(onshore FMR) × (% time onshore)]}/(% time at sea).
[c]Animal not recaptured.

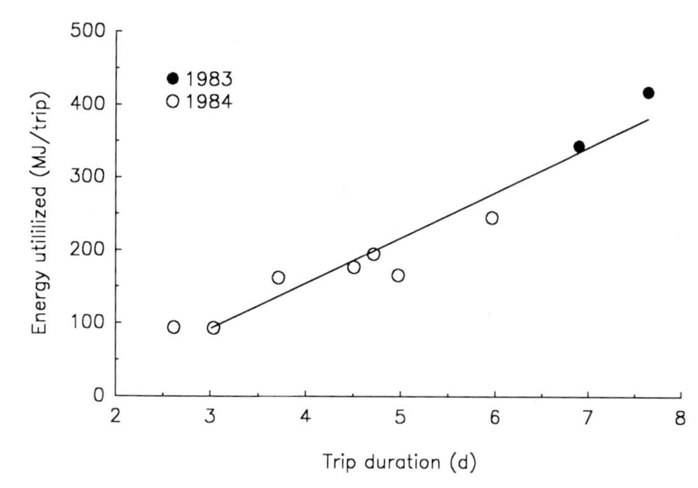

Fig. 2. Energy utilization of foraging California sea lions significantly increased as trip duration increased [r = 0.966, n = 9, $p < 0.01$; 1984 alone r = 0.947, $p < 0.01$, n = 7; y(MJ) = 6.27 x(d) – 1.66]. Total energy utilization was 2.3 times higher in 1983 than in 1984 (U = 14, $p = 0.05$)

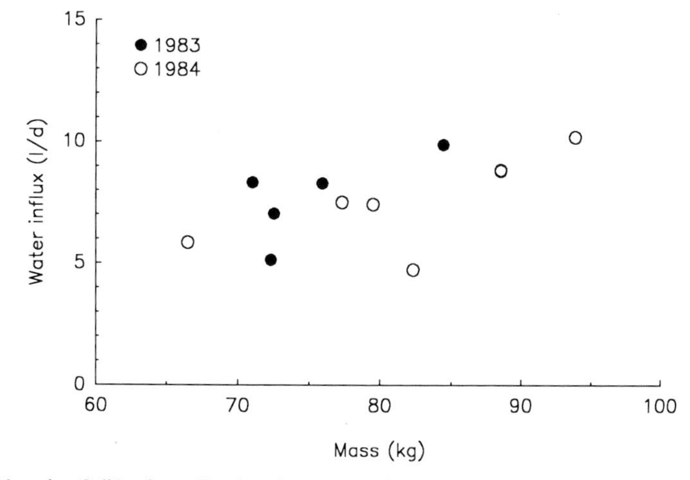

Fig. 3. Water influx of foraging California sea lion females as a function of body mass in July during the 1983 El Niño and in 1984. There was no relationship between body mass and water influx over both years combined (r = 0.620, $p > 0.05$) or treated separately (1983 n = 5, r = 0.706; 1984 n = 7, r 0.727)

bolic rate (FMR) and absolute FMR were significantly greater in 1983 than in 1984 (U = 14, N = 7, N = 2, $p = 0.05$, one-tailed test) (Tables 2, 3). At-sea FMR was from 1 to 18% higher than overall FMR and on average was 8% greater. Metabolic rate did not vary as a function of female body mass (r = 0.332, n = 2; $p > 0.10$; 1984 alone n = 7, r = 0.203). Energy utilization over a foraging trip was 2.4 times greater (U = 14, $p < 0.05$, one-tailed test) in 1983 (382 ± 37 MJ/trip, n = 2) than in 1984

(162 ± 21 MJ/trip, n = 7). Energy utilized over the foraging interval was significantly and positively correlated with the time spent at sea (r = 0.966, n = 9, $p < 0.01$; 1984 alone r = 0.947, n = 7, $p < 0.01$) (Fig. 2). However, differences in mass change over a trip (kg/trip) were not significantly related to energy expenditure (MJ/trip) (r = 0.581, n = 9, $p > 0.05$) nor was the rate of mass change (kg/day) correlated with the rate of energy expended over the trip (W) (r = 0.682, n = 9, $p > 0.05$).

Water influx rates (l/day or ml kg^{-1} day^{-1}) were not different between years (Tables 2, 3; Fig. 3; 1983 mean = 7.72 ± 0.79 l/day, n = 5; 1984 mean = 7.66 ± 0.72 l/day, n = 7; U = 18, $p > 0.05$). Water influx rate (l/day) was not correlated with body mass over both years combined (r = 0.620, $p > 0.05$) or treated separately (1983 n = 5, r = 0.727; 1984 n = 7, r = 0.706). Furthermore, there was no relationship between water influx rate and the rate of mass change between years or for all data combined ($p > 0.05$, 1983 n = 5, r = 0.804; 1984 n = 7, r = 0.184; combined r = 0.381, n = 12).

16.4 Discussion

16.4.1 Foraging Energetics

The at-sea FMR of female California sea lions foraging during 1983 was 6.9 times and during 1984, 4.8 times the predicted basal metabolic rate for a terrestrial mammal of equal size (Kleiber 1975). This difference is similar to the six to seven times the predicted BMR observed in smaller (37 kg) northern and Antarctic fur seal females (Costa and Gentry 1986; Costa et al. 1989).

Seasonal differences in food intake can be estimated from water influx rates if the diet and its composition are known and the animal does not drink (Shoemaker et al. 1976). Laboratory feeding studies have shown that California sea lions do not ingest sea water while feeding on either squid, herring, or mackerel (Costa 1987). Using the data on prey water content (WC) from Table 1 and total water influx rate (TWI) from Tables 2 and 3, food intake can be estimated as follows:

$$\text{Food intake (g/day)} = \frac{(\text{TWI ml } H_2O/\text{day}) - (\text{MWP ml } H_2O/\text{day})}{(\text{WC ml } H_2O/\text{g food})}$$

MWP is the metabolic water production calculated from metabolic rate data in Tables 2 and 3, using a constant of 0.0231 ml H_2O/kJ for 1983 and 0.0228 ml H_2O/kJ for 1984. These constants were derived from the composition of the diet assuming 0.0278 g H_2O are produced per kJ of fat oxidized and 0.0220 g H_2O per kJ protein (Costa 1987). Estimates of prey food intake are presented in Table 4. Even though there were marked differences in diet there were no significant differences in the rate of prey mass or prey energy consumed between years (U = 18, $p > 0.10$) (Table 4). This implies that sea lions foraging in 1983 expended energy at a higher rate to consume equivalent amounts of prey.

We have chosen not to estimate the rate of food energy intake from the data in Table 4, due to the considerable variation in energy content of prey and the uncertainty as to the actual prey taken by individual sea lions. Errors in the actual prey

Table 4. Rates of prey mass and energy intake are estimated from field metabolic rate (FMR), total water influx measured by HTO dilution, metabolic water production (MWP) estimated from FMR using the constants 0.0231 g H_2)/kJ for 1983 and 0.0228 g H_2)/kJ for 1984. Preformed water influx is the difference between total water influx and MWP. Mass of prey consumed was calculated by dividing preformed water influx by the water content of the diet, which was 74.9% ωατερ ιν 1983 ανδ 80.3% ωατερ ιν 1984

Animal No.	FMR W	Water influx (l/day)			Prey intake mass
		Total	MWP[a]	Preformed	(kg/day)
1983					
15	547	9.87	1.09	8.78	11.73
18	534	7.03	1.07	5.97	7.97
23		5.13	(1.08)	4.05	5.41
24		8.31	(1.08)	7.23	9.65
25		8.27	(1.08)	7.19	9.60
Mean =	541	7.72	1.08	6.64	8.87
SE =		0.79		0.79	0.89
1984					
184	475	7.42	0.94	6.48	8.10
284	405	5.84	0.80	5.04	6.30
484	380	4.74	0.75	3.99	4.99
584	335	8.84	0.66	8.18	10.22
684	465	8.81	0.92	7.89	9.86
784	366	7.49	0.72	6.77	8.45
884	443	10.21	0.87	9.34	11.67
Mean =	410	7.62	0.81	6.81	8.51
	20	0.71	0.04	0.70	0.88

[a]The MWP values for animals 23, 24, and 25 are assumed to be the mean for animals 15 and 18.

taken or its energy content greatly effect estimates of prey energy intake, while having only a minimal effect on estimates of prey mass intake (Costa 1987). For example, if animals in 1983 ate exclusively whiting, but we erroneously assumed they ate exclusively mackerel, the mass intake would be overestimated by only 13% (mackerel 9.3 kg/day; whiting 8.2 kg/day), whereas the energy intake would be overestimated by 82% (whiting 350 W; mackerel 640 W).

The foraging energetics data are in accord with changes in pup growth recorded by other investigators. For example, Ono et al. (1987) found that lactating sea lions on nearby San Nicolas Island increased the duration of their foraging trips in 1983 compared with 1982, 1984, and 1985, but found no difference in the amount of time mothers spent ashore with their pups. This implies that females remained at sea longer to acquire the same amount of milk energy to provision their pups. Isotopic dilution measurements of pup milk intake also support this trend. In 1983 male pups received 12% and female pups 3% less milk energy than in 1984. Such reductions in the rate of milk delivery are consistent with pups receiving the same amount of milk per shore visit because the visits are less frequent. Again, such differences are consistent with lower growth rates, lower pup mass, and higher mortalities that were observed in 1983 compared to other years (Boness et al., DeLong et al., this Vol.).

However, inconsistent with the above data is the fact that lactating sea lions gained less mass in 1984 than in 1983 (Tables 2 and 3). In our opinion this inconsistency reflects the difficulty of accurately measuring mass change in foraging sea lions. For example, study animals spent variable amounts of time onshore prior to being recaptured. If a female found her pup immediately, the pup could have consumed substantial amounts of milk (2 to 3 kg) prior to being recaptured and weighed, whereas a female that was recaptured prior to being reunited with her pup would not have given any milk to her pup and thus would have a greater body mass. Moreover, considering the difficulties of weighing a large struggling animals on a hanging spring scale, inconsistencies in female body mass are not surprising given the small mass changes measured (2–3 kg) relative to the total mass of the female (80 kg).

Our data indicate that while at sea, lactating sea lions increased both their rate of energy expenditure and their total time spent foraging in response to changes in food availability. This combination resulted in a greater energy expenditure per foraging trip in 1983 than in 1984 (Tables 2, 3). What possible changes in at-sea behavior may account for such differences in foraging energetics? Using time-depth recorders Feldkamp et al. (this Vol.) found that lactating sea lions worked harder in 1983 than in 1982. For example, during EN study animals spent a greater percentage of total trip time diving, exhibited diving bouts of longer duration, made dives of greater duration, and possibly dived to deeper depths as well as spending less time swimming and resting than pre-EN animals (Feldkamp et al., this Vol.). Such changes in foraging behavior would have obvious implications to time-energy budgets. For example swimming and diving require greater rates of energy consumption than resting (Feldkamp 1987). If sea lions spent less time resting and more time actively pursuing prey, they would increase their at-sea FMR. Furthermore, different foraging strategies may result in different FMRs (Costa 1988a, 1991). These observed changes in at-sea behavior are consistent with the increases in at-sea FMR observed in our study animals.

Although oceanographic conditions were more favorable during 1984, they had still not returned to normal as evidenced by longer than normal trip durations and lower than normal pup growth rates on San Nicolas Island (Boness et al. and Heath et al., this Vol.) during 1984, and lower than normal weights of pups at approximately 4 months of age on San Miguel Island (DeLong et al., this Vol.). The differences in foraging energetics observed between 1983 and 1984 are probably minimal and if data were available from a more "normal" year, the expected deviations may be even greater. Northern fur seal females foraging over 2 years in the Bering Sea exhibited significantly different FMRs between years without changes in trip duration or pup mortality (Costa and Gentry 1986). In northern fur seals, normal year to year variations in the rate of energy expenditure may be associated with changes in prey type or may be due to modifications in the at-sea time-energy budget (Costa 1988a; Costa et al. 1989). However, northern fur seals foraging off San Miguel Island also responded to the 1982/83 EN by remaining at sea longer (DeLong et al., this Vol.). In contrast, Antarctic fur seal females foraging over two seasons, one when prey was scarce and one when prey was normally available, responded to reduced prey availability by increasing trip duration without modifying FMR (Costa et al.

1989). Such differences were assumed to be due to several factors. First, it was suggested that Antarctic fur seals operate near their metabolic maximum and have a reduced ability to increase FMR. This is because most (95%) of their time at sea is spent actively, that is either swimming or diving. Therefore, they have a reduced ability to increase their overall metabolic effort. Second, and possibly more importantly, since they feed exclusively on Antarctic krill, they cannot benefit from differences in rates of energy expenditure and mass gain associated with foraging on different types of prey. The foraging behavior of California sea lion females during the 1982/83 EN incorporated both of the responses observed for Antarctic and northern fur seals. California sea lions are similar to Antarctic fur seals, in that while at sea they normally spend only 4.4% of their time resting (Feldkamp et al., this Vol.). However, California sea lions prey on a variety of species, whereas Antarctic fur seals prey only upon krill. This implies that variations in foraging energetics are more related to differences in diving pattern and prey species than reductions in time spent resting. Apparently, such modifications in diving pattern and the concomitant increases in FMR were not sufficient to compensate for the changes in prey abundance and distribution caused by the 1982/83 EN. It is likely that the sea lions' metabolic maximum was reached and like Antarctic fur seals, the remaining alternative was to increase trip duration.

One might speculate that during normal year to year variations in prey availability, California sea lions can compensate for alterations in prey distribution and abundance by switching the prey type (Bailey and Ainley 1982; Antonelis et al. 1984) or by modifying the time and energy spent actively foraging while at sea. Ideally, such changes would preclude increases in trip duration, since increasing the time at sea reduces the amount of time the mother can spend attending her pup. Therefore, longer foraging trips should be the last option. It is far more advantageous to alter foraging behavior and at-sea metabolism to maximize foraging efficiency and thereby assure that offspring growth is optimal. If, however, as was the case during the 1982/83 EN, modifications in the at-sea FMR and time-energy budgets are insufficient to compensate, females have no alternative but to increase the amount of time spent foraging at sea.

Acknowledgments. This work is the result of research sponsored in part by NOAA, National Sea Grant College Program, Department of Commerce, under grand #NA80AA-D-00120 and NA85AA-D-SG140, project No. R/U 92, through the California Sea Grant College Program, and in part by the California State Resources Agency. Additional funds were made available from the Office of Naval Research contract #N00014-87-K-0178. The U.S. Government is authorized to reproduce and distribute for governmental purposes. We thank S. Feldkamp and L. Higgins for assistance with fieldwork at San Miguel Island and M. Zavanelli and L. Higgins for help with laboratory analysis. D. Lavigne, M. Kretzmann, G. Worthy, M. Castellini, and the committee of "N" journal club provided valuable comments on the manuscript.

17 Effects of the 1982–83 El Niño on Several Population Parameters and Diet of California Sea Lions on the California Channel Islands

R.L. DeLong, G.A. Antonelis, C.W. Oliver, B.S. Stewart, M.C. Lowry, and P.K. Yochem

17.1 Introduction

Prior to the 1982–83 El Niño, California sea lion (*Zalophus californianus*) populations in the California Channel Islands were recovering from low abundance levels that resulted from unregulated exploitation and indiscriminate killing which began in the 1800s and continued through the late 1930s (Bonnot 1937; Bartholomew 1967; Stewart et al., in press). There has been an increasing trend in sea lion abundance on the California Channel Islands since 1958 (Bartholomew 1967; Odell 1971; Le Boeuf and Bonnell 1980; Stewart et al., in press). Based upon counts of pups born between 1971 and 1981, the sea lion population at San Miguel Island increased at about 5% annually (DeMaster et al. 1982). The population in the Channel Islands is estimated to number 87 000 animals (Boveng 1988).

The 1982–83 El Niño had a substantial influence on physical and biological aspects of the marine ecosystem in the Southern California Bight (Fahrback et al., Chap. 1 and Arntz et al., Chap. 2, this Vol.). In this chapter we describe the effects of the 1982–83 El Niño on California sea lion pup production, weight of pups at approximately 4 months of age, juvenile distribution and mortality, and summer diet of sea lions on rookeries in the California Channel Islands.

17.2 Methods

Counts were made of live and dead pups at the end of the breeding season in mid- to late July each year from 1982 through 1986 on San Miguel, Santa Barbara, San Nicolas, and San Clemente Islands. All counts were made during ground surveys using the techniques described by Stewart and Yochem (1984). Estimates of pup production (number of pups born) were obtained by summing the counts of live and dead pups. These counts did not include premature births (i.e., pups born prior to 15 May). Pup counts for 1982 from Santa Barbara Island are from Heath and Francis (1984). Dead pup counts are known to underrepresent neonatal mortality as pups are washed away, covered by blowing sand and disappear due to decomposition within 14 days (DeLong and Antonelis, unpubl. data; Francis and Heath, Chap. 12, this Vol.).

Pup weights were obtained in late September of each year from 1982 through 1986 when pups were tagged on San Miguel and San Clemente Islands. Pups from each cohort were approximately 4 months of age at the time of weighing. At each island groups of approximately 100 to 200 pups were herded and held for weighing and tagging. Individual pups were placed in hoop nets and weighed on spring scales with weights recorded to the nearest 0.5 kg. Weight data were compared using a three-way analysis of variance (Baker and Nelder 1978).

Sightings of previously tagged animals from both research programs designed to study marked animals and from opportunistic sightings by the general public provided information on interannual changes in juvenile distribution and mortality. Estimates of pup mortality (between birth and tagging) and of juvenile mortality (between tagging and age 1) on San Clemente Island were obtained from data collected during monthly surveys. On surveys during summer months dead and live pups were counted, and on surveys during the remainder of the year tags were recovered from dead, tagged sea lions.

Summer sea lion diets at San Miguel and San Clemente islands were documented by analyzing scats using methods described by Antonelis et al. (1984). Scats were collected from areas utilized primarily by adult females during July and August 1982 and 1983 at San Miguel Island, and from areas occupied by animals of all age/sex categories at San Clemente Island during June through August 1982 and 1983. A two-way analysis of deviance using binomial errors (McCullagh and Nelder 1983) was used to compare frequency of occurrence data of each prey species between islands and years. This analysis does not account for the possibility of interdependence among prey species (i.e., co-occurrence).

17.3 Results

Numbers of sea lion births declined at all rookeries in the Southern California Bight in 1983 (Fig. 1). The magnitude of the decline in births varied among rookeries with decreases of 30% at San Miguel Island, 43% at San Nicolas Island, 62% at San Clemente Island, and 71% at Santa Barbara Island. In 1984 births increased on San Miguel Island, decreased further on San Nicolas Island, and remained low on Santa Barbara and San Clemente islands. By 1986 numbers of births had recovered to the 1982 levels at San Miguel and Santa Barbara islands, but births remained low at San Nicolas and San Clemente islands. In 1986 numbers of pups born on all islands in 1986 were still 10% below the numbers born in 1982.

The weights of male and female pups showed similar annual trends declining about 25 to 35% at San Miguel and San Clemente Islands from 1982 to 1983 (Table 1). Pup weights remained low in 1984 and did not return to 1982 values until 1985. Females were significantly lighter than males on both islands during all years ($F_{1,1408} = 217.7$, $P < 0.001$). At each island weights of pups were significantly lighter in 1983 and 1984 than in all other years ($F_{4,1408} = 254.3$, $P < 0.001$). The only significant difference among islands and years occurred between 1983 and 1984 ($F_{4,1408} = 5.39$, $P < 0.001$), when pup weights increased at San Miguel Island but decreased slightly on San Clemente Island.

Counts of dead pups at San Miguel Island do not indicate a change in neonatal mortality for 1983 or 1984. In 1982 10% of pups born died by 1 August, whereas the proportion of dead pups was 6 and 4% in 1983 and 1984, respectively. Tags

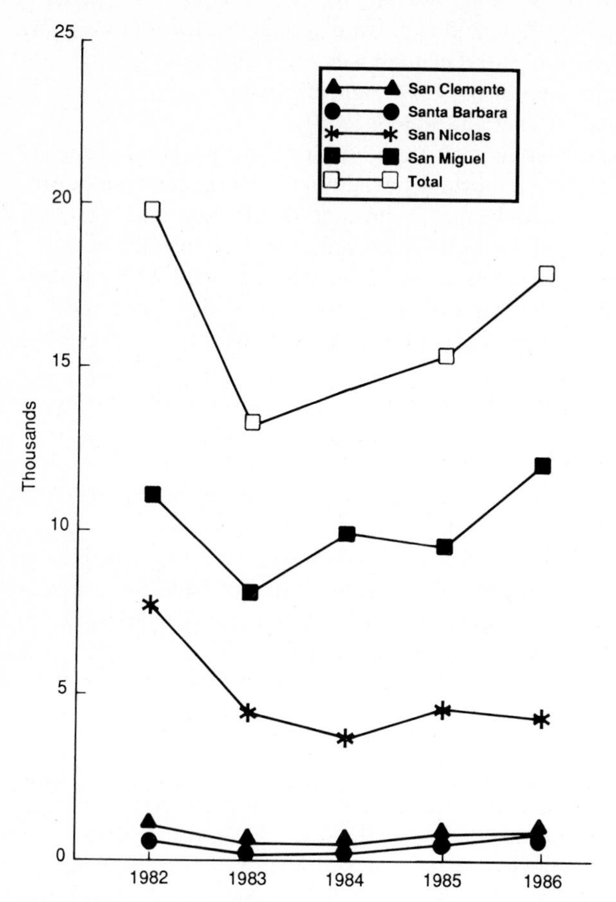

Fig. 1. Numbers of California sea lion pups born on San Miguel, San Nicolas, Santa Barbara, and San Clemente Islands, 1982 to 1986

recovered from animals marked as pups on San Clemente Island and subsequently found dead there through April 1984 indicate a high mortality of pups in 1983 (Table 2). Seven percent of the pups tagged in 1983 were subsequently found dead on San Clemente within 8 months of tagging, whereas up to 1% of pups were found dead in other years. The total number of dead pups recovered at San Clemente Island between August 1983 and May 1984 was five times the mean number for other cohorts from 1981 through 1986 (Table 2).

The distribution of resightings of tagged California sea lion pups did not indicate movement farther north than that observed before or after the El Niño. Yet, Huber (this Vol.) reported an increase in the number of resightings of tagged yearling and 2-year-old sea lions in 1983 and 1984 on the Farallon Islands.

Differences in the occurrence of prey species of adult females varied between San Miguel and San Clemente Islands, and between 1982 and 1983 (Fig. 2). Pacific

Table 1. Weights of California sea lion pups at approximately 4 months of age, San Miguel and San Clemente Island 1982–1986

Year	Male	n	SD	Female	n	SD
			San Miguel Island			
1982	21.7	66	3.94	18.9	83	3.47
1983	15.5	136	3.32	13.3	151	3.27
1984	17.0	55	3.44	13.7	44	2.83
1985	22.3	59	3.45	17.2	41	2.62
1986	21.5	81	4.29	18.8	83	3.47
			San Clemente Island			
1982	21.5	54	3.65	18.8	62	2.89
1983	16.1	66	4.30	13.6	76	4.00
1984	14.6	74	3.47	13.1	70	2.74
1985	21.7	74	3.30	18.1	57	2.13
1986	23.2	51	2.76	19.6	56	2.63

Table 2. California sea lion pups tagged on San Clemente Island and recovered dead on the island during the following 8 months, 1982–1986

		Tags recovered			
Cohort	No. tagged	Males	Females	Total	%
1981	71	0	0	0	0
1982	134	0	1	1	0.7
1983	145	7	5	12	8.3
1984	145	0	1	1	0.7
1985	132	0	0	0	0
1986	107	0	0	0	0

whiting (*Merluccius productus*) occurred more frequently in scats collected at San Miguel Island than at San Clemente Island ($P < 0.05$) during both years. Market squid (*Loligo opalescens*) and blacksmith (*Chromis punctipinnis*) were detected more frequently on San Clemente Island than San Miguel Island ($P < 0.05$) for both years. Pacific whiting was found more often in 1982 than in 1983 at both islands ($P < 0.05$), rockfish (*Sebastes* spp.) occurred more frequently in 1983 than in 1982 at both islands ($P < 0.05$). Significant interaction between island and year (ANOVA main effects) was detected for the occurrence of jack mackerel (*Trachurus symmetricus*), northern anchovy (*Engraulis mordax*), and pelagic red crab (*Pleuroncodes planipes*). No significant difference was detected in the occurrence of Pacific mackerel (*Scomber japonicus*) between islands or years.

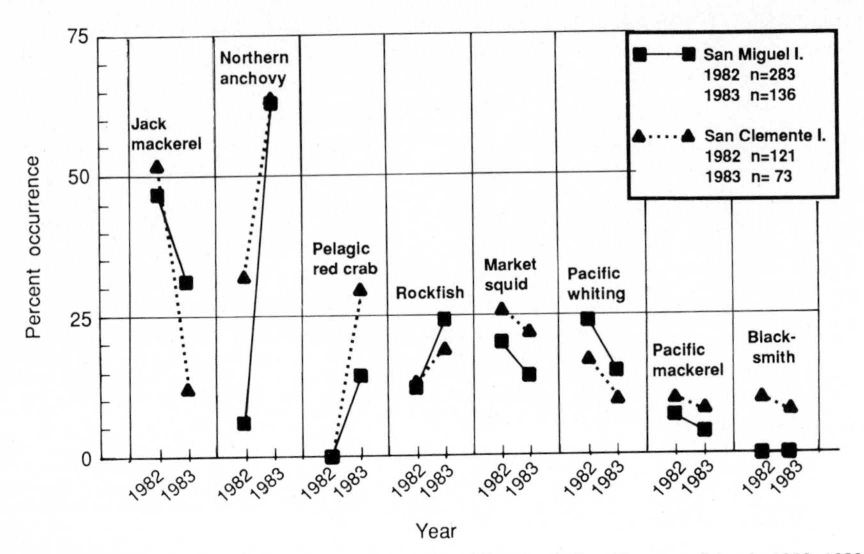

Fig. 2. Summer diet of California sea lions on San Miguel and San Clemente Islands, 1982–1983

17.4 Discussion

The correlation between oceanographic changes associated with the 1982–83 El Niño and the declines in numbers of pups and neonatal weights of sea lion pups on the California Channel Islands suggests that substantial environmentally induced changes occurred in the prey resources available to pregnant and lactating sea lions. Changes between 1982 and 1983 in diets, increased foraging efforts (Feldkamp et al., this Vol.), increased foraging energetics of lactating females (Costa et al., this Vol.) and lower pup weights all indicate that females foraged more actively and accumulated less energy as a result of reduced prey availability.

Differences in the magnitude of declines in births among rookeries indicate that the El Niño did not affect the Southern California Bight uniformly in 1983. Compared to other islands, births declined the least at the most northern island (San Miguel). This may be due to the clinal differences of the El Niño (Fiedler 1984a; Arntz et al., Chap. 2, this Vol.) where much warmer sea surface temperatures were recorded around San Clemente Island than around San Miguel Island. Thus, lactating and pregnant females found prey resources more available at higher latitudes where sea surface temperatures were less elevated.

It is not clear, however, whether the decline in births resulted from reduced pregnancy rates among females, increased female mortality, or both. The increases in pup numbers during years following the El Niño at San Miguel, Santa Barbara, and San Clemente Islands provide evidence for a decrease in pregnancy rates in 1983 and 1984. The further decline in births at San Nicolas Island in 1984 and the continued low counts through 1986 complicate this interpretation. The 43% decline in pup births at San Nicolas Island in 1983 undoubtedly resulted from environmental influences similar to those influencing other rookeries, but the lack of recovery in subsequent years may have been due to other factors such as human disturbance (Stewart et al., in press) which would have stimulated emigration. Whether females died or emigrated to other breeding colonies in the Southern California Bight or in Mexico remains unknown. Because total births on the California Channel Islands in 1986 were still less than in 1982, we believe that adult female and juvenile mortality may have increased in 1983, depressing the number of females in the population.

The significantly lower weights of sea lion pups at both San Miguel and San Clemente Islands in 1983 and 1984 compared to other years indicate that lactating females had reduced foraging success and consequently poorer milk production. From 1983 to 1984 the weights of pups on each island changed differently as indicated by the significant island-year interaction. Pups on San Miguel were heavier and pups on San Clemente were lighter in 1984 than 1983. Apparently, foraging conditions for females were somewhat better near San Miguel Island in 1984 , but were poorer near San Clemente Island in 1984 than 1983. At both locations residual physical and biological oceanographic effects of the El Niño were still apparent during 1984 (Norton et al. 1985; Cole and McLain 1989) and were reflected in lower pup weights. It was not until the return to more normal marine environmental conditions in 1985 that weights of pups returned to pre-El Niño values.

The number of dead pups found in 1983 suggests that neonatal mortality did not increase at San Miguel Island. Neonatal mortality did, however, increase in 1983 on San Nicolas Island (Ono et al. 1987; Francis and Heath, Chap. 12, this Vol.) and based on tagging records (Table 2) also increased on San Clemente Island. Although there are inconsistencies in the mortality data between islands, most suggest that neonatal and juvenile mortality increased at most islands in 1983.

Several studies have discussed the relationship between availability and abundance of prey and seasonal and annual variation in the diets of the California sea lion (e.g., Bailey and Ainley 1982; Antonelis et al. 1984; Lowry, unpubl. data). Similar correlations occurred in 1982 and 1983 as the distribution of sea lion prey assemblages changed in the Southern California Bight (Bailey and Incze 1985; Fiedler et al. 1986). During this period, differences in prey species occurrence were apparent between San Miguel and San Clemente Islands. Compared to other prey species, the occurrence of northern anchovy exhibited the most dramatic change from 1982 to 1983 with increases of 30 and 50% for San Clemente and San Miguel Islands, respectively. Several factors were likely causes of that increase, such as an expansion of the central stock to the north and farther offshore and an influx of the anchovy stocks from the southern population, as well as a corresponding decline in the availability of the other prey (e.g., jack mackerel).

When the Davidson and California Countercurrent systems are intensified and persist for long periods during El Niño conditions, pelagic red crabs are frequently transported northward from coastal waters of Baja California into the waters off California (Boyd 1967). Pelagic red crabs had not been previously recorded as sea lion prey, yet its presence apparently provided an alternate food source for sea lions during a time when the availability and abundance of more commonly occurring prey had declined. Pelagic red crabs were also found in the diets of northern fur seals (*Callorhinus ursinus*) (DeLong and Antonelis, this Vol.), harbor seals (*Phoca vitulina*) (DeLong and Antonelis, unpubl. data), and northern elephant seals

(*Mirounga angustirostris*) in the Southern California Bight during those years influenced by the 1982–83 El Niño (Hacker 1986; Antonelis et al. 1987).

17.5 Summary

The most conspicuous effects of the 1982–83 El Niño on the California sea lion population in the Southern California Bight were the large declines in numbers of pups and reduced pup weights in 1983 and 1984. These changes were evidently mediated by decreased prey availability which affected females most and resulted in lowered reproductive success and pup growth. Decreased pup production resulted from decreased pregnancy rates, and perhaps increased mortality of adult females in 1983. Since numbers of pups had not recovered to 1982 levels by 1986 it is probable that juvenile female mortality also increased.

Acknowledgments. We thank the many field biologists who participated in tagging and resighting sea lions at the various rookeries, especially E. Jameyson, D. Skilling, and A. Smith at San Miguel Island, L. Ferm and J. Wexler at San Clemente Island, and H. Huber at the Farallons. We thank the Channel Islands National Park and the Pacific Missile Test Center for permitting our access to San Miguel and San Nicolas Islands. The paper was reviewed and improved by the comments of R. Gentry, R. Merrick, M. Perez, H. Braham, W. Pearcy, K. Ono, an anonymous reviewer, and the Publications Unit, Alaska Fisheries Center. A. York provided statistical consultation and review. The work of BSS and PKY was supported by contracts from the United States Air Force.

18 The Effect of El Niño on Pup Development in the California Sea Lion (*Zalophus californianus*) I. Early Postnatal Growth

D.J. Boness, O.T. Oftedal, and K.A. Ono

18.1 Introduction

Environmental conditions that alter the nutritional status of lactating females, or the amount of foraging time needed to maintain good nutritional status, may secondarily affect the development of their suckling young although the evidence for this is equivocal. It has been suggested that the best single measure of development of an organism is postnatal mass change (Layne 1968). Among mammalian young that feed solely on maternal milk during lactation a net accumulation of mass is possible only if nutrient transfer from the mother exceeds the maintenance requirements of the young. Maternal food shortage can cause a reduction in milk transfer (Loudon and Kay 1984; Oftedal 1985) and thus in the nutrients available for gain in mass by suckling offspring.

Experimentally induced food shortage among females in controlled captive situations has been shown to reduce early postnatal growth of young in some species (McClure 1981; Labov et al. 1986; Lee 1987). Additionally, experimentally enhanced food resources in free-ranging opossums (*Didelphis marsupialis*) produced larger young in provisioned females than in unprovisioned females (Austad and Sunquist 1986). However, in two other experimental studies food shortage either had no effect on suckling young, or the effect was the opposite of that which was expected (Gosling et al. 1984; Wright et al. 1988).

The altered prey distribution off the southern California coast during the 1983 El Niño appears to have affected the foraging effort of lactating California sea lion females. Lactating females had significantly longer foraging trips during El Niño than in other years (Ono et al. 1987; Heath et al., this Vol.). In this chapter, we examine neonatal size and early postnatal growth of California sea lion pups in four consecutive years, encompassing the 1983 El Niño, to determine whether the altered maternal foraging effort associated with El Niño affected pup growth.

18.2 Methods

The study was carried out between 1982 and 1985 at San Nicolas Island, California (33°N, 119°W). Approximately 6000–8000 pups are born there yearly between late May and early July (Stewart and Yochem 1984). Our study sites consisted of an area of rocky shelves and boulders and several beaches (Areas 240 and 211–212, respectively in Bonnell et al. 1980).

El Niño began late in 1982, after our growth measurements were taken, and peaked in 1983. We thus refer to the 1982 cohort as PRE El Niño, the 1983 cohort as EN, and the 1984 and 1985 cohorts as POST1 and POST2, respectively.

Capture dates varied annually because many of the measurements of pups were collected as part of other studies. Pups were taken during the following dates: 1982 cohort, May to September 1982; 1983 cohort, May to December 1983; 1984 cohort, May to December 1984; 1985 cohort, June 1985 to January 1986.

Each year most pups at a behavioral study site (Area 211) were weighed, measured, and marked in June, and remeasured and tagged in August (Ono et al. 1987). Additional measurements were obtained from pups that were individually netted either on beaches adjacent to the behavioral study site, or in Area 240. These pups were typically selected from small groups of resting pups, and were weighed, measured, and individually marked with bleach, paint, and/or flipper tags prior to release. We refer to these data as cross-sectional because they involve single captures of individual pups at various times.

Young pups were weighed to the nearest 1 oz. using a Chatillon hanging spring scale of 60 lb. capacity and older pups to the nearest 0.5 lb. with a Chatillon Dynamometer of 100 lb. capacity. Measurements were made of nose to tail length, axillary girth, and right fore and rear flipper lengths (Am. Soc. Mammal. 1967). In 1982 and 1985 all measurements were not obtained on all animals, whereas in 1983 and 1984 all measurements were taken on most individuals.

The ages of most young pups (less than 2 months of age) were known exactly or to within 2 days since birth dates were determined by observation of:

1. Pups being born;
2. Fresh placentas and blood on females or pups; or
3. Remains of umbilical cords attached to pups.

The umbilical cord is usually lost by the third day postpartum (unpubl. data). The approximate ages of older pups for which birth dates were not known were estimated from the mean birth date for all 4 years (16 June).

Statistical analyses were performed using programs from either the Statistical Package for the Social Sciences (SPSS) (Nie et al. 1975) or SAS (Version 6; SAS Institute Inc.). Overall year effects were tested by ANOVAs that included sex as an independent variable in order to control known sex differences (unpubl. data). Means presented in tables are least squared means for year effects, reflecting an equal weighting of the sexes. Pairwise differences between years were tested using the sequential Bonferroni technique after randomly adjusting the samples in each year to include an equal number of each sex in the sample (Holm 1979). In cross-sectional analyses, only the last set of measurements for any individual was used to avoid unequal representation. Differences were considered statistically significant at $p \leq 0.05$.

18.3 Results

Neonate Measurements. Measurements of pups within 3 days of birth were analyzed to estimate size at birth. There were no significant year effects for any of the measurements taken (Table 1). Nonetheless, for all five body measurements the mean values in EN were slightly lower than those in other years.

Postnatal Growth. Early postnatal mass gain of California sea lion pups was not linear (Fig. 1). A polynomial regression analysis of cross-sectional measurements from each year was restricted to pups 200 days old or less because pups older than this may begin to ingest solid food (unpubl. data) and measurements beyond this age are not available from all 4 years. Body mass changes during this period of life were exponential for most year and sex cohorts, indicating that the rate of gain declined with age. Although quadratic terms were not significant in three of the eight year and sex groups (females in PRE and both males and females in POST1), combining the separate analyses for each year (Fisher's combined probability test, Sokal and Rohlf 1969), the exponential pattern of mass gain was significant for both male and female pups ($X^2 = 62.3$, df = 8, $p < 0.001$ and $X^2 = 22.8$, df = 8, $p < 0.005$, respectively).

A statistical comparison of the curves in Fig. 1 was not made because of the limitations in the data among years, as mentioned above. A visual inspection of the curves suggests that the decline in rate of mass gain as a function of age was greatest in EN. In fact, male pups appeared to show an average loss in body mass between about 120 and 180 days in this year. Female pups did not show a mass loss during this period, but they gained little. The size of both male and female pups at about 450 days of age in EN was high relative to expectations from the shape of the

Table 1. Mean size of neonate (< 3 days of age) California sea lion pups (mass is in kg and all other measures are in cm)

	PRE (1982)	EN (1983)	POST1 (1984)	POST2 (1985)	ANOVA	
Variable	Mean (SE, n)	Mean (SE, n)	Mean (SE, n)	Mean (SE, n)	F	P
Body mass	8.4 (0.146, 39)	8.1 (0.163, 31)	8.2 (0.229, 16)	8.6 (0.227, 16)	1.08	ns
Body length	73.7 (1.526, 7)	73.5 (0.730, 31)	74.2 (1.029, 16)	76.4 (2.034, 5)	1.02	ns
Axillary girth	50.0 (1.056, 7)	48.2 (0.505, 31)	48.7 (0.712, 16)	50.0 (1.408, 4)	1.54	ns
Fore flipper	24.1 (0.618, 7)	23.9 (0.295, 31)	24.0 (0.416, 16)	24.4 (0.427, 15)	0.30	ns
Rear flipper	16.3 (0.417, 5)	16.2 (0.168, 31)	16.3 (0.238, 16)	— (—, -)	0.72	ns

Note: Means have been adjusted by statistically removing the effect of sex since the proportion of each sex is not the same in each year cohort.

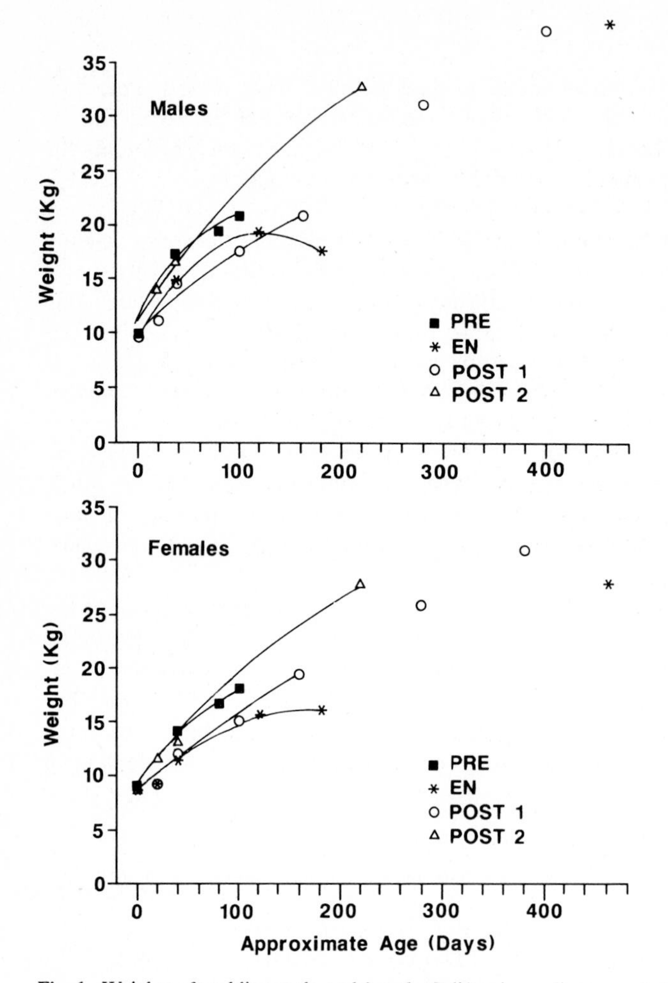

Fig. 1. Weights of suckling male and female California sea lion pups in four different year cohorts. Each point represents a mean for a 20-day period. The curves are derived from quadratic regression analyses. The total numbers of male pups weighed in the four cohorts [PRE (1982), EN (1983), POST1 (1984), and POST2 (1985)] were 92, 130, 93, and 125, respectively; the numbers of female pups were 91, 133, 89 and 133, respectively. Most pups are no longer suckling by 1 year of age

growth curves for the first 200 days of age. In POST1, the only other year for which we have data over a comparable period, both male and female pups showed sustained growth over the entire time interval.

Most pups are weaned by 1 year of age; only about 30% suckle into their first year (unpubl. data; Francis and Heath, Chap. 21, this Vol.). Most yearlings sighted at the rookery were observed suckling.

The longitudinal data obtained each year at the behavioral study site provide an opportunity for statistical analysis of yearly differences in growth rates of pups over the first 2 months of life. As there were significant differences in growth rates between the sexes and as larger animals might grow at a different rate than smaller ones, year effects on growth were assessed by analysis of covariance. The covariance model included year and sex as main effects and size at initial capture as a covariate. There was no significant interaction between year and sex effects. Significant differences among years were found in the rate of gain in body mass, fore flipper length, and rear flipper length (Table 2). Year effects for gain in body length and axillary girth were not significant. These two measures are probably more prone to measurement error in live animals than are other measures because movement by an individual may change these dimensions.

The general pattern which emerged revealed that pup growth was slowest in EN, intermediate in the POST1 year, and fastest in PRE and POST2 years. Pairwise comparisons of years indicated significantly lower rates of gain in body mass and fore flipper length in EN than in any of the other years (Table 2). Pairwise analyses were not done for body length, axillary girth, or rear flipper length because there were few samples or no samples in some years, or because the year effect for a measure was not significant. Although POST1 measurements did not differ significantly from either PRE or POST2, all body measures in POST1 were intermediate in value between those of EN and PRE or POST2.

Table 2. Mean growth rates (kg/day and cm/day) of California sea lion pups determined from longitudinal data on pups in the first 2 months of life

Variable	PRE (1982)[a] Mean (SE, n)	EN (1983) Mean (SE, n)	POST1 (1984) Mean (SE, n)	POST2 (1985)[b] Mean (SE, n)	ANOVA F	P	Significance[c] of yearly differences
Body mass	0.132 (0.007, 32)	0.101 (0.007, 30)	0.125 (0.007, 32)	0.130 (0.006, 38)	4.52	0.005	83 84 85 82
Body length	— (—, -)	0.249 (0.016, 30)	0.269 (0.016, 32)	0.350 (0.048, 7)	2.74	0.070	—
Axillary girth	— (—, -)	0.158 (0.014, 30)	0.165 (0.014, 31)	0.225 (0.040, 8)	2.81	0.070	—
Fore flipper	— (—, -)	0.043 (0.006, 30)	0.071 (0.006, 32)	0.073 (0.007, 31)	7.02	0.002	83 84 85
Rear flipper	— (—, -)	0.044 (0.005, 30)	0.063 (0.004, 32)	0.065 (0.014, 6)	5.06	0.009	—

Note: As in Table 1, the means are adjusted to remove the effects of sex differences.
[a]In PRE, measurements other than weights were not obtained.
[b]In POST2, logistical constraints allowed only weight and fore flipper measurements to be taken on most pups.
[c]Differences between years were determined by Holm's test. For this test each sex was equally represented in a given year. This was done by randomly deleting individuals from the analysis for the sex with the greater number of individuals. Years are arranged in ascending order of growth rates and underlining denotes years that are not significantly different ($p > 0.05$). This analysis was not performed for body length, axillary girth, and rear flipper length because sample sizes for these measurements in 1985 were so small and no measurements were available in 1982.

18.4 Discussion

The results of this study suggest that El Niño negatively affected the early postnatal growth of California sea lions from southern California. Surprisingly, however, El Niño did not affect prenatal growth, as suggested by a lack of difference in neonate size among years. Sea lion females that were inseminated during the PRE breeding season and gave birth in EN should have been affected by El Niño in the mid- to late stages of pregnancy since El Niño began along the California coast as early as October 1982 (McGowan 1984). For females giving birth in POST1, the impact on food supply should have affected the entire pregnancy. Indeed, the shorter duration of the first attendance period of lactating females in EN (Ono et al. 1987) would seem to indicate that maternal reserves were reduced under El Niño conditions. Perhaps pups *in utero* are partially protected from mild to moderate food shortage because females mobilize reserves to support fetal growth. Alternative explanations for the apparent lack of effect of El Niño on neonatal size include the fact that the cost of pregnancy is energetically less demanding than lactation (Millar 1977; Oftedal 1985; Loveridge 1986; Oftedal and Gittleman 1989) and females can forage full time before their pups are born. Thus, even during food-poor years females may be able to meet the demands of pregnancy. Comparison of changes in body composition of pregnant females under different nutritional planes and estimates of prenatal foraging times might help clarify this issue.

There was a clear impact of El Niño on postnatal growth in suckling California sea lions at San Nicolas Island. Pups born in EN were smaller throughout most of the period of measurement (Fig. 1) and growth rates up to 2 months of age were substantially less in the El Niño year than in other years (Table 2). The marked decline between about 120 and 200 days postpartum in the growth curves for 1983 (Fig. 1) might reflect the greater difficulty of females to meet increasing demands of their growing pups under conditions that require increased foraging efforts. Several facts support this hypothesis:

1. Growth curves from other years show a trend toward declining pup growth rates to about 200 days postpartum;
2. The male pups, which are larger than the female pups, in 1983 showed a net loss in mass between about 120 and 200 days, whereas the female pups only showed a leveling off in mass gain; and
3. At least some pups begin to feed on solid food (squid, *Gonatus* sp., and pelagic crab, *Pleuroncodes* sp.) at about 200 days postpartum (unpubl. data).

In EN the relatively large mean body mass of yearling pups (about 450 days old) in comparison to the relatively small mass at just under 200 days of age may reflect compensation for the earlier slow mass gain (or even mass loss). Such compensation could have been accomplished through increased solid food intake. We cannot, however, rule out the possibility that the relatively high mean weight at about 450 days simply reflects a biased sample of the hardiest pups which have survived.

The oceanographic evidence indicates that the 1983 El Niño had weakened by early 1984 (McGowan 1984). Studies of various aspects of the reproductive biology of the California sea lion (Ono et al. 1987; Heath et al., this Vol.; Iverson et al., this

Vol.) and the Galapagos fur seal (Trillmich 1986b), nonetheless, indicate that El Niño effects may be felt beyond the year in which an El Niño occurs. Our data are weakly suggestive of this extended effect. All measures of neonatal size and growth rates from POST1 are intermediate between those from EN and PRE, although the POST1 values are statistically indistinguishable from PRE and POST2 years.

Acknowledgments. Many people assisted in the collection of these data so that it is not possible to name them all here, but we are very grateful for their help. We particularly thank the following people for assisting us in the field during several seasons: S. Andrews, D. Baer, T. Beedy, L. Dabek, J. Francis, S. Gaines, C. Heath, S. Iverson, M. Klope, P. Lubchenco, L. Palermo, S. Pereira, W. Price, and W. Rice. L. Dabek also assisted in analysis of the data and W. Rice provided statistical advice. Helpful comments were made on earlier drafts of the paper by Drs. F. Trillmich and S. Iverson. We thank the U.S. Navy at the Pacific Missile Test Center, Point Mugu, CA, for permission to use San Nicolas Island (SNI) as our study site and for logistical support. R. Dow, the Ecology Coordinator at Point Mugu, was especially instrumental to the success of our work on SNI. Dr. K.A. Ono was supported during part of this study by a Postdoctoral Fellowship from the Friends of the National Zoo. The study was funded by grants from the Friends of the National Zoo and the Smithsonian Institution Fluid Research Fund.

19 The Effect of El Niño on Pup Development in the California Sea Lion (*Zalophus californianus*) II. Milk Intake

S.J. Iverson, O.T. Oftedal, and D.J. Boness

19.1 Introduction

Determination of milk yield is an essential component in the study of the effects of El Niño on pup development in the California sea lion. Reproductive effort and lactational performance of females directly influence growth of pups. The use of hydrogen isotope methods allows the direct measure of milk intake by pups in free-ranging animals with minimal disturbance of normal suckling patterns (Costa 1987; Oftedal and Iverson 1987). Pinnipeds are well suited for such studies. Parturition and lactation tend to occur in large aggregations, capture is relatively easy, and site fidelity permits repeated captures. Because cross-suckling is rare in otariids (Oftedal et al. 1987a), milk intake by the pup can be considered equivalent to milk production by the mother.

19.2 Methods

In this study we measured milk intake by deuterium oxide dilution in pups of the California sea lion during early lactation in both normal and El Niño years (see Oftedal et al. 1983; Oftedal and Iverson 1987; and Iverson 1988 for details of sample analysis). Pups were studied in 1982, 1983, and 1984, or before (PRE), during (EN), and in the first year after (POST1) the 1983 El Niño. Data were collected from a wild population of California sea lions on San Nicolas Island (see Boness et al., this Vol.). Isotope studies were conducted on 39 pups over 3-week periods beginning the first week of June or July. Given the constancy in yearly mean birth dates of sea lions on San Nicolas Island, as well as information on growth rates in early lactation (Boness et al., this Vol.), pups studied in June and July were considered to be first and second month postpartum, respectively.

The first month of lactation was studied in June of 1982 and 1984; the second month of lactation was studied in July of 1983 and 1984. As our initial objective was to measure differences in milk intake according to lactation stage, pups were not studied in the same months in 1982 and 1983. After we recognized the severe effect of El Niño in 1983, data were collected in both months in 1984 to allow comparison to prior years. Because milk intake is known to vary with stage of lactation in other species (Oftedal 1984), our comparisons are limited to PRE versus POST1 in the first month postpartum and to EN versus POST1 in the second month postpartum. While 1984 is referred to as post-El Niño, continued effects of El Niño were found to be present in 1984 and thus POST1 may be more similar to EN than originally expected (Ono et al. 1987).

Data were grouped by month of age and tested by two-way analysis of variance (sex × year) using Stat View 512+™ designed for the Macintosh™. Significance level was set at $P < 0.05$. Additional comparisons were made within 1984 between first and second month pups (two-way ANOVA, sex × age). Statistical interactions were not significant for any parameter for either age, and hence were not reported. Several pups in this study exhibited abnormal growth (below the first percentile of normal growth for pups on undisturbed sites of the island in each of those years), possibly indicating the effect of human or other disturbance. Because our intent here was to examine differences due to yearly effects (El Niño), these pups were excluded from the present analyses: two males and one female in June of 1982 and one male in June of 1984. Therefore, a different subset of pups was used in this analysis than in Oftedal et al. (1987b), although the initial data base was the same.

19.3 Results

The initial weight of pups used in isotope studies did not differ between non-El Niño and El Niño years in either month (Tables 1, 2). Initial weights in June (Table 1) represent pups of about 5 days of age, although precise birth dates were not known for all pups. The similarity in initial weight across years is consistent with the observation that El Niño did not have an effect on mean birth weight (Boness et al., this Vol.). By the second month of age, pups might be expected to be larger in non-El Niño years but our comparison is limited to POST1, a year during which growth was still affected by El Niño (Boness et al., this Vol.). Comparison of growth rates of isotope-labeled pups in the first month postpartum revealed significantly lower growth rates in POST1 than in PRE (Table 1), but no differences in the second month between EN and POST1 (Table 2). Growth rates of isotope-labeled pups did not change from the first to second month in POST1.

The percentage of body weight represented by body water was significantly lower in PRE than in POST1 in first month pups (Table 1), indicating that PRE pups had a higher body fat content. No differences were detected between EN and POST1 (Table 2). Body water percent declined with age from first to second month postpartum in POST1 ($P = 0.01$).

In the first month postpartum, fractional water turnover rate and daily milk intake were higher in PRE than in POST1 (Table 1). Milk intake expressed as a percent of body weight was also significantly higher in PRE compared to POST1. In the second month, no differences were found in water turnover or milk intake between EN and POST1, however (Table 2). In POST1, fractional water turnover and milk intake did not increase from the first to the second month postpartum. Because pups were larger in the second month yet consumed similar quantities of milk, milk intake as a percent of body weight declined with age in POST1 ($P = 0.03$, ANOVA).

Milk intakes were converted to daily gross energy intakes, based upon the caloric content of the milk (3.41 kcal/g; Oftedal et al. 1987b; Iverson 1988). Gross energy

Table 1. Growth, milk intake, and energy intake in suckling California sea lion pups in the first month of life during pre-(1982) and post-(1984) El Niño[a]

	June				
	PRE (1982)		POST1 (1984)		
	Male n = 7	Female n = 7	Male n = 4	Female n = 4	PRE vs POST1[b] (*P*)
Initial weight, kg	9.09 ±0.112	7.59 ±0.289	8.75 ±0.428	7.84 ±0.435	n.s.
Daily gain, g	144.1 ±16.13	120.0 ±8.16	107.5 ±5.74	95.0 ±11.75	*P* = 0.03
Body water, %	70.7 ±1.34	68.1 ±1.02	74.0 ±0.41	71.1 ±0.81	*P* = 0.02
Fractional water turnover	0.088 ±0.0030	0.084 ±0.0030	0.075 ±0.0047	0.076 ±0.0025	*P* = 0.01
Milk intake (MI), g/day	831 ±24.6	641 ±38.3	683 ±45.6	596 ±40.4	*P* = 0.02
MI as % of pup body weight[c]	8.00 ±0.260	7.37 ±0.242	7.06 ±0.441	6.82 ±0.207	*P* = 0.03
Energy intake (GE), kcal/day	2834 ±83.9	2187 ±130.6	2328 ±155.4	2032 ±137.8	*P* = 0.02
GE per MBS[d], kcal/$Wt^{0.83}$	406.3 ±12.78	362.8 ±13.20	353.9 ±21.71	335.9 ±9.73	P = 0.02

[a]Data from Oftedal et al. (1987b); values are means ± SEM.
[b]Data were tested by two-way ANOVA (sex × year); no significant interactions were present.
[c]Weight at mid point of study period for each pup, predicted by regression; all other values expressed on this basis.
[d]Metabolic body size (MBS) taken as $Wt^{0.83}$ for suckling young, GE, gross energy.

intakes of sea lion pups, both total and expressed per metabolic size ($kg^{0.83}$), were reduced in POST1 relative to PRE in the first month postpartum (Table 1), but did not differ between EN and POST1 in the second month postpartum (Table 2). Gross energy intakes did not differ with age in POST1 whether expressed as a total or on a metabolic size basis.

19.4 Discussion

The principles and reliability of isotope methods for the measurement of milk intake in pinnipeds are reviewed extensively elsewhere (Costa et al. 1986; Oftedal and Iverson 1987; Oftedal et al. 1987b). Every measured parameter (other than initial weight) was depressed in the year after El Niño as compared to the year before El Niño. The finding that pups had lower water content in early June in PRE than in POST1 may reflect greater fat content at birth in PRE. In both years pups averaged 5 days of age when body water measurements were made and thus were probably sampled prior to the first departure of the mother on a feeding trip. While females departed on their first foraging trip earlier in EN, there was no difference between PRE and POST1 (Ono et al. 1987), the years of our comparison. Greater fat stores

Table 2. Growth, milk intake, and energy intake in suckling California sea lion pups in the second month of life during El Niño (1983) and post-El Niño (1984)[a]

	July				
	EN (1983)		POST1 (1984)		
	Male n = 5	Female n = 4	Male n = 5	Female n = 3	EN vs POST1[b] (*P*)
Initial	10.32	9.12	11.52	9.26	n.s.
weight, kg	±0.715	±0.128	±0.533	±0.642	
Daily gain, g	109.2	75.8	117.8	39.7	n.s.
	±41.62	±62.56	±20.01	±42.60	
Body water, %	66.9	68.8	67.9	70.0	n.s.
	±3.57	±1.42	±0.59	±1.12	
Fractional water	0.073	0.071	0.072	0.073	n.s.
turnover	±0.0054	±0.0045	±0.0018	±0.0052	
Milk intake	681	600	769	582	n.s.
(MI), g/day	±87.9	±67.6	±47.2	±88.9	n.s.
MI as % of pup	5.96	5.93	6.11	6.00	n.s.
body weight[c]	±0.405	±0.467	±0.217	±0.533	
Energy intake	2323	2046	2624	1986	n.s.
(GE), kcal/day	±299.8	±230.6	±160.9	±303.0	
GE per MBS[d]	306.8	299.4	320.3	300.7	n.s.
kcal/$Wt^{0.83}$	±23.44	±25.30	±12.46	±29.91	

[a]Data from Oftedal et al. (1987b).
[b,c,d] See Footnotes, Table 1.

of pups in the first week in PRE than in POST1 could also be the result of a greater quantity of milk produced by females in the first few days postpartum, however. The decline in body water percent with age in POST1 is similar to that which occurs in most growing mammalian neonates (Moulton 1923; Spray and Widdowson 1950; Adolph and Heggeness 1971).

In POST1, male and female isotope-labeled pups exhibited 18 and 7% lower milk intakes, respectively, as compared to PRE. Energy intakes were likewise reduced. Energy intakes are best compared on a metabolic body size basis to compensate for effects of body mass on energetics (Brody 1945; Kleiber 1975). We used a scaling factor of body weight (kg) to the 0.83 power for suckling neonates (see Oftedal 1981, 1984; Oftedal et al. 1987b). Between PRE and POST1, energy intakes per metabolic body size decreased by 10%, coupled with a 23% lower growth rate for both sexes.

While we were unable to make a direct comparison of PRE to EN, the finding that parameters differed between PRE and POST1, but did not differ between EN and POST1, indicates that PRE and EN would likely have differed in the same manner.

Most studies attribute differences in reproductive effort during EN to a reduction in food supply available to lactating females along the California coast. During EN, lactating females departed earlier postpartum, while in both EN and POST1 duration of feeding trips was greater than during normal conditions, and less time was spent nursing pups (Ono et al. 1987). Our data are consistent with these findings.

Lactation represents the greatest proportion of the total energetic cost of reproduction, as much as 75–80% in many species (Oftedal 1985). In most species the onset of lactation is coupled with increased nutrient intakes by lactating females, while energy deficits are met by the mobilization of body fat stores (Bauman and Elliot 1983). Under nutritional or environmental constraints during El Niño years, it is possible that females were unable to support full milk yield due to limited body energy stores and limited availability of prey. Milk yield usually increases with age of the neonate during early lactation in most species (Bauman and Elliot 1983; Oftedal 1984). The finding that milk yield did not increase with age in POST1 suggests that normal milk yields were not maintained. In ungulates that are poorly nourished, lactating females exhibit a postpartum decline in milk yield rather than the characteristic rise (Oftedal 1985). An increase in duration of female absence may also have reduced the chances for the pup to obtain milk, if a female can only deliver a finite quantity of milk per period of onshore attendance. Ono et al. (1987) found that pups suckled less, as a percentage of maternal presence, in EN and POST1 as compared to PRE, suggesting that females may have had less milk to deliver.

Evidence from this study demonstrates the consistent and prolonged impact of El Niño conditions on reproductive performance in California sea lions. Longitudinal studies of milk intake by pups in non-El Niño years are necessary to determine whether milk yield normally rises after birth and to document the characteristic patterns of lactation and milk transfer in "normal" years.

Acknowledgments. Dr. K. Ono was instrumental in the ongoing success of field activities, and contributed valuable assistance and advice. M. Allen, D. Baer, M. Caspers, L. Dabek, J. Griffin, C. Halbert, L. Higgins, S. Pereira, and Dr. W. Rice also assisted in the field. We thank the U.S. Navy, and biologists at the Pacific Missile Test Center (PMTC), Point Mugu, CA for permission to use San Nicolas Island (SNI) as our study site and for logistical assistance. This study was supported by yearly grants from the Friends of the National Zoo, Washington D.C. and a grant from the Smithsonian Research Opportunities Fund.

20 The Influence of El Niño on Mother-Pup Behavior, Pup Ontogeny, and Sex Ratios in the California Sea Lion

K.A. Ono and D.J. Boness

20.1 Introduction

During an El Niño event, lactating females experience a lower food availability and, as a consequence, the milk and energy intakes of pups are reduced (Iverson et al., this Vol.). We may also expect changes in pup behavior reflecting this decreased energy intake.

Zalophus are highly gregarious, especially during the breeding season and interact with one another through a complex repertoire of social behavior. Pups swim poorly at birth, but slowly gain swimming skills throughout the season (Ono and Boness, unpubl. data). Both social and swimming skills are necessary for survival and appear to be learned through practice and experience in the first few months of life. The question addressed here is whether or not *Zalophus* pups, living in breeding colonies as far from the center of El Niño activity as the Channel Islands, exhibited changes in behavior similar to those previously observed in other species undergoing food shortages. We also ask whether the El Niño affected normal patterns of social development and swimming skills in *Zalophus*, and whether there were any changes in the sex ratio of pups born during this perturbation.

20.2 Methods

The study was conducted on San Nicolas Island, Channel Islands, California. Pups used for this analysis of behavior resided on Trailer Cove which is located on the northwest tip of the island (see Ono et al. 1987 for details of study area and methods).

Behavior, location, and association (see Sect. 20.3.5) of each member of a mother-pup pair was taken during hourly scan samples (Ono et al. 1987). Focal animals were searched for at the beginning of each hour scan, and recorded if and when they were found within that hour. The majority of animals were found within approximately the first 5 min of the scan. Consecutive data points for any given pair were used in the analysis if separated by at least 40 min. Data for pairs which did not have at least 250 observations were not used in this analysis. We did not always locate every focal animal each hour, and therefore each pup's behavior was analyzed individually by calculating the proportion of the total observations that were spent in each type of behavior. The mean of these proportions across individual pups in a given year, was subsequently calculated and is reported here. Sample sizes therefore reflect the number of pups or females used to calculate each mean and are not the total number of observations. Since male and female pups did not differ significantly in behavior (Mann-Whitney U-test, $P > 0.05$), data for both sexes have been pooled. Data for the few focal pups that died during the observation period, were not used for two reasons: (1) so that we could compare behavior of apparently healthy pups in all years; (2) to eliminate the bias posed by comparing data from pups of different ages, since pups died at different ages within and between years.

The statistical package SPSSPC (SPSS Inc.) was used to compute all statistical procedures. *P*-values for all multiple comparisons were adjusted using either the standard Bonferroni (Neter et al. 1985) or the sequential Bonferroni (Holm 1979) procedures. All X^2-analyses were carried out on the raw data frequencies, although the data are often expressed as percentages in the text. Statistical significance was accepted at the $P \leq 0.05$ level.

Since the 1982–83 El Niño did not affect *Zalophus* living in the Channel Islands until the 1983 breeding season, we will hereafter designate the 1982 breeding season as the pre-El Niño season, PRE, the 1983 season as the El Niño season, EN, and the 1984 season as the post-El Niño season, POST.

20.3 Results and Discussion

20.3.1 Suckling Behavior

We expected pups to attempt to maximize their energy intake during the El Niño year by spending a larger proportion of time suckling. However, pups spent a larger proportion of their total time (which includes observations taken both when the mother was absent and when she was present) suckling in PRE compared with the other 2 years. This unanticipated finding might have been due to the fact that females spent more time at sea during EN and POST than in PRE (Ono et al. 1987; Heath et al., this Vol.). If female availability was a limiting factor in the amount of time pups spent suckling during EN (and POST), we expected pups to compensate by spending a greater proportion of their time suckling when their mothers were *available* in EN and POST (i.e., only considering those observations taken when the mother was present). There were no differences between years in this measure, and pups spent only a small proportion of the time with their mothers suckling ($\bar{x}$ = 27.2% ± 9.3; Ono et al. 1987). In summary, female availability does not seem to be responsible for the observed decreases in suckling in EN and POST.

The reduction in the time spent suckling may reflect a decrease in the amount of milk consumed by pups, given that milk intake decreased during EN and POST in comparison to PRE (Oftedal et al. 1987b; Iverson et al., this Vol.). Although no data are available on the relationship between suckling duration and milk intake in *Zalophus*, Trillmich (1986b) found that suckling bout length was correlated with the amount of milk obtained by Galapagos fur seal pups, as determined by weighing pups before and after suckling. Lastly, there was a decrease in growth rate for pups during the El Niño year (Boness et al., this Vol.), again reflecting a lower milk intake.

20.3.2 Sneak Suckling

"Sneak suckling", defined here as an attempt to suckle from females other than a pup's mother, is common in some pinniped species such as elephant seals (*Mirounga angustirostris*, Reiter et al. 1978), but occurs only rarely in well-nourished (*Zalophus* pups (it is, however, more commonly observed in yearling and juvenile *Zalophus*; Ono and Boness, unpubl. data). We found that a larger percentage of our marked pups attempted to sneak suckle at least once in EN and POST compared to PRE (Ono et al. 1987). Most of the repeated sneak suckling attempts were made by very emaciated pups while their mothers were away at sea, although even apparently healthy pups attempted sneak suckling in EN (pers. obser.). It is important to note, however, that most sneak suckling attempts by pups were unsuccessful (90.5%), with the unrelated pup failing to contact the female's teat before it was threatened away. Those few attempts that did result in suckling were of short duration (1983; $\bar{x}$ = 52 s, SD = 86, n = 4; 1984: $\bar{x}$ = 39, SD = 31, n = 5). Therefore, sneak suckling does not appear to have substantially enhanced the milk intake of pups in any year.

20.3.3 Other Behavior

In order to conserve energy, we expected pups during EN to rest more and participate less frequently in high energy utilizing activities. This expectation was not fully realized. Pups did not spend more time resting during EN (Ono et al. 1987). Pups did, however, decrease the amount of time spent in active behaviors in both EN and POST. We defined "active" behaviors as those apparently requiring moderate to large expenditures of energy. These include play, interactive (other than suckle), and locomotory components. A fourth behavioral category, "low active", for which there were no a priori predictions for change, increased in EN and POST (PRE: $\bar{x}$ = 9.94, SD = 3.81; EN: $\bar{x}$ = 19.99, SD = 3.20; POST: $\bar{x}$ = 28.88, SD = 3.85 (n's as in Fig. 1); PRE vs EN, $P_{(1\text{-tailed})}$ = 0.0001; EN vs POST, $P_{(1\text{-tailed})}$ = 0.0001). Pups appeared to conserve energy in the EN and POST years by participating less in high energy utilizing behaviors, but not by resting more often.

We also asked which specific types of activities are curtailed in response to nutritional shortage. The components "play on land" (play) and "swim and play in water" (swim) are assumed to be important for several developmental processes: socialization, coordination, and muscle strengthening. Play decreased during the EN and POST years. However, the proportion of time spent by pups in swim did not differ significantly between years (Ono et al. 1987), suggesting that the development of swimming skills may be too important to pups to be curtailed for energy conservation purposes.

Long-term nutritional stress has been shown to cause an increase in aggressive behavior in young of other species (Dasmann and Taber 1956; Zimmermann et al. 1974, 1975), leading us to anticipate that pups would be involved in more aggressive encounters in EN. There were no differences, however, in the proportion of time spent in aggressive interactions (mostly pup-pup interactions) between years. Aggressive encounters are of very short duration, so that it is difficult to accurately

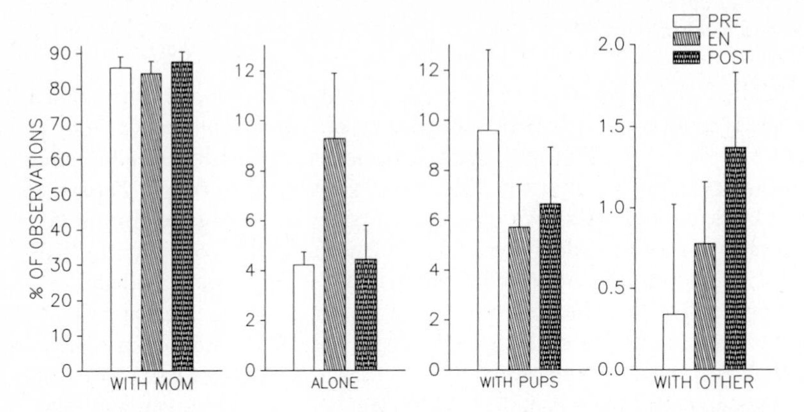

Fig. 1. The mean proportion of observations when the mothers were present on the study area in which pups were alone or associated with their mothers, pups, or others (see text for definitions), in the years PRE (1982), EN (1983) and POST (1984). *Vertical lines* above *columns* indicate two standard errors. Sample sizes were 15 for PRE, 17 for EN, and 23 for POST

assess their abundance using hourly scan samples. The nonsignificant ($P > 0.05$) trend was for an increase in EN and POST (Ono et al. 1987).

20.3.4 Behavioral Ontogeny

The ontogeny of behavior might be retarded when a pup experiences severe food restriction. The behaviors swim and play were analyzed to test for differences in the onset of these behaviors. The date of first occurrence of each behavior was found for all focal pups in all 3 years. There was no difference in the date or age (in days) of onset of either behavior between years (Table 1); pups began playing at about 13 days of age and began swimming at about 20 days in all years. The El Niño, therefore, did not appear to affect the normal ontogeny of these two developmentally important behaviors.

20.3.5 Associations

Along with the observed decrease in play, one might expect a decrease in the amount of time spent in association with other animals, i.e., one might expect the animals to exhibit less social contact and spend more time alone. Experimentally induced protein-calorie malnutrition in Rhesus monkeys has been shown to decrease tactual contact and approach-play behavior (Zimmermann et al. 1975). It appears that starving animals do not choose to expend energy in social interactions. Our measure of association was whether or not animals were in spacial proximity (less than approximately 30 cm between pups, and 60 cm between mothers and pups or between females) to one another, and not necessarily *actively* interacting. For gregarious pinnipeds which are not restricted by available space, spatial proximity

Table 1. Ontogeny of behaviors "swim and play in water" (swim) and "play on land" (play) for marked pups on San Nicolas Island, 1982–84. Means, standard deviations, and sample sizes given for Julian calendar dates and age in days. An ANOVA revealed no significant differences between years for any of the variables

	PRE = 1982	EN = 1983	POST = 1984
Date of 1st recorded (swim)	187.8 ± 12.4	186.4 ± 6.0	188.1 ± 9.4
n =	15	19	24
Date of 1st recorded (play)	179.6 ± 8.0	179.9 ± 8.6	181.5 ± 5.5
n =	16	19	23
Age of 1st recorded (swim)	21.1 ± 13.2	19.2 ± 6.0	21.2 ± 8.1
n =	15	14	24
Age of 1st recorded (play)	12.9 ± 7.3	11.9 ± 5.8	14.3 ± 6.5
n =	16	14	23

may reflect sociality whether for the purpose of interaction, protection, or communal heat conservation. It appears that individuals which prefer to be alone, such as dying pups and females giving birth, are able to maintain separateness, generally by exhibiting aggression to conspecifics which are too close.

In order to evaluate changes in the extent to which pups and their mothers associated with each other or with other sea lions, we divided the hourly observations into those taken when the mother was present and those taken when she was absent. Associations for pups were then categorized as "alone", "with mom", "with pups" or "with other" ("with other" category includes juveniles, other females, subadult males and adult males) (Figs. 1 and 2). Associations for mothers were divided into the categories "alone", "with pup", "with female(s)" and "with other" (including juveniles, other pups, subadult and adult males) (Fig. 3). The mother-pup association took precedence over all others, so that a pup was considered to be associated with its mother if it was within one pup body length of her, even if other pups or females were also within this distance. The same criterion was applied to the mothers. In order to simplify statistical analysis, an "association index" (AI) was calculated for each pup using the following equation:

$$AI = \% \text{ with pups} + \% \text{ with other})/(\% \text{ alone} + \% \text{ with pups} + \% \text{ with other}).$$

When their mother was present, there was a significant difference in pup association during EN compared with both PRE and POST (AI: $\bar{x}$ = 0.64, SD = 0.25 PRE; $\bar{x}$ = 0.44, SD = 0.23 EN; $\bar{x}$ = 0.64, SD = 0.20 POST; PRE vs EN Student's t = 2.37, df = 29, $P_{(1\text{-tailed})} = 0.025$; EN vs POST t = –4.24, df = 69, $P_{(1\text{-tailed})} = 0.0001$). Since there was no significant between-year difference in the proportion of time pups spent with their mothers (Fig. 1; $P > 0.05$ all comparisons), the association index is not biased by this category. Pups appeared to spend less time with other pups in EN and POST and more time alone in EN (Fig. 1). Pups were also in

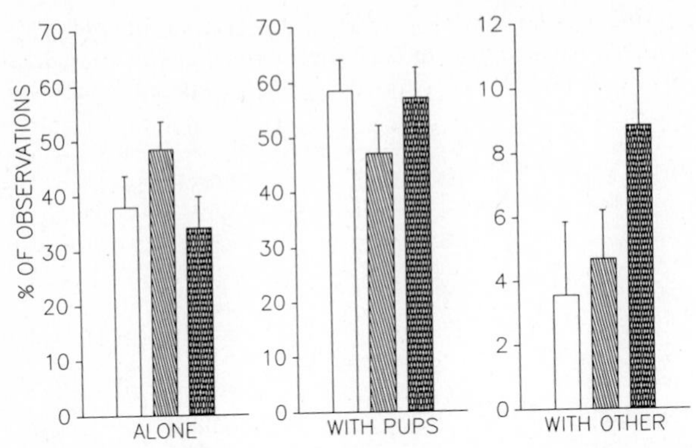

Fig. 2. The mean proportion of observations taken when the mother was absent from the study area in which pups were alone or associated with pups or others. Legends and sample sizes as in Fig. 1

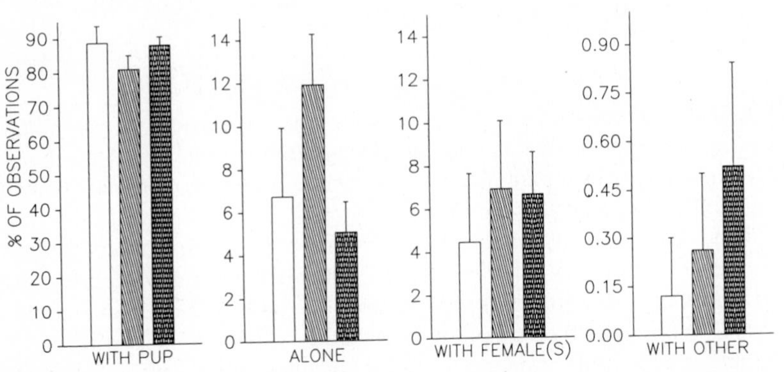

Fig. 3. The mean proportion of observations of females on the study area in which they were alone or associated with their pups, females, or others. Legends and sample sizes as in Fig. 1

associationless in EN than in PRE or POST when their mothers were absent (AI: $\bar{x} = 0.62$, SD = 0.12 PRE; $\bar{x} = 0.52$, SD = 0.11 EN; $\bar{x} = 0.66$, SD = 0.14 POST; PRE vs EN t = 2.60, df = 29, $P_{(1\text{-tailed})} = 0.015$; EN vs POST t = –4.24; df = 69, $P_{(1\text{-tailed})} = 0.0001$). Again, pups spent more time alone and less time with other pups (Fig. 2). During EN, pups had fewer opportunities to interact with other animals because of the lower level of associations. This might have affected the development of socialization in these young.

Mothers, while on the study area, spent significantly less time with their pups in EN (Fig. 3; PRE vs EN t = 2.41, df = 28, $P_{(1\text{-tailed})} = 0.025$; EN vs POST t = 2.85, df = 29, $P_{(1\text{-tailed})} = 0.008$). The different pattern observed for mothers and pups in the proportion of observations during which they were seen together (cf. Figs. 2 and 3) is statistically possible because both members of the focal pair were not always recorded each hour. In addition to the significant yearly differences in time spent with pup, some females were *always* observed with their pups. We were, therefore, unable to use the association index to evaluate yearly differences in patterns of female association, and instead analyzed the frequency of each category of associations directly. Mothers spent significantly more time alone during EN compared with the other 2 years (Fig. 3; PRE vs EN $t = -2.6$, $df = 26.4$, $P_{(1\text{-tailed})} = 0.03$; EN vs POST $t = 50$, $df = 27$, $P_{(1\text{-tailed})} = 0.001$). There was no difference in the amount of time females spent in association with other females or "with other" sea lions between years ($P > 0.05$).

The decrease in the proportion of time females were observed with their pups during EN may be partially attributable to lowered milk production that year, as evidenced by a lower milk intake by pups during EN (Iverson et al., this Vol.). Females may have needed less time for milk transfer during EN, and pups spent a smaller proportion of their time suckling (see Sect. 20.3.1). This would reduce the amount of time that females were required to stay with their pups. The observed differences in female association may also have been motivated by increased ambient air temperatures during EN (Ono et al. 1987), i.e., females may have left their pups more often to cool off in the water (pers. obser.). The finding that this pattern is significant in the female data, but not in the pup data makes us suspect that this a statistical artifact; however, the pattern of a lower pup association with mother during EN is also present in the pup data, but to a much smaller degree.

20.3.6 Sex Ratios

The sex ratio of pups born in the El Niño and in prior and subsequent years is a subject of theoretical and empirical interest. Previously reported analyses of the secondary sex ratio of otariid pups born during the El Niño and surrounding years have shown no significant differences (Trillmich 1986b; Ono et al. 1987), although either very large sample sizes or very large differences in ratio are needed to obtain a significant difference using a pairwise analysis. Another way to analyze these data is to utilize all the available information simultaneously in a regression.

To do this we first ranked the years by our a priori expectation of the sex ratio (expressed as percent males). We based our a priori expectations upon the theory of Trivers and Willard (1973). They hypothesized that mothers in superior condition will, on average, produce an excess of sons when males have the higher variance in mating success and mating success is contingent upon resources invested in offspring by their mothers. Conversely, females in poor condition should shift to the production of daughters since they are the safer investment. There may also be a simpler explanation to our expectation of a female-biased sex ratio during the EN. Since male pups are larger at birth (Boness et al., this Vol.), and therefore, presumably require more energy in utero, and since there appeared to be more aborted fetuses as well as fewer pups during EN (Francis and Heath, Chap. 12, this Vol.), we might expect more male pups to be aborted prior to birth.

Ranking the years by the potential for maternal investment, the El Niño year (EN) clearly ranks worst (rank = 1), as corroborated by the high pup mortality in this year as well as the emaciated state of many surviving offspring. The effects of El Niño were muted but still manifest in the year following (POST). For this reason we ranked this year second (rank = 2). Ranking of the other 2 years for which we have data, 1982 and 1985, is complicated by the fact that the PRE year was obviously unaffected by the impact of El Niño, while females in 1985 may have been able to invest in offspring to an unusually high degree. The high potential for maternal investment in 1985 is a consequence of many females not giving birth in 1983 and 1984 (see Trillmich and Dellinger, this Vol.; Francis and Heath, Chap. 12, this Vol.), and therefore being able to recoup from their prior pregnancy for an extended period of time, coupled with an apparent return to normal resource availability. Both 1982 and 1985 were therefore probably "good" years relative to the potential for maternal investment in offspring. The fact that females had recuperated from the impact of El Niño by 1985 is manifest in the high birth weights in this year (Table 4 in Ono et al. 1987; Boness et al. this Vol.). We therefore decided to pool these 2 years and assigned each a rank of 3.5. First, we used the weighted (by sample size) least squares regression of percent male offspring vs our a priori ranking of years to obtain the estimated slope and its standard deviation. A Student's t-test with two degrees of freedom was next used to test the null hypothesis of a zero slope versus the alternative of a slope greater than one. We observed a significant positive slope, ($b = 5.49$, $t = 3.34$, $df = 2$, $P = 0.039$). Based upon many indirect measures from 1982 and 1985, we felt that 1985 probably had the highest potential for female investment in offspring. Ranking 1985 as 4 and 1982 as 3, the above pattern is strengthened ($b = 5.01$, $t = 10.24$, $df = 2$, $P = 0.005$). We therefore conclude that while there was no significant deviation from an equal sex ratio in individual years, pooling the data across years supports the Trivers and Willard hypothesis.

Acknowledgments. We would like to thank our field assistants, L. Dabek, P. Fromhoff, E. Gimble, L. Maul, L. Osborn, S. Pereira, and Y. Yount, without whom much of the data would not have been collected. Logistic support from the Commander and personnel of the Pacific Missile Test Center, Naval Air Station, and San Nicolas Island was greatly appreciated. W. Rice provided statistical consultation, and K. Kovacs, F. Trillmich, and W. Rice contributed helpful comments on the manuscript. Support for this work was provided by the Friends of the National Zoo and the Lerner-Gray Fund for Marine Research of the American Museum of Natural History.

21 The Effects of El Niño on the Frequency and Sex Ratio of Suckling Yearlings in the California Sea Lion

J.M. Francis and C.B. Heath

21.1 Introduction

Gentry et al. (1986a) proposed that environmental unpredictability is responsible for the long duration of maternal care that has evolved among some temperate and tropical otariids. By extending maternal care and even simultaneously suckling young from two different years, mothers can "hedge their bets" against an uncertain environment (Stearns 1976; Gentry et al. 1986a). Female Galapagos fur seals, for example, spend 2 years or more rearing their pups in an environment subject to frequent fluctuations in prey availability resulting, in part, from the El Niño phenomenon (Trillmich 1986a).

The capability of some otariids to forgo reproduction in a given year and continue investment in a previous year's offspring sets them apart from phocids. Despite the important flexibility that this capacity confers, a full understanding of its relationship to environmental conditions awaits more adequate data on the duration of maternal care in otariids. While more than half of the 14 species of fur seals and sea lions have been observed suckling pups past 1 year of age (Stirling 1983), quantitative information on the proportion of pups suckling into their second year and beyond are, for the most part, unavailable. In one exception, Trillmich (1986b) presented data from two normal years, indicating that 44% of the 1- and 2-year-olds suckling were 2-year-olds, while in a particularly good year 2-year-olds comprised only 10% of those receiving extended care.

Variation in environmental quality may lead to changes, not only in the occurrence of extended maternal care, but also in the proportion of male versus female pups that receive it. Such measures are particularly important in testing or expanding sex allocation and parental investment theories (Fisher 1930; Trivers 1972; Trivers and Willard 1973; Maynard Smith 1980).

The following research on maternal investment in the California sea lion measures the proportion of yearlings suckling and, in particular, the variation in this proportion over a 3-year period encompassing the most severe EN event in recorded history. Further, it examines the sex ratios of yearlings receiving extended maternal care in each year for a more complete understanding of parental investment and sex allocation in this species.

21.2 Methods

A total of 929 pups were tagged on San Nicolas and Santa Barbara Islands, California in early August at the end of the 1981, 1982, and 1983 breeding seasons. The capture technique among involved three to eight people approaching groups of 20 to 100 pups and cutting off their access to sea. All pups were sexed and tagged following capture and those on San Nicolas were weighed to the nearest kilogram. Tags were subsequently read with the aid of binoculars and 15–45× spotting scopes from observation points above the rookeries. The behaviors of the tagged individuals at the time of their sighting were noted, including whether or not the animals were suckling.

Resights of tagged sea lions on San Nicolas Island during the 1982–1985 breeding seasons, late May through late July, included 2 months of daily observation at the beaches of tagging plus systematic biweekly searches over the entire island population. The majority of tag resights reported in this study were collected during this part of the year. Investigators also recorded tagged animals during periodic winter visits to San Nicolas. At Santa Barbara Island, resights occurred primarily during the tagging visits at the end of July and as a consequence were limited to less than 1 week each summer. Resights at other locations along the coast of California were provided as a by-product of ongoing investigations of pinnipeds at the Farallon Islands, Año Nuevo Island, and Monterey breakwater. Resight effort at these three locations was comparable between years. In addition, information on occasional sightings of animals on coastal beaches and at sea (comprising 6% of all yearlings resighted) were provided through various federal, state, and private agencies.

21.3 Results

21.3.1 Yearling Suckling

A substantial proportion of California sea lion pups suckled as yearlings (12–24 months old) and this proportion varied both annually and by sex of pup. First, we will present data on suckling frequency and then follow with data on intersexual and interannual variation in yearling suckling behavior. Of the 929 sea lion pups tagged over the course of this study, 218 (23%) were resighted as yearlings and 73 (8%) were observed suckling at least once as yearlings. Since some yearlings remain at sea or are not resighted while on land, and because tag loss occurs, 23% is undoubtedly a low estimate of yearling survival. Nevertheless, these data indicate that as much as 33% (73/218) of surviving sea lion pups may suckle into their second year.

Of those yearlings resighted on the island of tagging, nearly half (73/149 = 49%) were observed suckling. Many of these were regular visitors and were seen suckling frequently, on average 1 day of every 5 days observed. Because over 90% of the tagged yearlings present on the island on any given day were regular visitors (sighted more than 5 days during the breeding season) and 85% of the regular visitors sampled were observed suckling, most yearlings present on the island on any given day were sucklers.

21.3.2 Sex Bias in Suckling and Migration

Not only was a yearling likely to suckle if on the island of tagging, but it was equally likely to suckle regardless of sex (Table 1). However, since more female than male yearlings were sighted on the island where they were tagged, significantly more female yearlings were observed suckling (43% of females resighted vs 25% of males resighted; $X^2 = 5.269$, df = 2, $p < 0.05$).

Female yearlings were more likely to be resighted on the island of tagging while males were more often found at other locations. Of yearlings resighted on the island of tagging, 59% (88/149) were female while females comprised only 22% (15/69) of the yearlings resighted at locations other than where they were tagged (2 × 2 contingency table, sex by location, $X^2 = 27.96$, df = 1, $p < 0.001$).

Yearlings were observed suckling exclusively on the island of tagging. There have been no reports of yearling suckling from the other nonrookery haulouts, including the Farallon Islands where there have been over 10 years of extensive observations (H. Huber, pers. comm.). Further, the island of tagging can be assumed to be the island of birth. Only one pup out of 929 tagged was observed on an island other than where it was tagged in the month following tagging, indicating negligible interisland movement of pups at the age of tagging.

Maternal care in the California sea lion can occasionally extend beyond a pup's second summer. Of the five sea lions (3% of the resighted population) seen suckling beyond 15 months of age, two were male. Three of these individuals were 24 months old (two males, one female) and we have seen larger juveniles of unknown age suckling, in addition to one adult female suckling from another (see Pitcher and Calkins 1981, for a similar observation in the Steller sea lion).

Table 1. Sex differences in percentage of California sea lion yearlings observed suckling as compared to those (A) resighted on their island of birth (San Nicolas and Santa Barbara Islands) and (B) resighted at all locations (numbers of individuals are in parentheses)

A	Male	Female	n
Suckling	48% (29)	50% (44)	(73)
Not suckling, but resighted on birth island	52% (32)	50% (44)	(76)
	(61)	(88) $X^2 = 0.090$	(149) $p < 0.75$

B	Male	Female	n
Suckling	25% (29)	43% (44)	(73)
Not suckling, but resighted	75% (86)	57% (59)	(145)
	(115)	(103) $X^2 = 7.46$	(218) $p < 0.01$

21.3.3 Interannual Differences

The proportion of male yearlings observed suckling was significantly higher in the 1983 cohort as compared to the 1981 and 1982 cohorts with no significant difference between 1981 and 1982. Corresponding to the increase in suckling observed, the proportion of the male cohort resighted as yearlings on their island of birth more than quadrupled from 1982 and 1983 to 1984 (1981–83 cohorts, respectively, Table 2). In contrast, neither the proportion of the female cohort resighted as yearlings nor the proportion observed suckling varied significantly between years. As their attendance on the island of birth increased, resights of males at locations away from the island of birth showed a marked and significant decline from 1982 to 1983 (1981 vs 1982 cohorts) and a tendency toward decline from 1983 to 1984 (1982 vs 1983 cohorts, nonsignificant $p = 0.054$). Females, again, showed no significant changes between years.

Pup weights at 2 months of age decreased significantly from 1981–1983 (Table 3) corresponding to the increases in yearling suckling and attendance on the islands of birth reported above. The average male weighed 5.5 kg (28%) less in 1983 as compared to 1981 while female weights dropped 3.9 kg (24%) over these 3 years. The pups were tagged between August 6–9 in all 3 years and there appears to be no difference in the timing of births between seasons (see pup census curves in Heath and Francis 1984).

These trends suggest that lighter pups are more likely to suckle as yearlings. However, only in the 1981 cohort, both the heaviest and the only pre-EN cohort,

Table 2. Between year, by sex comparison of percentage of California sea lion cohorts[a] resighted as yearlings on and off the island where they were born, including the percentage observed suckling as yearlings

	Cohort 1981	1982	1983	X^2 tests
Suckling on island of birth (%)				
Male	2.0	5.3	12.7	81,82 < 83
Female	11.5	8.6	10.6	NS
Resighted only on island of birth (%)				
Male	4.6	11.1	26.4	81 < 82 < 83
Female	20.6	16.7	24.4	NS
Resighted off island of birth (%)				
Male	18.3	9.7	3.6	81 > 82,83[b]
Female	3.8	4.8	0.8	NS
Total resighted (%)				
Male	22.9	20.8	30.0	NS
Female	24.4	21.5	25.2	NS

X^2 goodness of fit analyses compare number resighted or suckling to those not resighted or suckling within each cohort. Sequential Bonferroni method (Holms 1979) was used to adjust for multiple comparisons. Null hypotheses were rejected at the $p < 0.05$ level.

[a]See Table 4 for cohort sample sizes.

[b]Fisher's exact text for pairwise 1984 comparisons.

Table 3. Yearly weights (kg) of California sea lion pups at approximately 2 months of age on San Nicolas Island[a]

	1981			1982			1983		
	Mean	SD	n	Mean	SD	n	Mean	SD	n
M	18.1	3.4	107	15.7	2.9	153	13.0	3.1	110
F	14.9	3.2	82	13.6	2.4	135	11.3	2.2	123

[a]All pairwise t-tests between years, within sex and between sexes within years significant at $p < 0.002$). Tests corrected for unequal variances where necessary and for multiple comparisons (sequential Bonferroni test, Holms 1979).

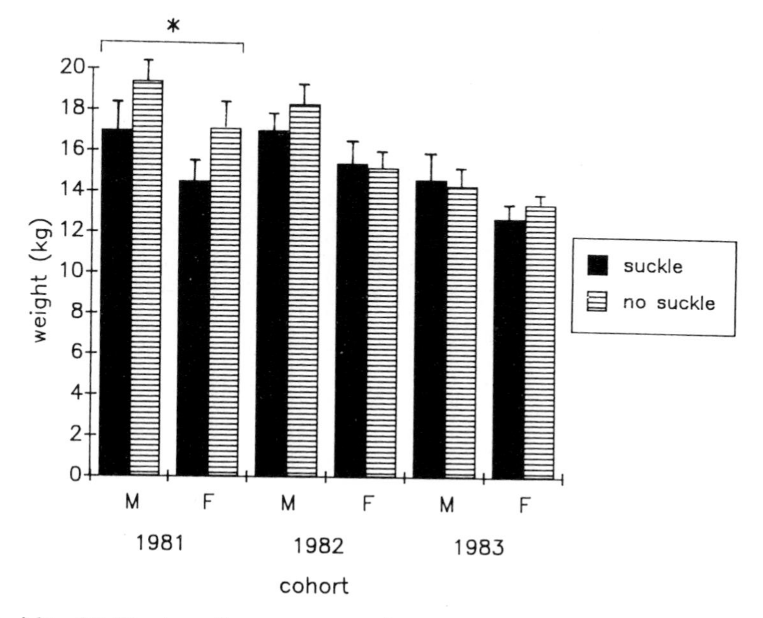

Fig. 1. Yearly weights of California sea lion pups at approximately 2 months of age on San Nicolas Island among individuals resighted and suckling vs resighted but not suckling as yearlings. *Error bars* represent 1 SE. *Asterisk* denotes significant ($p < 0.05$) combined probability (Fisher 1954) of male (*M*) and female (*F*) t-tests within year

were suckling yearlings significantly lighter (as pups) than those which were resighted but not observed suckling (Fig. 1).

Lighter pups were less likely to be seen again as yearlings, but only in the 2 years during which yearlings were subjected to the environmental conditions on EN. Yearlings of the 1982 and 1983 (but not 1981) cohorts who were not resighted were significantly lighter than their cohort mates who were resighted (Fig. 2).

The sex ratio of pups at tagging (2 months of age) was significantly male biased in the pre-EN years (1981 and 1982) combined as compared to that of the 1983 EN year [$X^2 = 3.674$ $p < 0.05$ (1-tailed); Table 4]. No data on sex-specific pup or fetal

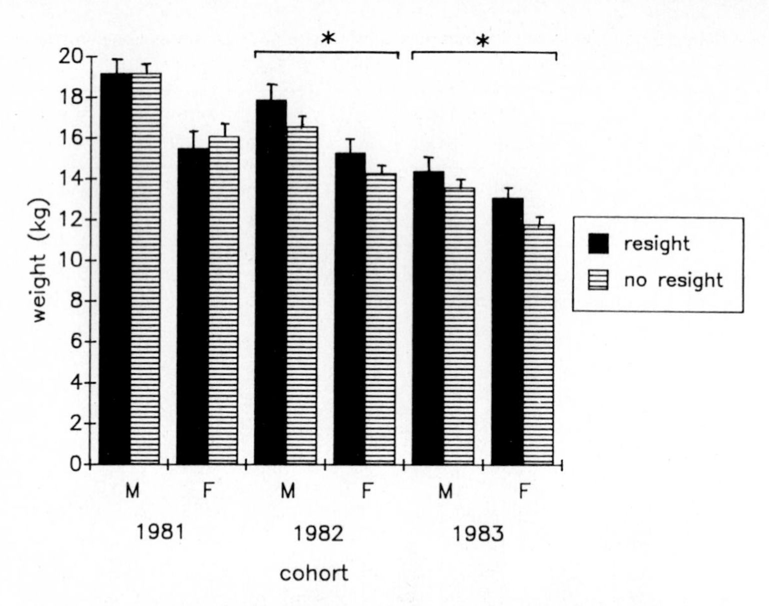

Fig. 2. Yearly weights of California sea lion pups at approximately 2 months of age on San Nicolas Island among individuals resighted vs individuals not resighted either on or off the island of tagging. *Error bars* represent 1 SE. *Asterisk* denotes significant ($p < 0.05$) combined probability (Fisher 1954) of male (*M*) and female (*F*) t-tests within year

mortality rates are available for this species. Thus, we are unable to determine whether this bias in sex ratios is a product of differential postnatal or prenatal mortality or a reflection of sex ratios at conception.

21.4 Discussion

21.4.1 Interannual Differences

The increase in yearling suckling in 1984 resulted from increased male attendance on the island of birth; the percentage of tagged female yearlings resighted varied insignificantly over the 3 years studied. Since resights of tagged male yearlings at their primarily northern haulout locations declined in 1983 and there was a tendency toward decline in 1984 with comparable observation effort between years, it may be that the increase in males on their island of birth resulted from individuals remaining in the south during the usual winter migration. Whether the males appeared on the island of birth with higher frequency in 1984 in order to suckle, or the suckling was a by-product of an otherwise motivated proximity to their birthsite is unclear at this time.

The factors promoting the suckling behavior of yearlings can be viewed from two perspectives, that of the yearling and of the mother. Yearlings, as they grow, become increasingly capable of foraging for themselves. Yearlings satisfy nutritional

Table 4. Percentages of male and female California sea lion pups tagged at 2 months of age on San Nicolas and Santa Barbara Islands (number tagged in each cohort is in parentheses)

	Cohort			
	1981	1982	1983	N
Male	53.9 (153)	54.9 (226)	47.2 (110)	(489)
Female	46.1 (131)	45.1 (186)	52.8 (123)	(440)
	(284)	(412)	(233)	(929)

needs through consuming either milk or prey, and their decision to suckle or forage is probably based on the availability and quality of these resources. If environmental conditions or poor physical condition make foraging difficult, or if the opportunities for suckling are readily available, total or partial dependence on maternal milk may be beneficial. From the maternal perspective, extended care will be promoted when a female either loses or does not produce a pup and the previous year's offspring is present. Females may also suckle a yearling and pup simultaneously, as has been reported for the Galapagos fur seal (Trillmich 1984) although this is very rare in the California sea lion. Typically, only the yearling survives in such circumstances. In all cases, care for a pup should extend into the second year as long as the benefit, in terms of maternal reproductive success, exceeds that to be gained through investment in future offspring (Trivers 1972).

During the 1983 EN, changes in prey distribution correlated with a decline in pup condition, high pup mortality both in utero and postpartum, and reduction in copulation frequency, which may account, singularly or in combination, for the increased yearling suckling behavior observed in the California sea lion the following year. Numerous observations of prey species off the coast of California indicated shifts in distribution and abundance during the EN event (Arntz et al., this Vol.). Lower pup production and higher pup mortality during the first 2 months of life (Francis and Heath 1983, and Chap. 12, this Vol.), higher mortality in utero (Francis and Heath, Chap. 12, this Vol.), lower pup weights at 2 months of age (Ono et al. 1987; this study), and slower growth rates (Ono et al. 1987; Boness et al., this Vol.) indicate poor nutrition to the fetuses and newborns during the 1983 EN year and to some extent in the year following. That these changes were probably a result of reduced prey availability among mothers is indicated, at least for the 1983 breeding season, by increased duration of time spent at sea foraging (Ono et al. 1987; Heath et al., Feldkamp et al., Costa et al., this Vol.).

Pup condition appeared to affect, at least in some years, the probability of resight and suckling among yearlings. Lighter pups were less likely to be resighted as yearlings in the 1982 and 1983 cohorts and more likely to be observed suckling as yearlings in the 1981 cohort. Since the 1982 and 1983 cohorts were subject to the EN conditions, it is possible that the lack of resight more directly reflects the mortality of underweight pups than in the 1981 cohort. A larger proportion of those not resighted from the 1981 cohort, as compared to the other 2 years, may have been heavy, healthy, and feeding at sea. Data on survival to age 2 and beyond for these cohorts will address this issue. As for pup weight and yearling suckling, the data suggest that some threshold weight exists above which yearlings are unlikely to seek extended care. Thus, only in the 1981 cohort, having a larger proportion of heavy pups, was a relationship between suckling and weight apparent.

Not only did changes in prey availability directly affect pup condition, which could result in a lengthened dependence on maternal care, but it also affected the probability that a female was available as a source of milk for her yearling. More females were available for extended care as a result of increased numbers of abortions and increased pup mortality in 1983. A high number of females without pups were also available in 1984, due to high pup mortality (no data are available on abortion rate) as well as low fecundity resulting from low copulation rates in the preceding year.

It is important to note that the proportion of yearlings suckling in 1984, but not in 1983, was significantly higher than the 1982 pre-EN year. This may be an effect of the lower copulation rates and reduced fecundity of 1984 females already mentioned. It also may be a product of pup condition – the 1983 cohort, suckling as yearlings in 1984, suffered the double handicap of reduced prey availability during gestation and lactation while the pups of the 1982 cohort were in utero during what was, by all measures, a year of abundant resources (Arntz et al., Chap. 2, this Vol.). Higher pup weights in 1982 as compared to 1983 reflect this difference.

21.4.2 Sex Differences in Suckling and Migration

In order to interpret the differences observed between sexes in the migratory and suckling behavior of yearlings, one must consider the relative value for each sex in remaining near their birthsite. Adult females are forced to forage close to the island of birth, due to the relative immobility of their pups during the first year of care (few pups less than 1 year of age were resighted at areas other than the island where they were born). Thus, females probably depend upon a good knowledge of prey distribution surrounding the islands where they pup. Female yearlings, by remaining near the island of birth as they grow up, may become more acquainted with the foraging grounds upon which they will, for their lifetime, depend. Male yearlings and males in general migrate north during the winter while females apparently remain in the south near the Channel Islands (Mate 1975; and this Chap.). Since adult males fast during the breeding season (Schusterman and Gentry 1971), it is reasonable to assume that they are not as dependent as females on a well-learned knowledge of the fisheries surrounding the islands where they hold territories.

By the above reasoning one might expect a female bias among suckling yearlings, as was observed in this study. Since the bias occurred only in 1982 the year preceding EN and presumably the most "normal," it is likely that this represents the normal pattern for the California sea lion. Since male pups are consistently heavier than female pups at birth (Boness et al., this Vol.) and at 2 months of age (this Chap.; Boness et al., this Vol.) and data show greater milk intake by male pups during the first month of maternal care (Oftedal et al. 1987b), it is reasonable to assume that male pups in the California sea lion receive greater parental investment in the first few months of life. But the female bias in the sex ratio of suckling yearlings presented in this study suggests that, on average, weaning may occur later in females. Thus, while males may receive greater investment in early lactation, females, through a longer investment period, could actually be more costly than males. This may only be determined through careful measurement of investment over the entire course of maternal care.

Acknowledgments. We thank Lt. Cdr. Eugene Giffin and Mr. Ron Dow of the U.S. Navy Pacific Missile Test Center for logistic support on San Nicolas Island and Chuck and Patricia Scott, and the National Park Service for support on Santa Barbara Island. Wyatt Decker, Charles Deutsch, Jane King and Mark Lowry collected tag resights on San Nicolas Island and Harriet Huber, Rick Condit, Ken Nicolson, and associates provided resights for the Farallon Islands, Año Nuevo Island, and the Monterey breakwater, respectively. Tags were applied by many field assistants too numerous to name, but no less gratefully acknowledged. Charles Oliver compiled and forwarded tag resight information reported by numerous individuals and agencies both public and private. Drs. Sheila Anderson, Steven Albon, Claudio Campagna, Burney Le Boeuf, and Robert Trivers commented on earlier versions and Drs. Daryl Boness, Kathy Ono and Ms. Harriet Huber commented on more recent versions of this manuscript. Drs. Bill Rice and Kathy Ono provided statistical advice. Funding was provided by the Southwest Fisheries Center, Coastal Marine Mammals Program under the direction of Dr. Doug DeMaster and by the Biology Board of the University of California, Santa Cruz.

Part IV
The Elephant Seal

22 The Natural History of the Northern Elephant Seal

B.J. Le Boeuf

The largest phocids in the Pacific ocean are northern elephant seals, *Mirounga angustirostris*. Elephant seals breed on islands and peninsulas along the western coast of North America from mid-Baja California, Mexico to the Farallon Islands near San Francisco, California. The two largest rookeries are Isla de Guadalupe in Mexico and San Miguel Island in southern California; these two islands account for approximately 75% of the population's annual pup production. Three long-term studies during the breeding season have been in progress on this species over the period during which El Niño 1982–83 occurred. Stewart and Yochem (Chap. 25) present data from their studies on San Miguel and nearby San Nicolas Islands in the southern California Bight which span the period 1981–1988. Huber, Beckham and Nisbet (Chap. 24) present data from their study on the Farallon Islands near San Francisco that has been ongoing since 1974. Le Boeuf and Reiter (Chap. 23) summarize findings from Año Nuevo, California, collected during the period 1968–1988.

Because the effects of El Niño were primarily on the marine environment, it is useful to consider the context and period that elephant seals of both sexes and various age groups are at sea. Elephant seals lead a pelagic existence at sea and apparently forage individually. They are deep divers that appear to feed off the continental shelf along the west coast of North America from mid-Baja California to the southern Aleutians in Alaska.

Females spend about 34 days onshore from December to March giving birth, nursing their pups, and copulating. After weaning their pups, females spend approximately 2.5 months at sea feeding before coming back to land to molt, a process that takes about 1 month. Females spend the rest of the year, an 8-month gestation period that extends from about May to December, feeding at sea before returning to land to give birth.

Elephant seal pups are weaned in January, February, and March at 1 month of age. They remain on the rookery for an additional 2.5 months fasting and learning to swim and dive until about mid-May. By this time, all of them have gone to sea where they remain until the following fall or spring. During this first trip to sea, individuals feed only so much as to maintain their departure mass (Mason, Morris and Le Boeuf, unpubl. data).

Males present on the rookery during the breeding season, ranging in age from 5 to 14 years, spend March to July and September to November feeding at sea. Juveniles are at sea feeding from December to March and from about May to September.

23 Biological Effects Associated with El Niño Southern Oscillation, 1982–83, on Northern Elephant Seals Breeding at Año Nuevo, California

B.J. Le Boeuf and J. Reiter

23.1 Introduction

This chapter presents vital statistics on northern elephant seals, *Mirounga angustirostris*, inhabiting the rookery at Año Nuevo, California over a period that included EN 1982–83. Our approach involves comparisons of baseline data collected routinely in the years before, during, and after EN 1982–83. Specifically, we examine:

1. Direct effects on mortality and distribution of adults caused by a rise in sea surface level, high coastal winds, high surf, and winter storms associated with El Niño (Fahrbach et al., this Vol.), and
2. Indirect or long-term effects on mortality, reproduction, and foraging effort associated with fluctuations in water currents, upwelling, and water temperature – perturbations that might have caused changes in the composition, distribution, abundance, and availability of the food base of these phocids (Arntz et al., this Vol.).

Northern elephant seals are deep-diving pelagic seals that forage along the continental slope from mid-Baja California, Mexico to southern Alaskan waters. Depending on sex and age, individuals remain at sea for 2.5 to 8 months at a time. During the 2.5-month period at sea following lactation, adult females dive virtually continuously to mean depths in the range, 400 to 700 m (Le Boeuf et al. 1986, 1988, 1989b; Naito et al. 1989). Females average 2.5 to 3 dives per hour around the clock; 99% of these dives are to depths greater than 200 m. It is clear that much of the time spent diving is spent feeding because all females studied increased their mass over the period at sea. Seven females monitored in 1985 and 1986 gained a mean of 76.5 ± 13.9 kg during a mean of 72.5 ± 5.0 days at sea (Le Boeuf et al. 1988).

Given their deep-diving habits, one might expect elephant seals to be less affected by fluctuations in sea temperature, sea level, and currents than shallow water feeding sea lions. EN 1982–83 seemed to have less of an effect on deep-water organisms, which comprises most elephant seal prey, than on organisms living close to the surface that sea lions exploit. Thus, one might predict that the weather and sea perturbations that influenced the distribution of marine life during EN 1982–83 had a less deleterious effect on elephant seals than on Pacific coast sea lions and fur seals. We test this prediction.

23.2 Methods and Background Information

We were in a good position to assess the effects of EN 1982–83 on the elephant seals at Año Nuevo because the colony has been under close observation every breeding season since 1968. In 1975, when elephant seals began breeding on the mainland, the study was enlarged to cover both the mainland and the island. The long-term study involves routine monitoring of data such as the following: number of pups born, weaned and pup mortality rate; number and distribution of adults and juveniles; fecundity and reproductive success of marked known-age females and the sex ratio of pups produced; and the length of the foraging trip to sea by marked females following breeding.

In general, observations were continuous throughout daylight hours on both the island and the mainland during the entire breeding season from mid-December to mid-March, and observations were conducted approximately once a week during the rest of the year. Censuses and monitoring of pup deaths were conducted daily during each breeding season.

For most analyses, we attempt to compare the data from El Niño years to those of preceding and succeeding years.

23.3 Direct Immediate Effects

Mortality Due to Storms. On three consecutive days, from 27 to 29 January 1983, giant storm-driven waves cresting above the year's highest tides smashed the coast of California. The surf pounded Año Nuevo island and the mainland at the peak of the breeding season, a time when the maximum number of females were giving birth and nursing their pups. The scene on Año Nuevo Island (Fig. 1) was described as follows (Le Boeuf and Condit 1983, p. 14):

"The Año Nuevo Island rookery was packed with animals when giant waves curling over a 6.6-foot tide inundated the harems on the morning of 27 January. Wedged together on the large sandy Point Beach on the lee side of the island were 95 males, 936 nursing or pregnant females, and at least 510 pups. There were also 20 females, 20 suckling pups, and a few males in a small cove on the island's weather side. For three hours the unusually high surf pounded the island. Waves flooded the beaches and sent logs, boulders, and flotsam careening against animals, over cliffs, and onto the highest reaches of the island. Piers, catwalks, and retaining walls erected by the U.S. Coast Guard in the 1930s were demolished and washed away. The tight group of over a thousand seals was wrenched apart and dispersed. With each incoming surge bodies were hurled pell mell over others and against cliffs and were then swept back to sea in the strong aftertow. As hundreds of pups were swept away hundreds of mother seals bobbed about in the white foam, calling to their helpless offspring. About two hundred females with pups managed to cling to higher ground at the base of the cliffs in about a foot of water."

By 30 January, no pups remained on the exposed Cove Beach. The sandy Point Beach, the prime breeding area for the seals, was reduced to 10% of its normal size and only 290 pups remained in the harem. Scores of pregnant females and at least

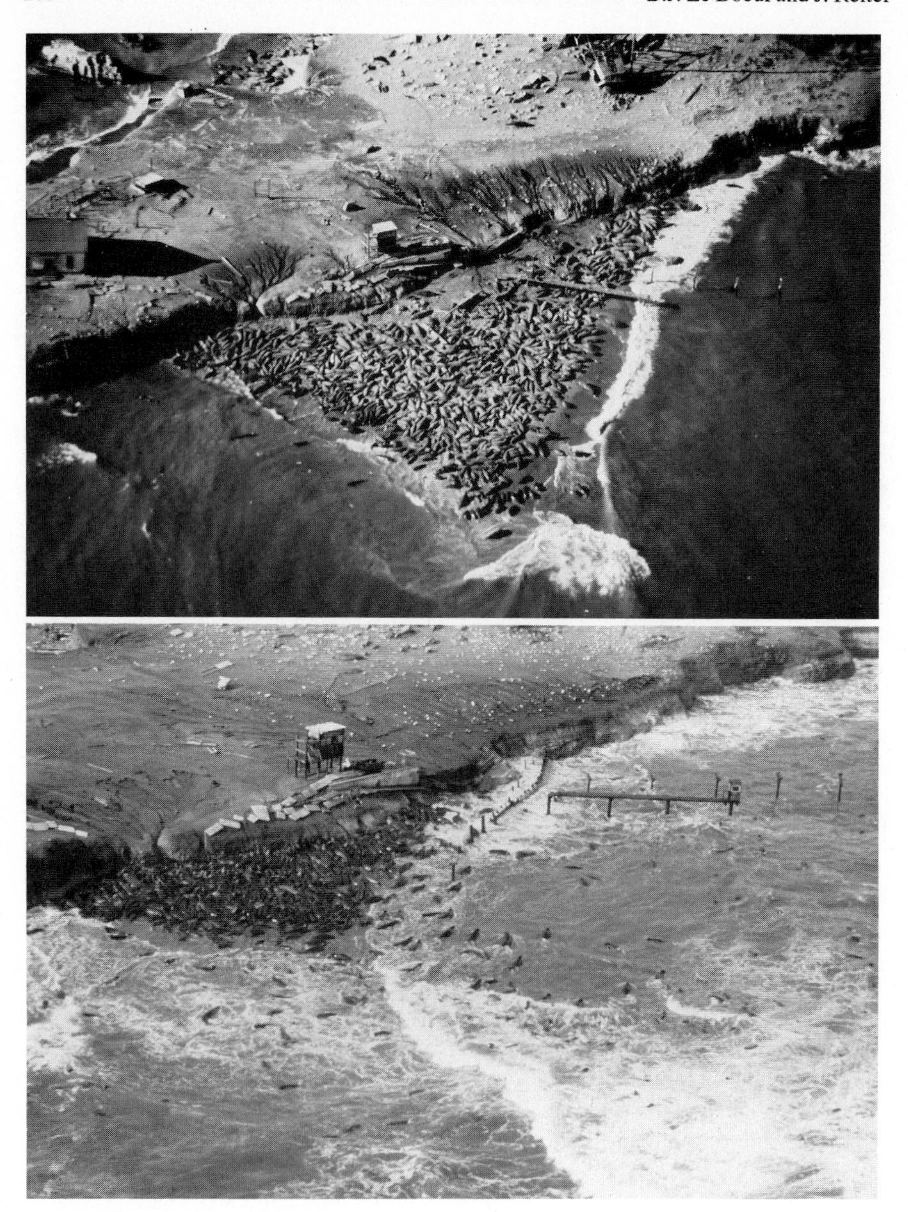

Fig. 1. Photograph of the Point Harem on Año Nuevo Island at peak season on a calm day at high tide (*top*). The same harem is shown during a storm with attendant high surf at the peak of the breeding season in 1983 (*bottom*)

Table 1. Number of pups born and pup mortality rate at Año Nuevo Island and mainland from 1977 to 1987

	Island		Mainland	
Year	No. Pups born	Percent died	No. Pups born	Percent died
1977	798	20.0	16	0
1978	908	44.0	86	9.3
1979	1072	36.4	101	20.8
1980	1194	36.0	159	3.1
1981	900	40.0	325	11.1
1982	1100	25.0	369	7.6
1983	975	70.0	591	11.3
1984	956	48.5	729	7.8
1985	1016	37.8	685	8.3
1986	889	40.2	853	6.2
1987	716	22.6	808	8.9

125 mothers that lost their pups abandoned the ravaged island for the relative safety of adjacent mainland beaches.

Pup mortality on the island, from December 1982 to March 1983, was estimated as 70% of pups born; 683 pups died out of 975 pups produced. The mortality rate in 1983 far exceeded that observed in preceding and succeeding years (Table 1). Since the island colony reached carrying capacity in 1977, variation in the pup mortality rate has been highly correlated with storms at peak season. Before 1977, when colony number was increasing, the entire breeding beaches were not fully occupied and storms had a less deleterious effect. During the years, 1968 to 1976, the pup mortality rate ranged from 13 to 26% of pups born (Le Boeuf and Briggs 1977).

Many pups died as a result of being washed out to sea. Many evidently drowned but some may have been eaten by white sharks, *Carcharadon carcharias*. Over 250 dead pups washed ashore during the few days following the storms. However, the majority of pups died a weak or two later because of separation from their mothers caused by the physical disruption of the harems (Le Boeuf and Briggs 1977). Storms and high surf increase crowding in harems, heighten aggression between females, and make mother-pup separation more likely. Orphans not reunited with their mothers within 2 days must attempt to steal milk from other nursing females or they die of starvation. Most females nursing their own pups bite milk thieves. Starvation and trauma caused by female bites is the major cause of pup death in most years (Le Boeuf and Briggs 1977).

The high mortality rate on Año Nuevo Island in 1983 was due to the coincidence of several events:

1. Higher than normal sea level and the highest tides of the year coupled with high coastal winds and storms created unusually high surf that inundated traditional breeding areas.

2. The storms struck when the number of females and suckling pups were at their highest peak.
3. The island colony was at carrying capacity.
4. All space on the large breeding area on the Island, the Point Beach, was taken up by the animals, and high cliffs prevented the animals from retreating to higher ground. Flooding of the breeding beaches on the island was the worst that we have ever observed.

In contrast to the island, the pup mortality rate on the adjacent mainland, separated from the island by a 200-m channel, was 11% of pups born, only slightly higher than in previous years (Table 1). Although mainland beaches also took a physical pounding that eroded cliffs and caused large areas of dunes to be washed away, females and pups on the mainland had one advantage. They were widely dispersed along the shore, breeding in 10 small harems containing 10 to 150 females, and most importantly, most females and pups could move inland to high and dry areas away from the dangerous surf. However, where high cliffs inhibited retreat, pup mortality was high. For example, one mainland beach being used for the first time, housed 12 suckling pups before the storm. All of them were swept away when the first high seas washed over the beach. Even when females moved inland, beaches behind them were often cut away, leaving meter-high vertical banks that prevented lagging pups from following their mothers inland to safety.

Other exposed elephant seal rookeries in California also incurred high pup mortality rates. All 20 pups born in the newly established breeding colony at Cape San Martin in Big Sur were killed. Numerous starving or dead pups washed ashore in southern California, evidently from the offshore rookeries on San Miguel or San Nicolas islands, where 2000 pups were estimated to have died (Le Boeuf and Condit 1983).

Only nursing pups were endangered by the high surf. No deaths among juveniles or adults of both sexes was noted.

23.4 Indirect Effects

If the distribution of elephant seal prey was changed significantly by El Niño, this might be reflected by a decrease in survival or reproduction of adult females, pups, and adult males.

23.4.1 Females

Temporal Patterning of Reproduction. If foraging was more difficult for pregnant females during the El Niño years, they did not respond by remaining at sea longer and delaying parturition. The normally distributed curve of females present on the rookery during the breeding season (Fig. 3 in Le Boeuf and Briggs 1977), which also shows the onset and rate of female arrivals and the rate at which females return to sea, did not change significantly from 1968 to 1988 despite a sevenfold increase in the population of females (from 247 giving birth in 1968 to 1742 giving birth in 1986) and despite storms and bad weather at peak season in the years, 1973, 1978, 1981 and 1983. The time and slope of female arrivals and departures varied little over the years; females began arriving in early December and the last breeding female departed on about 10 March. In all years, female numbers peaked between 26 January and 2 February. The timing of parturition, nursing, and mating is evidently deeply fixed and unresponsive to environmental fluctuations.

Postlactation Foraging Duration. At the end of lactation, adult females go to sea to feed for about 70 days. This post-reproductive foraging period would appear to be a critical time for females. They have just lost approximately 42% of their mass during a 4-week nursing period during which they do not feed or drink (Costa et al. 1986) and, being fertilized, they are beginning another pregnancy. It is likely that they go directly to food. Much of their time at sea appears to be spent feeding since females increase their mass, on average, by 1 kg/day, which would require that they consume about 20 kg of prey per day (Le Boeuf et al. 1988). Upon their return from a period at sea lasting approximately 70 days, they are 25% heavier than at the end of lactation during initial water entry. If foraging was difficult during the El Niño years, one might expect that the females would have stayed at sea longer.

We determined the duration of the period at sea between the end of lactation and the beginning of the molt for females that weaned their pups successfully during the years 1976 to 1987. We simply noted the departure and return dates of marked, tagged females. The sample sizes for these years were 31, 18, 17, 12, 4, 38, 21, 23, 16, 20, 55, and 33, respectively. Figure 2 shows that the mean duration at sea was significantly higher by 5–15% in the years 1982 to 1985 than in preceding or

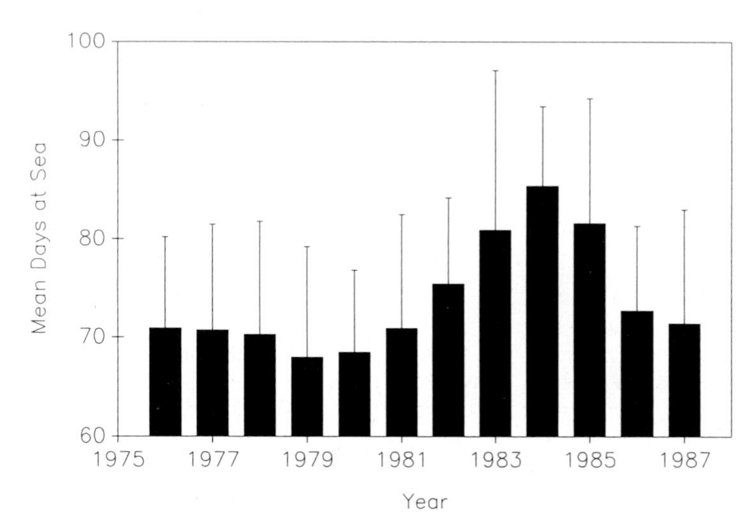

Fig. 2. Mean duration at sea during the period between lactation and molting as a function of breeding year among female elephant seals that weaned pups successfully

succeeding years ($F = 5.32$, $df = 11, 275$, $P < 0.05$). The highest mean duration at sea was in 1984 but the longest periods at sea immediately following lactation were recorded in 1983. One female bearing a diving instrument (not included in the sample) was at sea for 127 days (Le Boeuf et al. 1986); one female, unencumbered by an instrument, was at sea 104 days.

Survival to the Next Year Following Reproduction. If the increased time at sea reflects a more difficult time obtaining food or more energy spent securing it, we might expect this to be reflected in a lower rate of survival to the next year or decreased probability of producing or weaning a pup in the next year, possibly because of a reduced amount of energy transferred to the pup.

We calculated the percentage of tagged females that were resighted 1 year after producing a pup by conducting a search in each of the 3 years following the breeding episode. Searches were made daily during the breeding season and once a week during the rest of the year. Virtually all females resighted were observed during the breeding season at Año Nuevo or the rookery at southeast Farallon Island.

Table 2 shows the minimum number of females that survived from one breeding season to the next from 1981 to 1986. It is clear that the minimum survival rates did not vary substantially over the sample period whether one makes the comparison 1, 2, or 3 years after breeding (Chi-square = 1.34, $df = 4$, $P > 0.05$). The minimum survival rate was as high in El Niño years as in non-El Niño years. Moreover, adjusting for an estimated 6% tag loss per year (Le Boeuf and Reiter 1988), the best estimate of annual survival during this period is in the range, 60 to 69%.

Reproductive Success: Pup Production Following Reproduction. Because 97% or more of the females observed on a rookery give birth to a pup, observation of a female on a rookery can be virtually equated with giving birth. Thus, it follows that reproductive success, as measured by pups produced, also did not vary significantly across the sample years indicated in Table 2. In any case, we did not observe a decrease in natality during the EN years.

Table 2. Percent annual survivorship of female northern elephant seals from one breeding season to the next during the years 1981 to 1985. The percentage of females surviving to the year after breeding was calculated by searching for tagged females during 1, 2, and 3 years after breeding

		Percent females surviving to next year based on observation in year:		
Breeding season	n	1	2	3
1981	151	47.1	55.6	58.9
1982	204	53.9	61.8	63.2
1983	183	53.0	62.3	62.3
1984	193	44.6	56.5	58.5
1985	162	47.5	53.1	No data

23.4.2 Pups

Mass. If foraging conditions were poorer in El Niño years, females might have transferred less energy to their pups and weaned them at reduced weights compared to other years. This was clearly not the case in 1982 and 1984. In a study by Le Boeuf (1989a), using annual sample sizes in the range 28 to 98, mean mass of pups at weaning (both sexes combined) was 127 ± 36 in the years 1978, 1982, 1986, and 1987. Mean weaning mass was higher in 1980 (146 ± 31), 1984 (134 ± 19), and 1985 (137 ± 23). Pups were not weighed in 1983.

Survivorship. Although pups appear to have started life in as good condition in 1982 and 1984 as in other years (as reflected by mass), was pup survival over the first period at sea as high as in other years? This is a vulnerable time for pups for they must avoid predators and feed on their own after a 2.5-month fast. On average, only 45% survive the first period at sea to 1 year of age (Le Boeuf and Reiter 1988). Preliminary data indicate that they are shallower divers than adult females (P. Thorson and B. Le Boeuf, unpubl. data) and hence, may exploit different prey. Their prey distribution may have been changed during the El Niño years making it more difficult for them to feed. If so, one would predict low mass and poor condition when the animals return from sea, or in the extreme, a lower survivorship rate during the El Niño years than in other years.

From annual tagging studies and tag recovery observations, we calculated the survivorship curves shown in Fig. 3 for the years 1981 to 1984. The survivorship curve is lowest for the peak El Niño year, 1983, with the adjacent years, 1982 and 1984, being intermediate to the non-El Niño year, 1981. The survivorship curve in 1981 is similar to the survivorship curves of 1976 to 1980 (B. Le Boeuf, unpubl. data).

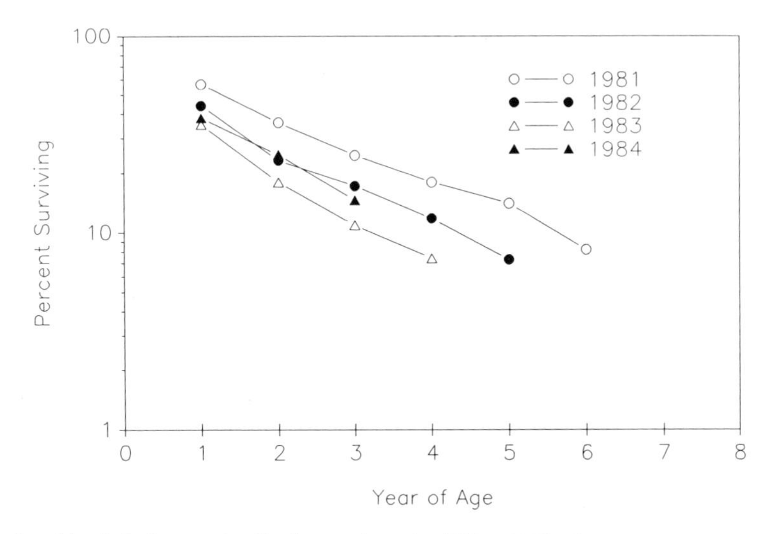

Fig. 3. Log survivorship of elephant seals of both sexes from the 1981 to 1984 cohorts

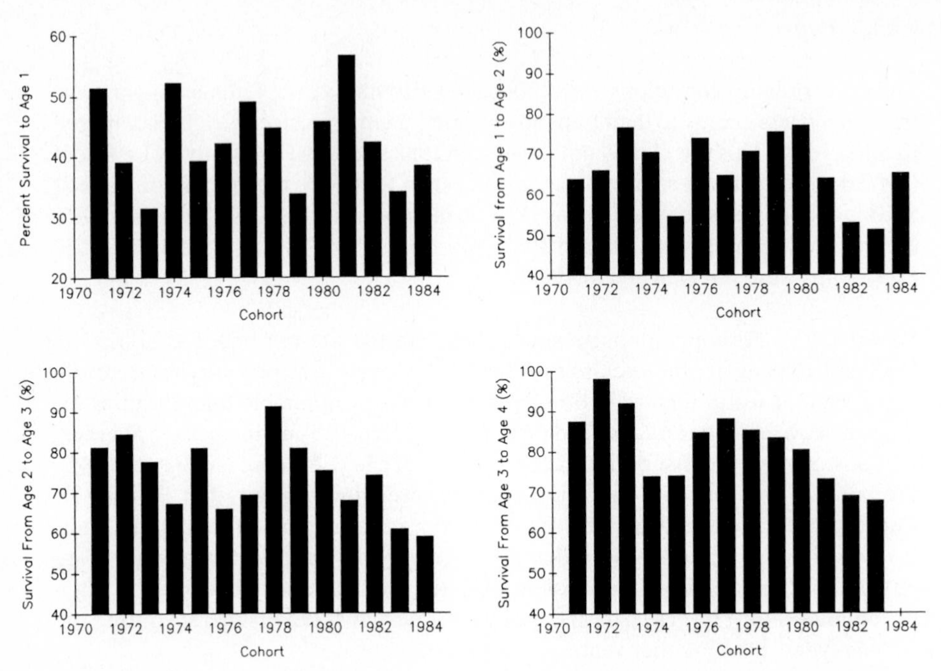

Fig. 4. Cohort survival of elephant seals of both sexes to age 1, from 1 to 2, 2 to 3, and 3 to 4

Moreover, the discrepancy in survivorship between the years 1982–1984 and 1981 increases with advancing age. This effect is shown more clearly in Fig. 4. First, it is clear that survivorship to 1 year of age varied significantly across cohorts (Chi-square = 112, df = 13, $P < 0.05$). The 1983 cohort has low survivorship to age 1 but no more so than the years 1973 and 1979. However, survivorship during the intervals, age 1 to 2, 2 to 3, and 3 to 4, reveals that the 1983 cohort is always significantly lower than other cohorts (Chi-square goodness of fit = 11, df = 1, $P < 0.05$). Other significant differences include adjacent years: 1984 to years 1, 3, and 4, and 1982 to years 2 and 4. By the time pups reached 4 years of age, only 7.5% of the 1983 cohort was still living. Next in line were the cohorts of 1982 (11.8%), 1975 (13.0%), 1973, and 1976 (about 17%). Survivorship to age 4 was highest in 1971, 1979, 1980, being slightly over 20%.

Age at Primiparity. The last point we address is cohort variation in age at primiparity. If body condition is positively correlated with onset of reproductive age (see Boyd 1984), and the female cohort from 1983 had a more difficult time finding food, the onset of reproduction might be later relative to other cohorts.

Table 3 shows that 1982, 1983, and 1984 were poor cohorts in terms of the low percentage of females that were primiparous at age 3, 4, or 5. The 1983 cohort had the poorest record. The 1982 and 1984 cohorts were somewhat more successful but less so than in other years. It is not clear whether this effect is due to reduced survival or deferrment of reproduction or a combination of both variables. The aberrant

Table 3. Percentage of females from various cohorts that were primiparous at age 3, 4, or 5 and older

		Percent females primiparous at age:			
Cohort year	N	3	4	5+	3, 4 or 5+
1977	293	3.8	9.9	5.5	19.1
1978	476	5.5	9.0	4.8	19.3
1979	97	5.2	8.2	11.3	24.7
1980	113	8.8	10.6	6.2	25.7
1981	227	4.8	15.9	5.3	26.0
1982	332	1.8	8.4	4.2	14.5
1983	277	1.1	5.4	2.9	9.4
1984	471	1.5	11.7	2.3	15.5
1985	221	3.6	8.6	n.a.[a]	n.a.
1986	231	3.5	n.a.	n.a.	n.a.

[a]Not available.

distribution for the 1979 cohort, with a higher percentage of females being primiparous at age 5 than age 4, may reflect difficult foraging for 4-year-olds in 1983 that resulted in the postponement of reproduction.

23.4.3 Males

We determined the percentage of identifiable males that returned to Año Nuevo from one breeding season to the next, i.e., the minimum annual survival rate (Fig. 5). The lowest minimum survival rate from 1 year to the next was from 1983 to 1984 (40.2%) followed by 1984 to 1985 (44.4%). However, differences across years are not statistically significant (Chi-square = 2.84, df = 6, $P > 0.05$). The sample size for a single year is also too small to determine what age class had the lowest survival rate. However, summing all years across age classes confirmed an effect reported earlier (Le Boeuf 1974). Annual survival rate decreased systematically with age, from 63 to 48% in 4–7-year-olds to 45% or below in adult males 8 years of age or older.

23.5 Conclusions

The year, 1983, and to a lesser extent, 1982 and 1984, were poor years for elephant seals breeding at Año Nuevo, California. As a large-scale natural experiment that caused changes in weather, sea temperatures, and currents, EN 1982–83 exerted diverse, depressing effects on survival and reproduction.

The principal effects of EN 1982–83 on the Año Nuevo colony can be summarized as follows:

1. A 44% increase in pup mortality in 1983 relative to other years, due to inundation of breeding beaches at peak season.

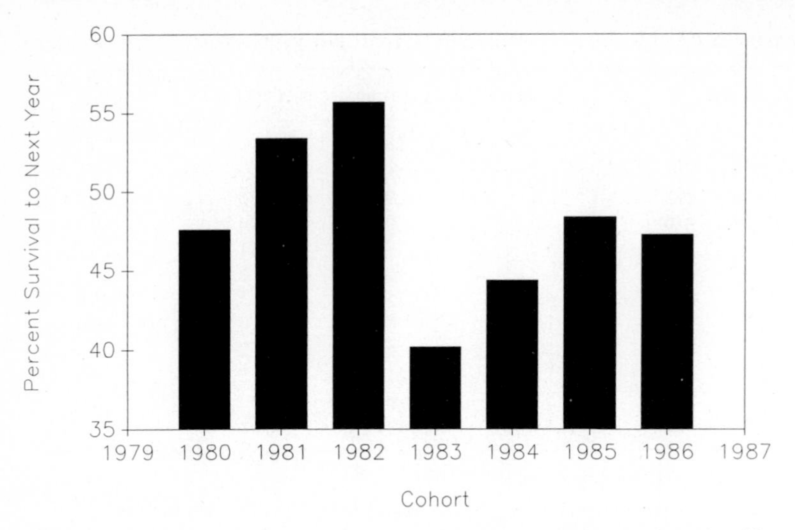

Fig. 5. Percent annual survivorship of tagged male northern elephant seals on Año Nuevo from one breeding season to the next during the years 1980 to 1986. For example, only 40% of the males present in 1983 were observed in 1984. All males were known-age adults, 5 to 14 years of age, that were born on Año Nuevo. The sample sizes for the years 1980 to 1986 were 124, 86, 79, 87, 81, 124, and 129, respectively

2. Reduced survivorship of pups born in 1983 to age 2, 3, and 4, relative to other cohorts.
3. Lower reproductive success of female pups born in 1983 relative to other cohorts, as measured by proportion of females primiparous at ages 3 and 4.
4. An increase in the length of the foraging period at sea of females following reproduction during the years, 1982–1985.
5. Reduced survival of breeding males from 1983 to 1984 relative to other years.

We found *no* evidence for predictions that during the El Niño years:

1. Females would be late arriving on the rookery to reproduce (arrival to give birth, nurse, and mate);
2. Breeding females would exhibit a lower survival rate to the next breeding season;
3. Breeding females would produce fewer pups the following breeding season;
4. The mean mass of weaned pups would decline (tested in 1984 but not in 1983); and that
5. Survivorship of pups to 1 year of age would be lower than usual.

Consideration of the causes of these effects, and the sex and age class affected, is illuminating, especially for comparisons with other species.

The devastating effect of the winter storms of 1983 on elephant seal colonies along the coast of California was only partly due to El Niño. The high pup mortality on Año Nuevo Island and on some other rookeries in central and southern California was due to the concomitance of several events – bad weather with high winds and high surf, high tides, peak number of females and suckling pups present, crowded conditions allowing little room for escape to dry land, and higher than normal water level. Only the latter condition was unique to El Niño. The bad weather and storm conditions were associated with El Niño, but storms and high surf occur in any year. The timing of the storm, 19–21 January, was the key to the high mortality in 1983. Even a milder storm at the same time of year would have caused numerous pup deaths on Año Nuevo Island, given the number of pups present and the lack of access to safety. Equally bad storms in previous years (1969, 1978, and 1981) caused less pup mortality because the storm struck before peak season or because the storm struck at peak season before the colony was at carrying capacity. The effect of the 1983 storms on pup mortality was also a matter of location. Where females and their pups could retreat inland from the dangerous surf, such as on the Año Nuevo mainland, pup mortality was no higher than usual. Lastly, the storm effect on elephant seal pup mortality had no parallel among sea lions and fur seals that breed in the Pacific during the summer months.

Of equal importance is how possible changes in the distribution of prey or abundance and condition of prey, associated with perturbations in sea temperature and currents, affected foraging and subsequent survival and reproduction. This discussion is speculative because we do not know precisely what northern elephant seals from Año Nuevo were eating and where they were foraging. In general, we know that seals associated with central California rookeries are most often observed north of their rookeries off California, Oregon, Washington, and Vancouver Island, British Columbia; a few individuals have been sighted as far north as the southern Aleutian Islands in Alaska (Condit and Le Boeuf 1984).

Studies of gut contents and observations of feeding indicate that elephant seals feed on a variety of prey that includes cephalopods, crustaceans, tunicates, and fish (Condit and Le Boeuf 1984; Antonelis et al. 1987). Several species of pelagic squid and octopus make up the largest number of prey species found in seal stomachs. Pacific hake, *Merluccius productus*, is the most frequently occurring teleost fish and several species of sharks, rays, and ratfish are common cartilaginous fish prey. The prey of elephant seals suggest that they are capable of foraging in a variety of marine environments from epipelagic to benthic. However, the diving pattern of adult females indicates that most foraging is in the mesopelagic zone between 400 to 700 m (Le Boeuf et al. 1988).

The effect of the aberrant sea temperature and currents on elephant seal foraging was not as visible and dramatic as it was on some otariids. The sea lion and fur seal pattern consists of females alternately foraging for 2–7 days and feeding their pups for about 2 days (Gentry et al. 1986a). Clearly, good foraging must be available close to the rookery. In contrast, female elephant seals wean their pups after 27 days of nursing during which they do not feed, drink, or enter the water (Le Boeuf et al. 1972; Reiter et al. 1981; Costa et al. 1986). Although the tax on the body is great, when females return to the water to feed, they are not constrained to remain near their pups as are otariid females. They have more freedom to move to where the prey are located. We do not know whether females traveled further afield in 1983, but they spent more days at sea than usual, a result also observed in California sea lions breeding at San Miguel and San Nicolas Islands in southern California

(Boness et al. 1985; Costa et al., this Vol.; Heath et al., this Vol.). Indeed, the seals spent more days at sea in 1984 than in any other year. This may have been caused by the necessity of traveling further north to feed on known prey such as market squid, *Loligo opalescens*, which decreased in catches off California and Oregon but increased off the coast of Washington in 1983 and 1984 (CalCoFi Report 1985; Pearcy et al. 1985). We have no evidence that spending more days at sea than usual had a deleterious effect on female survival or reproduction the following year. Adult males breeding in 1983 showed a reduced survival rate to the next breeding season. This sex difference suggests that males may have had a more difficult time finding or catching prey than females. This could have arisen if males feed on different prey, consume more prey per unit time away, or forage in different locations than females. At present, we have no information on the diving or foraging behavior of males that bears on this speculation.

Our most significant result is the poor survival and reproductive record of the 1983 cohort. This was due to the combined direct and indirect effects associated with El Niño. First, bad weather caused high pup mortality prior to weaning and, second, some unknown mechanism (possibly poor forage) caused reduced survivorship to reproductive age of females. The decreased survivorship of the 1983 cohort was not apparent at age 1 but became more and more pronounced with age. Decreased survival to reproductive age, a major component of reproductive success in female elephant seals (Le Boeuf and Reiter 1988), coupled with a reduced percentage of females giving birth for the first time at ages 3 and 4, indicates that the 1983 cohort will leave significantly fewer progeny than adjacent cohorts. Moreover, the 1983 cohort will not have the substantial advantages – the "compound genetic interest" effect – that females accrue from reproducing early (Lewontin 1965). Regardless of the underlying mechanisms, it is clear that significantly fewer progeny will be produced by the 1983 cohort than by others that immediately preceded or succeeded it. In the long run, the reproductive performance of the elephant seals born at Año Nuevo, California, in 1983 may be as low as some of the sea lions and fur seals whose first year class was decimated (Limberger et al. 1983). The underlying mechanisms might be different but the end results are similar.

Cohort variation in reproductive success is widespread in nature and the factors that cause this variation are significant forces in selection (Clutton-Brock 1988). Our data show that the 1982–83 El Niño Southern Oscillation produced a complex of physical perturbations that apparently caused extensive cohort variation in survival of elephant seals breeding in central California.

Acknowledgments. Selina Gleason was a great help analyzing data and we thank her. Supported by National Science Foundation grants. Marking solutions were provided by Clairol, Inc.

24 Effects of the 1982–83 El Niño on Northern Elephant Seals on the South Farallon Islands, California

H.R. Huber, C. Beckham, and J. Nisbet

24.1 Introduction

The colony of northern elephant seals (*Mirounga angustirostris*) on the Farallons has grown with remarkable speed since 1959 when the first elephant seal was seen there after an absence of more than 150 years. Most of this growth is due to immigration from southern rookeries (Cooper and Stewart 1983; Huber et al. in press).

A recent oceanic perturbation, the 1982–83 El Niño (EN), had profound biological effects on the Farallon northern elephant seal colony. Most of these effects lasted only a short time, but even the long-term effects, which include increased juvenile mortality for the 1983 and 1984 cohorts and apparently increased mortality among subadult males (age 5–6), have had much less of an effect on the elephant seal population than historic exploitation by sealers.

24.2 Methods and Study Area

The North, Middle, and South Farallon Islands are a series of small granitic islands 43 km west of San Francisco, California, USA, on the northern edge of the northern elephant seal breeding range. Elephant seals haul out only on Southeast Farallon and West End, the two islands which compose the South Farallon Islands. These islands are separated by a narrow surge channel and there is considerable interchange of elephant seals between them.

Observations began in December 1974 and continued year-round until April 1986. Daily observations of tagged and naturally scarred elephant seals during the breeding season (late November to mid-March) included the arrival, pupping, and departure dates for females; arrival and departure dates for males; and causes of pup mortality. We had two categories of individually recognizable animals: known-age animals (those tagged as pups on the Farallon, San Miguel, San Nicolas, and Año Nuevo islands) and known-history animals (those tagged as adults on the Farallons or those identified by unique natural scars). On the Farallons, from 90 to 100% of all pups surviving to weaning were tagged each year, except in 1979 when only 56% were tagged. Daily observations of tagged animals were made during the spring molt and fall haul out to determine survival of juveniles and of known-age and known-history adults.

Return rates were based on the proportion of tagged or naturally scarred animals returning the following year. Decreased return rates are not synonymous with increased mortality, since these animals may return in subsequent years.

We defined natality as the proportion of tagged females that gave birth in a year relative to the tagged females alive in that year. Age-specific natality was based on observations of females tagged as pups on South Farallon, Año Nuevo, San Miguel, and San Nicolas islands and giving birth on the South Farallon Islands. Age-specific natality rates are not independent but are based on females of known age seen over time, some individuals for up to 12 years. In calculating age-specific natality, we considered a female to have skipped a breeding season if (1) she was present but did not give birth, or (2) she was not present at the Farallons, nor was she observed at another rookery during that season, but she was known to be alive because she was seen subsequently at the Farallons either in the spring or fall haul outs or during the following breeding season (Huber 1987).

Immature survival is based on all resightings of an animal up to age 4; however, 90% of all individuals resighted are seen for the first time within 2 years after tagging as pups. The 1978 cohort consists of pups born between December 1977 and February 1978, etc. We did not attempt to account for tag loss in this study; therefore, all survival and return rate estimates are minimum estimates. We also assumed that there was no annual variation in tag loss, thus the minimum survival and return rate estimates are comparable between years.

For analysis of immature survival and male and female adult return rates, we used the exact test of binomial probabilities at the 0.05 significance level. As a result, for return rates, since the sets used were not always independent (some animals were measured year after year), the error probabilities may be underestimated. We used a one-way ANOVA to determine whether differences in breeding phenology for all males and females present from 1982 to 1986 were significant and then used the Tukey multiple range test to determine which years were different. For analysis of continuous breeding females over years, we used repeated measures ANOVA and, if differences existed, used the Tukey multiple range test to determine which years were different.

Sea surface temperature (SST) at Southeast Farallon Island was taken daily, weather permitting, and averaged into monthly means.

Stillbirth was determined as the cause of pup mortality when necropsy showed that the lungs had never been inflated.

Since age of first reproduction in elephant seals on the Farallons varies from ages 3 to 6 (Huber 1987), we considered females to be adults once they were known to have given birth. During the breeding season, males of unknown age (i.e., those not tagged as pups or not retaining tags) were categorized based on a combination of qualities including overall body size, chest shield development, and length of nose as described in Le Boeuf (1974). These categories (SA^2, SA^3, SA^4, bull; Table 2) were assigned an estimated age by comparison with known-age animals. Males were assigned to a single category for the entire breeding season.

In the early years of this study, sample sizes were very small and confidence intervals very large, therefore, we eliminated early years from analysis. For female return rates we analyzed years when more than 100 females were present (1977–86,

Table 1). For survival rates we analyzed years when more than 100 pups were born (1978–86, Table 1) and we also excepted 1977 when an experimental tag was used. For male return rates we analyzed years when at least 20 males were present in each age category (1982–86, Table 2) and for differences in male and female chronology we chose years before, during, and after the effects of the 1982–83 EN event (1982–86).

Table 1. Number of northern elephant seal pups born and pup mortality on the South Farallon Islands, 1975–1986

Year	Number of adult females present	Number of pups born	Change over previous year (%)	Pup mortality to weaning (%)
1975	36	35	+105	20
1976	62	60	+71	7
1977	107	104	+73	19
1978	138	133	+28	26
1979	192	184	+38	30
1980	236	232	+26	41
1981	300	292	+26	42
1982	375	367	+26	31
1983	487	475	+29	32
1984	479	447	−6	33
1985	390	360	−19	19
1986	447	434	+20	36

Table 2. Age classes of male elephant seals present during the breeding season on South Farallon Islands, 1976–1986

Age class	SA^2 (5 years)	SA^3 (6–7 years)	SA^4 (7–8 years)	Bull (8+ years)	Total males	Change over previous year (%)
1976	10 (46%)	6 (27%)	0	6 (27%)	22	
1977	18 (50%)	15 (42%)	1 (3%)	2 (6%)	36	+63
1978	26 (54%)	16 (33%)	4 (8%)	2 (4%)	48	+33
1979	34 (47%)	19 (26%)	13 (18%)	6 (8%)	72	+50
1980	39 (41%)	32 (34%)	13 (14%)	11 (12%)	95	+32
1981	62 (47%)	33 (25%)	21 (16%)	15 (11%)	131	+38
1982	65 (40%)	38 (23%)	35 (21%)	23 (15%)	161	+23
1983	89 (46%)	44 (23%)	37 (19%)	24 (12%)	194	+20
1984	63 (33%)	60 (30%)	32 (16%)	43 (21%)	198	+2
1985	31 (27%)	43 (28%)	35 (23%)	34 (22%)	143	−27
1986	25 (18%)	35 (25%)	31 (22%)	48 (35%)	139	−3

24.3 Results

Monthly mean SSTs at Southeast Farallon Island in 1983 followed the pattern of previous years (Fig. 1), but from January 1983 to April 1984 were higher than the monthly means of 1973 to 1982 (Fig. 1). By October 1984, SSTs returned to previous levels (Point Reyes Bird Observatory, unpubl. data). The SSTs were indicative of the geographical and temporal extent of the 1982–83 EN (see Arntz et al., this Vol.). Although 1984 was not an EN year, the effects of the 1982–83 EN were still prevalent around the Farallons in 1984.

Significant changes in the population dynamics of elephant seals on the Farallons coincided with the warm waters of the 1982–83 EN. Compared to previous years, pup production, survival of immatures, and return rates of males and females decreased. That these changes were not merely the beginning of a downward trend is shown by increases in number of pups born after 1985, increased survival for the 1985 cohort as well as increases in return rates of males and females after 1985. There were also significant changes in the chronology of the breeding season for males in 1984 which returned to previous timing in 1986.

24.3.1 Female Return Rates

From 1977 to 1982 (the pre-EN years), the proportion of tagged females returning to give birth the following year ranged from 0.64 to 0.78, whereas in 1983 and 1984

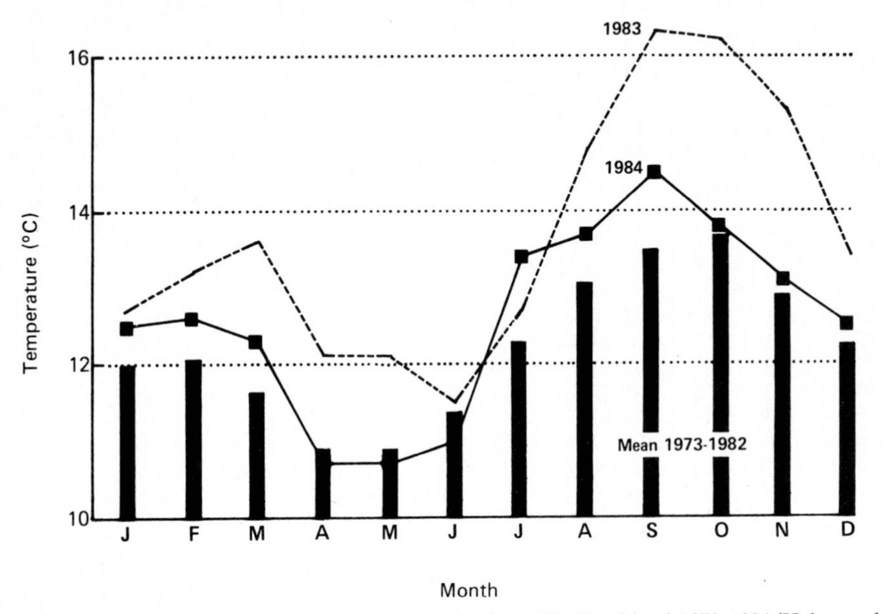

Fig. 1. Monthly mean sea surface temperature at Southeast Farallon Island, 1973–1984 (Huber et al. in press)

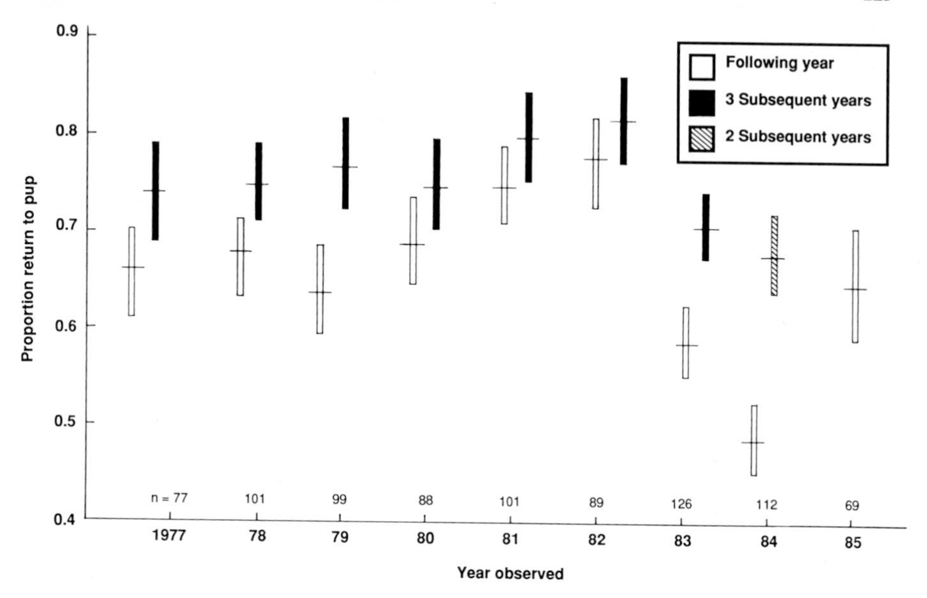

Fig. 2. Proportion of adult females (with 95% confidence intervals) returning to pup on the South Farallon Islands in the following year and within 3 subsequent years from initial observations 1977 to 1985. *n* Number of tagged or scarred females present in initial year

the proportion decreased to 0.58 and 0.49, respectively (Fig. 2). In 1984 the proportion of females returning the following year was significantly lower than all other years and in 1983 the proportion of females returning the following year was significantly lower than in all other years except 1979 and 1984 (Fig. 2). However, by 1985, the proportion of females returning the following year rose again to 0.65 which was not significantly different from the pre-EN years (Fig. 2). Every year some adult females fail to reproduce, thus the proportion of tagged females that return at least once within 3 years after their initial observation is higher than those returning just in the following year. The proportion of females returning within 3 years ranged from 0.74 to 0.82 between 1977 and 1982 (Fig. 2). In 1983, the 3-year return rate was 0.71, significantly lower than in 1981 and 1982 but not different from any other pre-EN years (Fig. 2). The proportion of tagged females that returned after 1984 (within 2 subsequent years) was 0.68. In examining Fig. 2 it is apparent that the proportion of females returning the following year and the proportion of females returning within 3 years were significantly different in 1979, 1983, and 1984.

Each year an average of 14% of adult females does not reproduce (Huber 1987), but in 1984 and 1985 a higher proportion of tagged adult females failed to give birth than in any other consecutive 2-year period during this study. However, adult female mortality was not a factor, because most of these females returned to give birth in 1986. The percent of nonpregnant adult females observed during the breeding season rose to 7 and 8%, respectively, for 1984 and 1985. This is significantly higher than the 2% observed for the previous 4 years (1980–1983) and the 3% in 1986 ($X^2 = 30.67$, df = 5, $P < 0.001$; Table 1). At least three females suppressed reproduction for three consecutive years, two from 1983 to 1985 and one from 1984 to 1986. We had not observed this on the Farallons in the previous 10 years.

Tagged adult females not seen within 3 subsequent years were assumed to have died (or lost tags). We found that between 1977 and 1984, the proportion never observed again (and presumed dead), ranged from 0.18 to 0.32 (Fig. 3). The only significant difference in female mortality from 1977 through 1984 was that mortality rates in 1981 and 1982 were significantly lower than in 1983 and 1984; there was no difference between mortality rates in 1983 and 1984 and mortality rates from 1977 to 1980 (Fig. 3).

24.3.2 Pup Production

The number of pups born on the Farallons increased annually until 1983, declined in 1984 and again in 1985, but by 1986 had risen to nearly 1983 levels (Huber et al. in press; Table 1). Pup mortality before weaning varied from 7 to 42% between 1975 and 1982 (Table 1). In 1983 and 1984 mortality before weaning did not increase, but was similar to mortality in 1982 and lower than in 1980 and 1981 despite the increased number of pups born in the later years (Table 1). However, the percent of stillborn pups doubled from less than 2% in the years 1978 to 1983 to 4% in 1984 and doubled again to 9% in 1985. By 1986, the percent of pup mortality due to stillbirths returned to 2%. The proportion of pup mortality due to stillbirths in 1984 and 1985 was significantly higher than all other years between 1981 and 1986 ($X^2 = 11.46$, df = 5, $P = 0.043$).

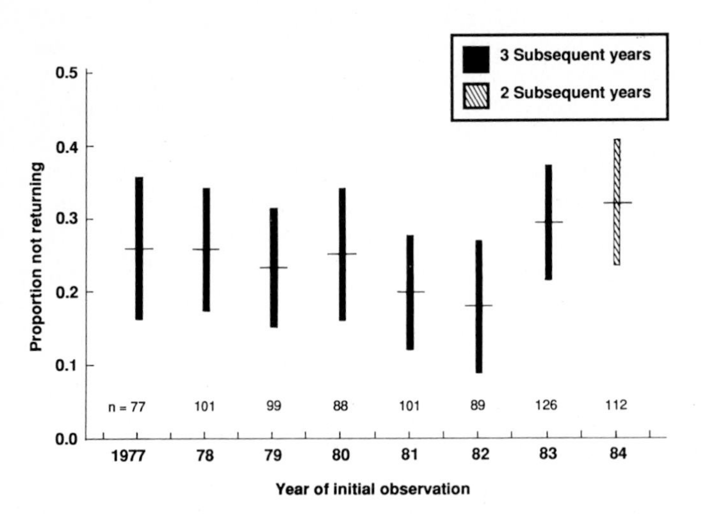

Fig. 3. Proportion of adult female elephant seals (with 95% confidence intervals) not observed in 3 subsequent years after initial observations 1977 to 1984. These females are presumed to have died or to have lost tags. *n* Number of females present in initial year

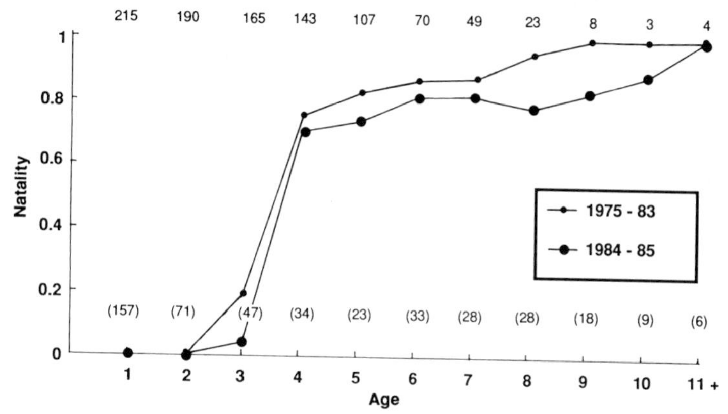

Fig. 4. Comparison of age-specific natality of elephant seals on South Farallon Islands between 1975 to 1983 and 1984 to 1985. *Numbers at top* of graph indicate sample sizes for 1975 to 1983 and *numbers in parentheses* indicate sample sizes for 1984 and 1985

24.3.3 Age-Specific Natality

For known-age females in the years 1975 to 1983, all females age 9 years and older gave birth (Fig. 4). However, in 1984 and 1985, all ages up to 11 years suffered reduced natality (Fig. 4). As a group, for females less than 11 years old, natality in 1984–1985 was significantly lower than from 1975 to 1983 ($P < 0.05$, signed rank test). Because of small sample sizes, particularly after age 8, the change between the two periods was significantly different only for 3-year-olds (exact test of binomial probabilities, $P < 0.05$).

24.3.4 Survival

Minimum survival of the 1983 and 1984 cohorts was significantly lower than for all other cohorts at every stage that we measured survival, to first fall (age 9 months), to second spring (age 14 months), third spring (age 26 months), and fourth spring (age 38 months) (Fig. 5A-D). Survival to first fall was 28 to 29% for the 1983 and 1984 cohorts compared to 44 to 56% for the 1978 to 1983 cohorts and 43% for the 1985 cohort (Fig. 5A).

Survival of the 1982 cohort was comparable to survival of the 1978 to 1981 cohorts through the second spring (1983). A significant decrease in survival of the 1982 cohort compared to previous cohorts began after the spring of 1983 which coincided with the warm water period of the 1982–83 EN (Fig. 5C,D).

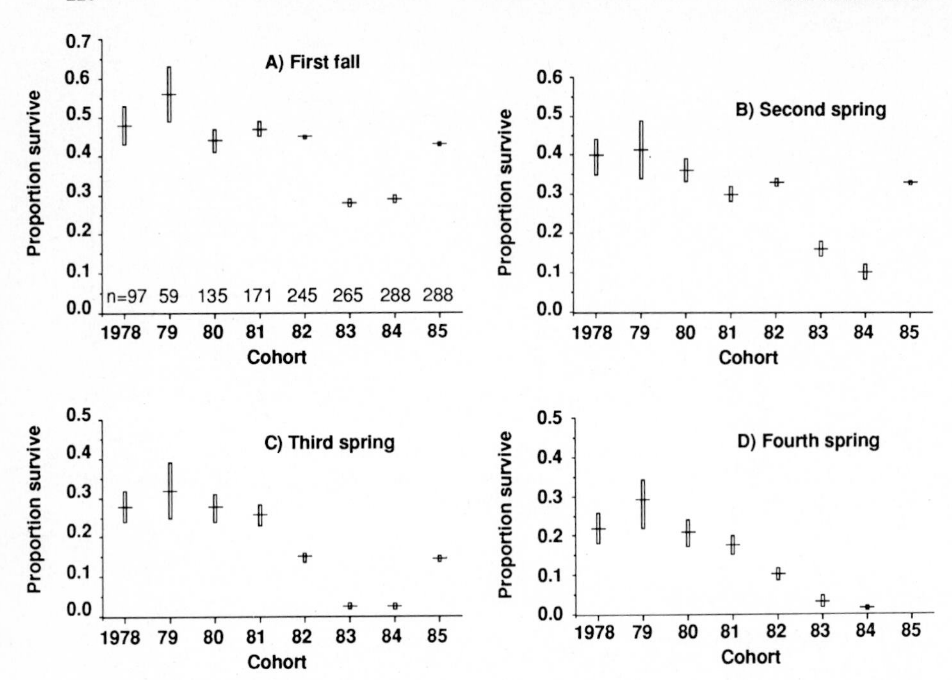

Fig. 5A-D. Minimum survival (with 95% confidence intervals) to fourth spring (age 3) for northern elephant seals tagged as pups on the South Farallon Islands, 1978 to 1985. **A** First fall (9 months); *B* second spring (14 months); **C** third spring (26 months); **D** fourth spring (38 months)

24.3.5 Male Return Rates

From 1976 to 1983 the total number of males age 5 to 14 present on the Farallons during the breeding season had an annual rate of increase of 30%. In 1984, the rate of increase of males dipped to 2% (Table 2). The following year, the number of males decreased dramatically by 27% (Table 2). By 1986, the rate of decline in number of males present decreased: there were only 3% fewer males present (Table 2).

From 1982 to 1985 the proportion of individually identified adult and subadult males returning the following year changed significantly (Fig. 6). In 1982, half of all tagged or scarred males present returned the following year (Fig. 6E). In 1983 the return rate for all age classes was down slightly to 0.47, but in 1984 the return rate decreased significantly to 0.32 (Fig. 6E). The return rate in 1985 for all age classes of males was significantly higher that in 1984, but also significantly lower than 1982 and 1983 (Fig. 6E).

For further analysis, we separated males into four estimated age categories: bull (8+ years), SA^4 (7–8 years), SA^3 (6–7 years), and SA^2 (5 years). There was no difference among the return rates of bulls in any year (Fig. 6A); however, the return rates of the younger SA^2's, SA^3's, and SA^4's were all significantly lower in 1984 than in 1982 and 1983 (Fig. 6B-D) and the return rates in 1985, although higher than in 1984, were not significantly different from either 1984 or from 1982 and 1983.

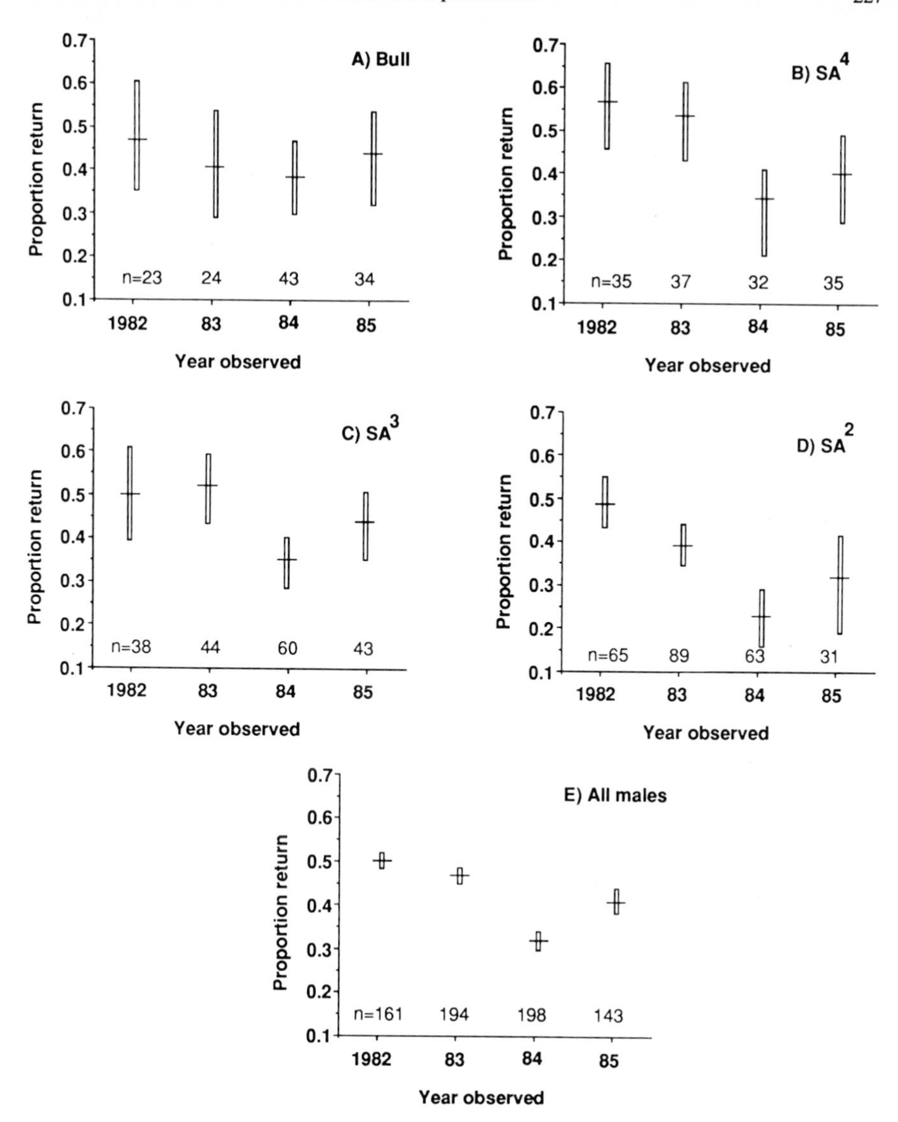

Fig. 6A-E. Proportion of adult and subadult males (with 95% confidence intervals) returning to the South Farallon Islands the following breeding season from initial observations in the 1982 to 1985 breeding seasons (**A** Bulls, 8+ years old; **B** SA^4's, 7–8 years old; **C** SA^3's, 6–7 years old; **D** SA^2's, 5 years old; **E** all age classes combined

Three males present in the 1984 breeding season, but not observed on the Farallons or elsewhere during 1985, returned to the Farallons during the 1986 breeding season. In the previous 10 years of observations we had not observed any males skipping a breeding season.

Table 3. Mean arrival dates, departure dates, and number of days ashore during the breeding season for elephant seal males on South Farallon Islands, 1982–86 (D = December, J = January, F = February, M = March)

	Mean arrival dates					*Mean departure dates*					*Mean days ashore*				
	Bull	SA4	SA3	SA2	All	Bull	SA4	SA3	SA2	All	Bull	SA4	SA3	SA2	All
1982	14D	19D	23D	2J	23D	6M	8M	9M	16F	2M	82.4	79.1	75.3	45.6	68.7
1983	20D	24D	1J	10J	31D	6M	11M	9M	28F	6M	75.8	76.9	67.2	48.0	65.3
1984	22D	31D	2J	9J	30D	9M	3M	6M	13F	4M	77.6	63.1	63.6	34.9	64.5
1985	24D	25D	30D	2J	27D	8M	4M	1M	14F	2M	73.2	68.8	60.9	43.6	65.4
1986	18D	19D	24D	6J	22D	11M	1M	28F	18F	4M	82.8	72.1	66.8	43.2	71.6

Table 4. Results of ANOVA for male elephant seal breeding season chronology on the Farallon Islands, 1982–86

Hypothesis	Statistic	*P* value[a]	Tukey multiple range test
All males			
No difference in arrival	$F_{4,615} = 7.94$	0.0*	83, 84, 85 later than 82 or 86
No difference in departure	$F_{4,615} = 1.25$	0.289	
No difference in days ashore	$F_{4,615} = 2.16$	0.07	
Bull			
No difference in arrival	$F_{4,153} = 2.23$	0.068	
No difference in departure	$F_{4,153} = 0.85$	0.496	
No difference in days ashore	$F_{4,153} = 1.88$	0.117	
SA4			
No difference in arrival	$F_{4,164} = 5.37$	0.0*	84 later than 82 or 86
No difference in departure	$F_{4,164} = 1.48$	0.211	
No difference in days ashore	$F_{4,164} = 3.80$	0.006*	84 less time ashore than 82 or 83
SA3			
No difference in arrival	$F_{4,174} = 4.48$	0.002*	84 later than 82 or 86
No difference in departure	$F_{4,174} = 2.61$	0.037*	82, 83, 84 later than 85, 86
No difference in days ashore	$F_{4,174} = 3.05$	0.018*	84, 85 less time ashore than 82
SA2			
No difference in arrival	$F_{4,109} = 1.01$	0.406	
No difference in departure	$F_{4,109} = 1.90$	0.116	
No difference in days ashore	$F_{4,109} = 0.89$	0.473	

[a]*Indicates significant difference.

24.3.6 Chronology

We observed significant changes in the chronology of the breeding season for male elephant seals between 1982 and 1986 (Tables 3 and 4). For all males combined we found mean arrival dates significantly later in 1983, 1984, and 1985 compared to 1982 and 1986, but no differences in departure dates or total days spent ashore (Table 4). To further analyze changes in the male breeding chronology, we separated males into four age categories: bull (8+ years), SA^4 (7–8 years), SA^3 (6–7 years), and SA^2 (5 years). For the oldest animals, the bulls, we found no significant differences in arrival dates, departure dates, or total days ashore; nor were there differences for the youngest animals, the SA^2's (Tables 3 and 4). However, the SA^4's arrived significantly later in 1984 than in 1982 or 1986, and spent significantly less time ashore in 1984 compared to 1982 or 1983, although departure dates were not significantly different (Table 4). SA^3's also arrived significantly later in 1984 than in 1982 or 1986, and remained ashore significantly fewer days in 1984 and 1985 than in 1982 (Table 4). In 1985 and 1986 SA^3's departed earlier than in other years (Table 4).

In comparing arrival, pupping, and departure dates for all female elephant seals arriving on Sand Flat, the major pupping area on Southeast Farallon, 1982 to 1986, we found no differences in the chronology of different years (Tables 5 and 6). In 1984 the mean arrival dates, mean pupping dates, and mean departure dates for pregnant females were later than in other years, but not significantly (Table 6). In comparing the number of days females spent on the island, we found no differences in amount of time females spent on the island postpartum, nor total time spent on the island (Table 6). We found that females spent a longer time ashore prepartum in 1986 than in 1982 and 1983 and, although the difference was statistically significant (Table 6), since the difference between the highest and lowest means was less than 1 day, we do not think the difference is biologically important.

For further analysis of female chronology we compared arrival, pupping, and departure dates for continuous breeders (females which pupped all 5 years between 1982 and 1986). These females arrived, pupped, and departed significantly later in 1984 and in 1986 than in other years (repeated measures ANOVA: arrival $F = 6.26$, $P = 0.003$; pupping $F = 7.86$, $P = 0.001$; departure $F = 6.07$, $P = 0.004$), but there was no difference in amount of time spent on the island (repeated measures ANOVA prepartum: $F = 0.581$, $P = 0.681$; postpartum $F = 0.575$, $P = 0.685$; total days ashore $F = 0.748$, $P = 0.574$). Because so many females suppressed reproduction in 1984 and 1985, we compared females that skipped breeding seasons but were present in both 1982 and 1986 with those continuous breeders present all 5 years, 1982 to 1986. Arrival, pupping, and departure dates were significantly later in 1986 for continuous breeders than for females that suppressed reproduction for 1 or 2 years (Mann-Whitney U-test: arrival $U = 340$, $P = 0.001$; pupping $U = 298$, $P = 0.003$; departure $U = 203.5$, $P = 0.024$). There was no difference between the two groups in 1982. We also found no difference between the two groups in number of days spent prepartum or postpartum or in total time ashore.

24.4 Discussion

Since year-round observations of the marine mammals and birds on the South Farallons began in 1972, there have been a number of "warm water" years. These normal perturbations of the oceanic environment have affected marine species differently depending on how warm the waters are, how long the warm waters remain, and how tolerant various species are to changes in their environment. The 1982–83 EN was

Table 5. Mean arrival, pupping, and departure dates and mean number of days prepartum, postpartum and total days ashore during the breeding season for elephant seal females on South Farallon Islands, 1982–1986 (J = January, F = February)

	Arrival			Pupping			Departure			Prepartum			Postpartum			Ashore		
	$\bar{x}$	SD	n	$\bar{x}$	SD	n	$\bar{x}$	SD	n	$\bar{x}$	SD	n	$\bar{x}$	SD	n	$\bar{x}$	SD	n
1982	11J	±9.37	137	16J	±10.07	90	12F	±10.02	96	4.89	±2.03	89	26.51	±3.43	89	31.47	±3.99	95
1983	13J	±10.24	178	18J	±9.88	173	14F	±10.83	171	4.72	±1.92	173	27.37	±4.26	166	32.08	±4.91	171
1984	13J	±9.22	127	19J	±9.06	127	15F	±9.70	127	5.22	±2.02	127	27.24	±2.96	127	32.46	±3.30	127
1985	12J	±10.03	103	17J	±9.46	103	14F	±9.52	103	5.27	±2.51	103	27.07	±3.01	103	32.34	±3.79	103
1986	13J	±8.89	95	18J	±8.68	95	14F	±9.49	94	5.84	±2.14	95	26.68	±3.55	94	32.5	±4.04	94

Table 6. Results of ANOVA for female elephant seal breeding season chronology on the Farallon Islands, 1982–86

Hypothesis	Statistic	*P* value[a]	Tukey multiple range test
No difference in arrival	$F_{4,635} = 0.91$	0.458	
No difference in pup date	$F_{4,583} = 1.67$	0.155	
No difference in departure	$F_{4,586} = 1.57$	0.180	
No difference in days prepartum	$F_{4,582} = 4.81$	0.002*	86 greater than 83 or 82
No difference in days postpartum	$F_{4,574} = 1.20$	0.310	
No difference in days ashore	$F_{4,585} = 1.05$	0.381	

[a]*Indicates significant difference.

stronger in severity and longer in extent than any of the other warm water phenomena since 1970 (Brinton 1981; McLain et al. 1985) and it had profound effects on the marine mammals (Huber et al. in press) and marine birds (Ainley and Boekelheide 1989) of the Farallons. Another warm water year, 1978, although not an El Niño year, appears to have had an effect on the Farallon elephant seal colony as well. Changes that may have occurred at the Farallon elephant seal colony during weaker EN years could have been masked by the very rapid growth of the colony due to immigration prior to 1978 (Huber et al. in press).

Decreased primary productivity affected the distribution of marine organisms during the 1982–83 EN (Barber and Chávez 1983, 1986; Pearcy and Schoener 1987; Adams and Samiere 1987, Samiere and Adams 1987; see Arntz et al. this Vol.). Since it is not yet known where elephant seals from the Farallon colony feed, or on what, it is difficult to assess just how modifications in prey abundance and distribution affected their feeding habits, but it is possible that variations in Farallon elephant seal biology in 1984 and 1985 are a result of changes in food availability due to alterations in the abundance and distribution of prey species or even the nutritive value of the available prey.

Because elephant seals fast during the winter breeding season, they would not be affected by changes in prey availability until spring, when pups first search for food on their own and adults are feeding to replenish fat supplies depleted during their 1- to 3-month fasts. Thus, elephant seals were not affected by the 1982–83 EN until spring of 1983. The warm water temperatures persisted well into 1984. If alterations in prey availability were present in spring 1983 and continued into 1984, then adult elephant seals would be most affected during the 1984 and 1985 breeding seasons and pups would be most affected in the springs of 1983 and 1984. This was the case with elephant seals at the Farallons.

Several studies indicate that female mammals in poor body condition suppress reproduction (Wasser and Barash 1983). Evidence for reproductive suppression in Farallon elephant seals in 1984 and 1985 are reduced numbers of pups born compared to previous and subsequent years, reduced natality among 3- to 10-year-olds, and increased incidence of stillbirths. Most adult females did not suppress reproduction, however, and although continuous breeders did arrive, pup, and depart 4 to 5 days later in 1984 than in other years, their total time ashore during the breeding season remained constant. We hypothesize that females may have arrived later in 1984 either because it was more difficult to find food before their extended fast and lactation period or because they were feeding further away from their usual feeding areas and it took them more time to get back to the Farallons. Interestingly enough, the differences in phenology were significant only for the subset of continuous breeders, there was no difference when we considered all females breeding at the Sand Flat. We also found, in comparing continuous breeders with females which suppressed reproduction for 1 or 2 years, that although there was no difference between the two groups in 1982, by 1986 continuous breeders arrived, pupped, and departed 4 to 5 days later than females which skipped years, implying some cost involved in breeding continuously.

Delayed arrivals among subadult males in 1984 may also be a result of complications involved in finding food in a warm water year. Although, for males, the mean arrival dates were up to 2 weeks later in 1984 compared to other years, departure dates were not significantly different over the years. Males usually leave a few days after the last estrous female departs, and, since the female phenology varies little, male departure dates were similar throughout the study. More variable than between-year variation was the variation for males of different ages. In general, the older the male, the longer he remained on the island, with 5-year-old males spending about half as much time ashore as the older age classes.

Both 1984 and 1985 were unusual years for the Farallon elephant seals during the breeding season, but for different reasons. In 1984 both males and continuously breeding females arrived significantly later and males spent significantly less time ashore; however, in 1985 many females suppressed reproduction completely and some males also skipped the breeding season. The depression in numbers of breeding animals present on the Farallons in 1985 cannot be explained completely by suppressed reproduction. Much of the growth in the Farallon colony is due to immigration, probably a result of crowding on rookeries at southern colonies. If crowding were lessened at the southern rookeries due to higher juvenile mortality in EN years, emigration would lessen. In that situation, numbers would not decrease on the larger rookeries such as San Miguel or Año Nuevo but would decrease at peripheral areas such as the Farallons which are not self-sustaining but rely upon immigration for their growth.

The numbers of male elephant seals and the proportion of each age class which makes up the Farallon colony has changed between 1976 and 1986. In the early years prior to 1983, the largest segment of the male population was young SA^2's (Table 2) and, based on tag returns, most of these were immigrants. After 1983 the number (and proportion) of young males decreased, which we attribute to the combined effects of lower rates of immigration and increased mortality among young animals. In the same way the number (and proportion) of bulls increased: few adult males move to other islands and unchanged return rates after 1983 and 1984 indicate no increase in mortality for adults in those years. The decline in the male population thus is due to lower numbers of young animals either because of death, emigration, or high juvenile mortality. A similar situation may be occurring among females in which the decrease in numbers present may be the result of fewer immigrants from the southern rookeries.

Acknowledgments. The elephant seal study on South Farallon Islands was supported by National Marine Fisheries Service (National Marine Mammal Laboratory and Southeast Fisheries Center), Marine Mammal Commission, and Point Reyes Bird Observatory (PRBO). Permission to work on the Farallons was given by U.S. Fish and Wildlife Service, San Francisco Bay Refuge, and U.S. Coast Guard, Twelfth District. Data analysis and write-up were supported by the National Marine Mammal Laboratory and PRBO.

L. Fry, S. Johnston, D. Judell, J. Nusbaum, A. Rovetta, and J. Swenson as well as other Farallon biologists and volunteers provided assistance and companionship in the field; Anne York provided statistical expertise. Ages of animals tagged on Año Nuevo, San Nicolas, and San Miguel islands were supplied by B.J. Le Boeuf and J. Reiter from the University of California, Santa Cruz. Bleach and dye used in marking animals were donated by Clairol, Inc., Stamford Conn. Earlier versions of this manuscript were improved upon by R. Merrick, E. Sinclair, P. Yochem, and A. York. Particular thanks are extended to Bob DeLong for advice and encouragement.

This is PRBO contribution No. 389.

25 Northern Elephant Seals on the Southern California Channel Islands and El Niño

B.S. STEWART and P.K. YOCHEM

25.1 Introduction

Northern elephant seals (*Mirounga angustirostris*) hauled out seasonally, and perhaps bred, on the southern California Channel Islands historically but they were locally extinct by the late 1800s (Stewart et al. in press). Several of the islands were colonized in the 1950s and births increased rapidly there through the early 1980s (Cooper and Stewart 1983), particularly on San Miguel Island which is now the largest rookery in the species' range and accounts for about half of all northern elephant seal births (Stewart 1989).

The Southern California Bight is a transition zone between the cold temperate Oregonian and warm temperate/subtropical Californian biogeographic provinces and, consequently, contains marine flora and fauna from both regions. Its most important oceanographic features are the surface currents, complex eddies, and related upwellings generated by the cool, southward flowing, low salinity, subarctic water of the California Current and the inshore, northward flowing California Countercurrent (Sverdrup et al. 1942; Reid et al. 1958).

Periodically, ocean warming events occur along the California Coast: California El Niño events occur coincidentally with tropical El Niño events and midlatitude warm events occur independently of tropical El Niño events (Norton et al. 1985).

Below, we discuss the influences of El Niño of 1982/83 (an intense tropical El Niño event) on several biological parameters of northern elephant seals at San Miguel and San Nicolas Islands. San Miguel Island (34°02′N, 120°23′W), about 38 km south of Point Conception, is the westernmost of the Channel Islands and San Nicolas Island (33°15′N, 119°30′W) lies about 100 km south of San Miguel Island and is about 92 km from the nearest mainland.

25.2 Methods

We observed the behaviors of breeding and molting elephant seals daily at selected sites and at least weekly at all others on San Nicolas Island from 15 December through 15 March, 15 April through 15 May and 1 June through 31 July each year from 1981/82 through 1986/87. During these observations we counted pregnant, lactating, and nonparous females and nursing, weaned, and dead pups. We spent between 2 and 6 weeks on San Miguel Island in January and February each year to document births and pup mortality, and to tag weaned pups. We used these data for the following analyses. Additional details of the study sites and methods can be found in Stewart and Yochem (1984) and Stewart (1989).

25.2.1 *Female Breeding Season Attendance and Spring Foraging Periods*

We determined the median date of arrival of adult females on San Nicolas Island during each breeding season using the "direct method" described for median birth date by Caughley (1977, p. 74).

We tested for the significance of skewness (Sokal and Rohlf 1981, p. 174–175) of female arrivals during each season and then used a multisample median test (i.e., a G-test of R × C independence; Zar 1974) to test the null hypothesis that the median arrival date did not differ among years.

We determined durations of pre- and postpartum periods ashore of tagged or marked (with bleach or hair dye) females during each breeding season and used a one-way ANOVA (Zar 1974) to test the null hypothesis that these periods did not differ among years. When the null hypothesis was rejected, we used a Tukey multiple range test (Zar 1974) to determine which years differed (significant at the 5% level).

We determined the durations of spring foraging periods of marked females by documenting their departure dates in January and February and their return dates in April and May.

25.2.2 *Physical Condition of Weaned Pups*

We tagged between 1500 and 2000 newly weaned pups each year at San Nicolas and San Miguel Islands and subjectively classified their physical condition as: 1 = every poor (emaciated), 2 = poor (thin), 3 = good (apparently well nourished and healthy), or 4 = superior condition (obese, "superweaner"; Reiter et al. 1978).

Using contingency table (chi-square) analysis we tested the following null hypotheses regarding the body conditions of weaned pups during each of the breeding seasons from 1981/82 through 1986/87:

1. The proportions of male and female pups in each condition category did not differ in each year;
2. The proportions of pups of each sex in each condition category did not differ among years; and
3. The proportions of individuals in each condition category did not differ between islands.

We used regression analysis to examine trends if the null hypotheses were rejected.

25.3 Results

25.3.1 Female Breeding Season Arrival and Attendance and Spring Foraging Periods

Pregnant females began arriving on the islands in mid-December of each year; female abundance ashore increased rapidly until about 18 January when it peaked, remained relatively constant for about 1 week, and then declined steadily during the first week of February as females weaned their pups and returned to sea (Stewart 1989). Patterns of female arrivals were significantly negatively skewed in all years (Stewart 1989). However, they arrived 5 to 8 days later in 1982/83 than in other years (Table 1). Females gave birth sooner after coming ashore, spent fewer days nursing their pups, and were ashore fewer days overall in 1982/83 and 1983/84 than in other years (Table 2).

Females that successfully weaned their pups spent more time at sea in spring each year than did unsuccessful females but both were at sea significantly longer in spring 1983 than they were in spring of 1982 and 1984 (Table 3).

Table 1. Median arrival dates of females on San Nicolas Island during the breeding season[a] (SE_{med} = standard error of the median, n = sample size)

Year	Median date of arrival	SE_{med} (days)	n
1981/82	13 January	0.7	234
1982/83	18 January	0.8	282
1983/84	13 January	0.8	234
1984/85	10 January	0.6	317
1985/86	11 January	0.7	341

[a]Median date of arrival in 1982/83 is significantly later than all other years (G = 5.43, df = 3, $0.25 > p > 0.10$).

Table 2. Duration of presence of parous northern elephant seal females on San Nicolas Island (SD = standard deviation, n = sample size)[a]

	Prepartum period ashore (days)			Parturition to departure (days)			Arrival to departure (days)		
Year	Mean	SD	n	Mean	SD	n	Mean	SD	n
1981/82	6.8	1.2	76	26.8	2.5	68	34.0	3.7	63
1982/83	5.9	1.2	62	25.1	2.0	56	31.7	2.9	51
1983/84	6.0	1.1	59	24.9	2.0	57	31.2	2.9	50
1984/85	6.8	1.3	67	26.9	2.4	61	34.0	3.4	60
1985/86	7.0	1.4	46	26.7	2.3	35	34.9	2.8	31

[a]1982/83 and 1983/84 differ significantly from other years for each parameter (Tukey multiple range test at $p < 0.05$): prepartum period, F = 9.176, df = 308, $p < 0.001$; parturition to departure, F = 10.828, $p < 0.001$, df = 272; arrival to departure, F = 11.055, df = 248, $p < 0.001$).

Table 3. Spring foraging periods (days) of northern elephant seal females from San Nicolas Island (SD = standard deviation, n = sample size)[a]

Year	Successful females[b]			Unsuccessful females[c]		
	Mean	SD	n	Mean	SD	n
1982	75.3	8.5	33	56.8	7.8	16
1983	85.7	10.3	28	65.3	9.5	15
1984	76.6	7.9	26	56.9	5.8	13

[a]Differences between successful and unsuccessful females were significant for each year (1982, $F = 29.59$, $df_1 = 1$, $df_2 = 48$, $p < 0.001$; 1983, $F = 24.09$, $df_1 = 1$, $df_2 = 42$, $p < 0.001$; 1984, $F = 65.31$, $df_1 = 1$, $df_2 = 38$, $p < 0.001$).
[b]Females that successfully weaned their own offspring; foraging periods were significantly longer in 1983 than in 1982 and 1984 although those in 1984 did not differ from those in 1982 ($F = 5.689$, $df_1 = 2$, $df_2 = 86$, $p = 0.007$; Tukey multiple range test at $p < 0.05$).
[c]Females whose pups died or were permanently separated from them during the breeding season; foraging periods were longer in 1983 than in 1982 or 1984 ($F = 3.695$, $df_1 = 2$, $df_2 = 43$, $p = 0.04$; Tukey multiple range test at $p < 0.05$).

25.3.2 Births

In 1982/83 births declined about 18% at San Nicolas Island but increased about 12% at San Miguel Island (Fig. 1). They then increased at both islands in 1984 and 1985, declined in 1986, and then increased slightly in 1987 (Fig. 1).

25.3.3 Pup Mortality

Preweaning pup mortality on San Nicolas Island was less than 4% in all years except in 1982/83 (Stewart 1989) when record high tides on 27–31 January (coinciding with peak female and pup abundance) were intensified by heavy storm surf and higher sea level, resulting in greater pup mortality and altered adult male and female distributions on many breeding areas. About 20% of the pups alive then died during that 4-day period; many pups were washed out to sea and drowned while others were permanently separated from their mothers and died within days from trauma and starvation. Storm-related mortality varied among sites (Fig. 2) depending on beach topography. The effects were most dramatic at site 231 (see Stewart and Yochem 1984), a beach that is backed by a steep bluff preventing access to higher ground. No pups died there before 27 January but after the storm tides from 27 to 31 January, when seals were frequently washed about in 1 to 2 m deep surf and beaten on rocks, 93% were dead; 9 pups drowned immediately, 84 were washed out to sea and presumably drowned, and 9 others were separated from their mothers, beaten against nearshore rocks and later died from trauma, starvation, or both. Eight weaned pups survived despite being constantly beaten against rocks by the surf. By 29 January, all females had deserted the beach and, evidently, the island; none of the tagged or marked females was later seen at another site.

Storm-related pup mortality was also high (81%) at site 232 (Stewart and Yochem 1984) were 91 pups were washed out to sea and 3 others were separated

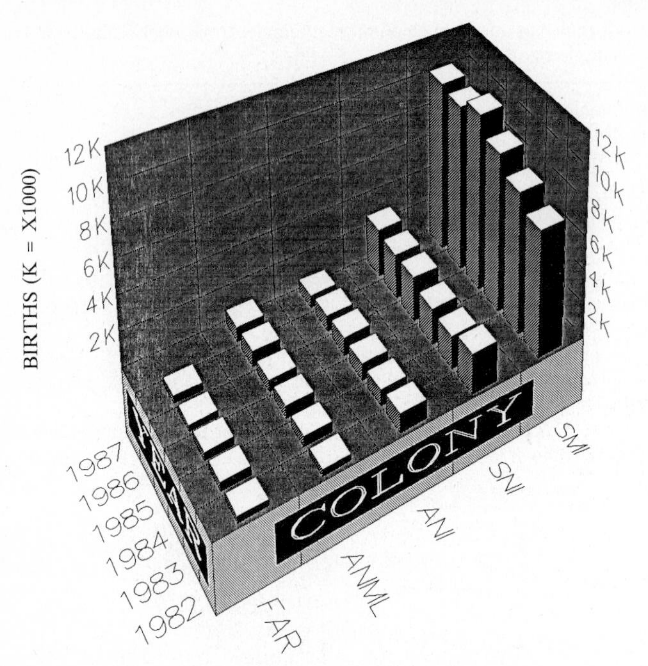

Fig. 1. Northern elephant seal births on the California Islands. *FAR* Farallon Islands (Huber et al., Chap. 24); *ANI* Año Nuevo Island, *ANML* Año Nuevo mainland (Le Boeuf and Reiter, Chap. 23); *SNI* San Nicolas Island, *SMI* San Miguel Island (Stewart 1989)

from their mothers. Only a few females and their pups managed to escape to higher ground.

Seals on other beaches were affected differently depending on the availability of and access to areas above the beaches. At sites where females and pups could move to higher ground (e.g., 221, 233, 262), pup mortality was much less (Fig. 2). These relocations resulted in fragmentation of large breeding groups, many isolated from each other, which in some cases led to redistribution of breeding males. Previously subordinate males became dominant in some small groups of females and were able to breed.

Preweaning pup mortality on San Miguel Island was generally less than 4% each year from 1981 through 1987 but in 1983 it was about twice as high (about 6.2%) as in other years; mortality was greatest at narrow, steep-backed beaches (Stewart 1989).

25.3.4 Physical Condition of Weaned Pups

Physical condition of pups differed between sexes in some years ($p < 0.003$), among years for each sex ($p < 0.05$), and between islands for some years ($p < 0.05$; Stewart 1989). Overall, there were declines in the proportions of males and females in very

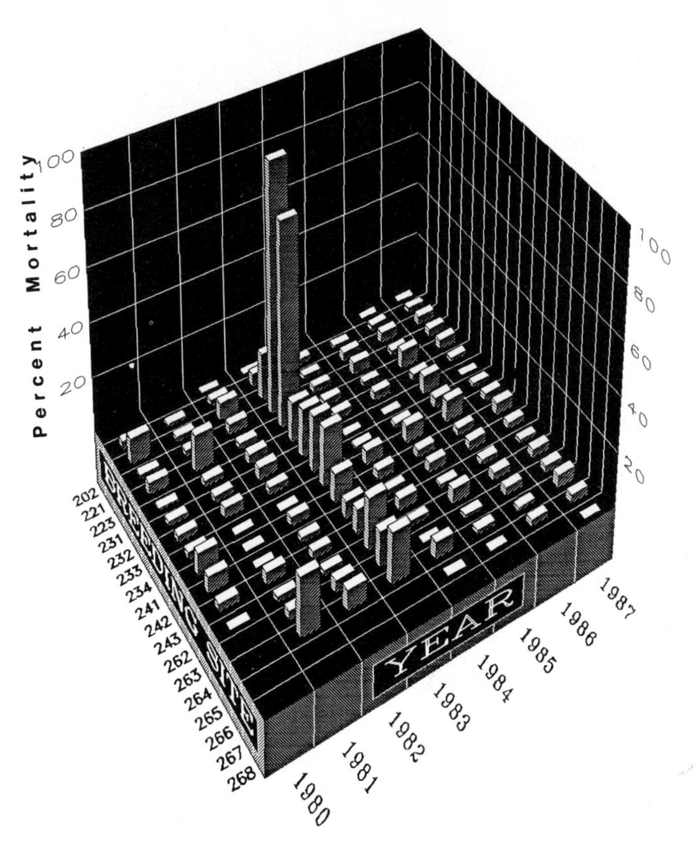

Fig. 2. Preweaning pup mortality of northern elephant seals at various breeding sites on San Nicolas Island

poor ($r^2 > 0.69$ in each case) and superior ($r^2 > 0.59$ in each case) condition, little change in the proportions in poor condition ($r^2 < 0.31$ in each case) and increases in the proportions in normal or good ($r^2 > 0.69$ in each case) condition (Figs. 3, 4, 5, 6). An exception was at San Miguel Island in 1983 when the proportions of male and female pups in poor condition were greater than in other years ($p < 0.01$ in each case).

25.4 Discussion

Pregnant females arrived about 1 week later than usual at San Nicolas Island in 1983, nursed their pups fewer days in 1983 and 1984, and were at sea feeding longer in spring 1983 than in other years. Evidently, females were in poorer physical condi-tion in 1983 and 1984 because of decreased food abundance or altered prey distribution, which may have required them to range farther to feed from late 1982 through early 1984. The higher preweaning pup mortality in 1983 was mostly attributable to

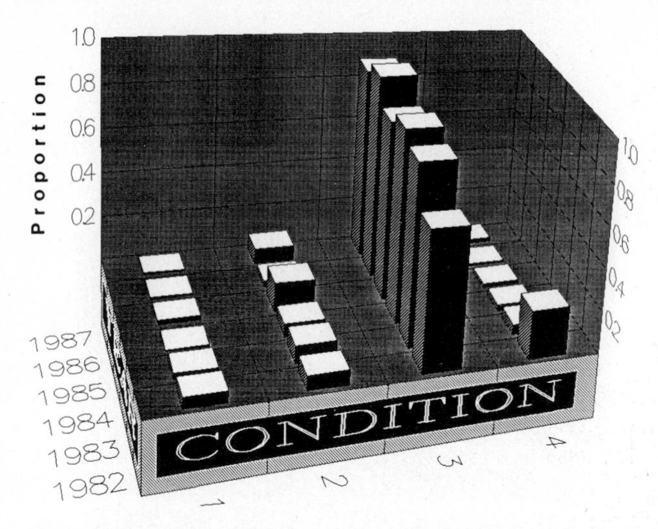

Fig. 3. Body condition of male weaned pups on *SNI* (San Nicolas Island): *1* very poor; *2* poor, *3* healthy; *4* obese

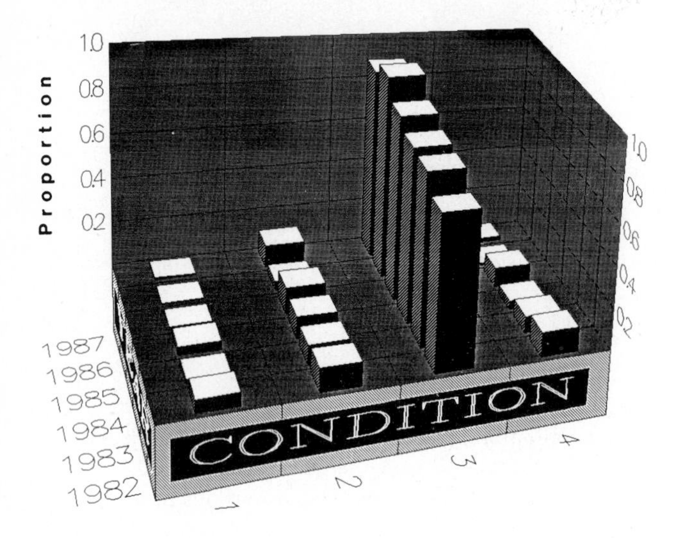

Fig. 4. Body condition of female weaned pups on *SNI* (San Nicolas Island): *1* very poor; *2* poor; *3* healthy; *4* obese

intense storms and astronomically high tides that occurred during peak breeding season. Similarly, the slightly poorer condition of weaned pups at San Miguel Island in 1983 was due to repeated disturbance to females and young pups during frequent storm-tide flooding of the large rookeries at Point Bennett. Even so, the condition of weaned pups was not substantially poorer in 1983 and 1984 to suggest that their subsequent survival would be less than that of other cohorts.

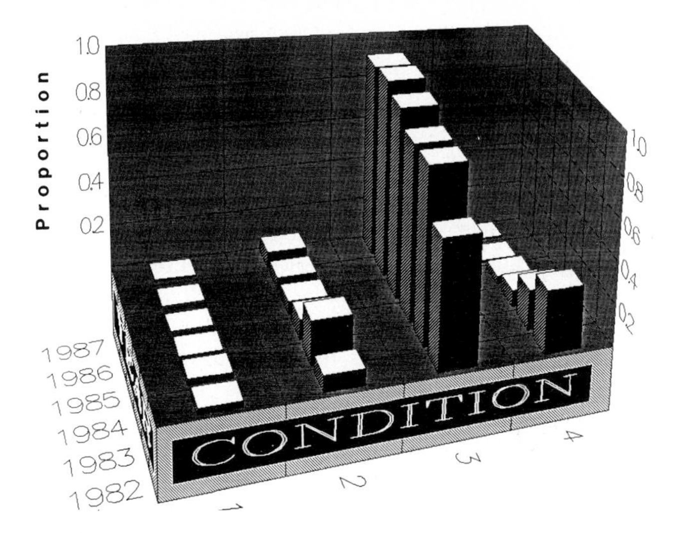

Fig. 5. Body condition of male weaned pups on *SMI* (San Miguel Island): *1* very poor; *2* poor; *3* healthy; *4* obese

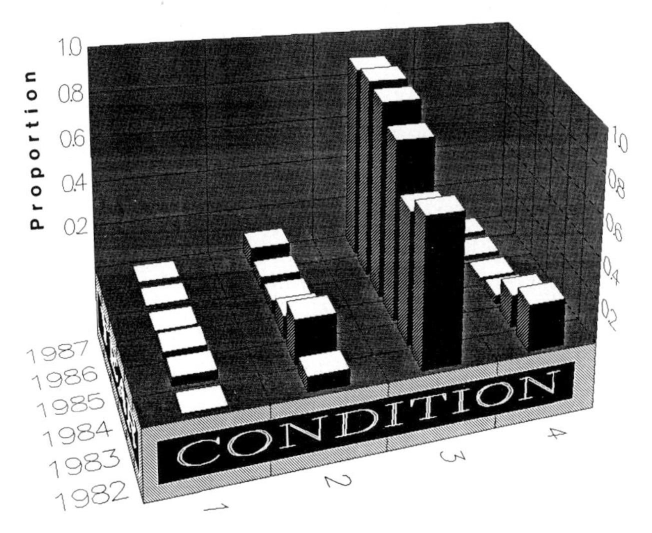

Fig. 6. Body condition of female weaned pups on *SMI* (San Miguel Island): *1* very poor; *2* poor; *3* healthy, *4* obese

By early 1983, sea surface temperatures were anomalously high from California to the Gulf of Alaska and into the eastern Bering Sea (Cannon et al. 1985; Niebauer 1985; Norton et al. 1985; Royer 1985; Tabata 1985). Upwelling and secondary productivity decreased in some areas and some large marine vertebrates other than northern elephant seals showed evidence of food shortages in 1983. These included:

1. Reproductive failure and high adult mortality of some seabirds that feed on planktivorous fish (e.g., cormorants, murres) were high in 1983 (e.g., Hodder and Graybill 1985), but species that fed on epibenthic fish (e.g., guillemots; Hodder and Graybill 1985) or had more varied diets (e.g., Western Gulls; Stewart et al. 1984) were less influenced;
2. Premature California sea lion births increased greatly (Stewart and Yochem, unpubl. data);
3. Births of California sea lions and northern fur seals decreased significantly (DeLong and Antonelis, this Vol.; Stewart et al. in press);
4. The diet of harbor seals at San Nicolas Island changed (Stewart and Yochem 1985) as did, evidently, the amount of time seals were at sea foraging (Yochem and Stewart 1985; Yochem 1987); and
5. Feeding trip durations of postpartum California sea lions and northern fur seals lengthened, and female California sea lions evidently dove deeper than usual to find prey (Feldkamp et al., this Vol.).

The increased sea surface temperatures, depressed thermoclines, and changes in nearshore current patterns in 1982 and 1983 had significant effects on the primary and secondary productivities in areas where many of these predators forage (e.g., Wooster and Fluharty 1985; Fahrback et al., this Vol.; Arntz et al., this Vol.).

Elephant seals feed at relatively great depths (female $\bar{X}$ = 333 m, Le Boeuf et al. 1986; male $\bar{X}$ = 389 m, DeLong and Stewart in press) away from the continental shelf (and therefore away from coastal upwelling systems; Stewart and De Long 1989), where their prey (pelagic squids, elasmobranchs and myctophid fish; Antonelis et al. 1987; DeLong and Stewart in press.) may have been less influenced by fluctuations in upper water column productivity. However, there were evidently changes in either the distribution or abundance of prey, or both, from late 1982 through mid-1984 that were sufficient to influence the physical condition of pregnant females. Births at San Nicolas and San Miguel Islands, combined, increased only 5% from 1982 to 1983 (Stewart 1989), substantially less than the average annual rate of increase of about 14% prior to 1982 (Cooper and Stewart 1983), suggesting that females were less productive in 1983 than in earlier years. The large increases in births from 1983 to 1984 (12%) and from 1984 to 1985 (14%) suggest, however, that the interruption in 1983 was due to lower pregnancy rates and, perhaps, increased abortions among females in 1982 rather than to increased mortality of adults. Indeed, Huber et al. (this Vol.) found no evidence for increased mortality of adult females at the Farallon Islands after 1983 although pregnancy rates evidently declined. Births declined about 4.5% on the Channel Islands and about 3% overall in California waters in 1986 (Fig. 1) and then increased only slightly (about 4%) on the Channel Islands in 1987 (Fig. 1; Stewart 1989); these trends suggest poorer than usual recruitment or survival, or both, of the 1982 through 1984 cohorts (see Huber et al., this Vol.; Le Boeuf and Reiter, this Vol.). Nonetheless, births have continued to increase on the Channel Islands in recent years (45% more pups were born there in 1989 than in 1982), indicating that any retarded growth or greater mortality of juveniles caused by the unusually strong 1982/83 El Niño had relatively short-lived demographic consequences. The greatest effect of periodic ocean-warming events appears to be disruption of the nutrient-rich upwelling systems that are critical to some shallow-feeding pinnipeds (e.g., Galapagos fur seals and sea lions, California sea lions). Even major disruptions like the one that occurred in 1982/83 evidently have only mild influences on the demography of northern elephant seals which appear to be relatively independent of such upwelling systems.

Acknowledgments. W. Ehorn and his staff at Channel Islands National Park facilitated access to San Miguel Island and R. Dow, his staff, and the Naval Command at Point Mugu Naval Air Station and at San Nicolas Island Outlying Landing Field permitted access to San Nicolas Island; we thank them for their support. Clairol Research Laboratories (Stamford, CT) generously supplied us with hair dye and bleach supplies. We thank J.R. Jehl, J.D. Hall, C.E. Taylor, G.A. Bartholomew, M.L. Cody, W.F. Perrin, N. Blurton-Jones, R.L. DeLong and two anonymous reviewers for their comments on the manuscript. The research was conducted under authorization of the Marine Mammal Protection Act of 1972 by Marine Mammal Permits Nos. 341, 367, and 579 and supported by contracts from the United States Air Force and by a University of California Research Fellowship to BSS.

Part V
Summary and Conclusion

26 The Effects of El Niño on Pinniped Populations in the Eastern Pacific

F. TRILLMICH, K.A. ONO, D.P. COSTA, R.L. DELONG, S.D. FELDKAMP, J.M. FRANCIS, R.L. GENTRY, C.B. HEATH, B.J. LEBOEUF, P. MAJLUF, and A.E. YORK

26.1 Introduction

Most hypotheses on adaptation stem from comparisons among species. The comparative method offers a way to study questions of ecological mechanisms and evolutionary adaptations, particularly in situations where experimentation is near impossible. Comparisons among species and areas in itself can advance the perspective greatly, but this approach suffers from unknown and uncontrollable variation among species and regions. Adding another dimension of comparison, a large ecological disturbance, to the system is extremely illuminating because it acts independently of preexisting species level and spatial variation and, therefore, simplifies comparisons by eliminating confounding variables. Thus, the chapters in this book are of interest because they not only chronicle the impact of El Niño (EN) on eastern Pacific pinnipeds, but they also provide new perspectives on the ecology, life history, and evolution of these species which could not otherwise have been obtained.

In the preceding chapters we described and analyzed the effects of EN on pinniped populations distributed over the Pacific east coast of the Americas from the Bering Sea (ca. 57°N) through California, Mexico, Ecuador (Galapagos), and Peru to northern Chile (ca. 23°S). We presented data on six species of pinnipeds, the northern fur seal (*Callorhinus ursinus*), the Galapagos fur seal (*Arctocephalus galapagoensis*), the South American fur seal (*Arctocephalus australis*), the South American sea lion (*Otaria byronia*), the California sea lion (*Zalophus californianus*), and the northern elephant seal (*Mirounga angustirostris*). Not only can we compare the effects of EN between two families of pinnipeds, the otariids and phocids, and between species of otariid seals, but we are also fortunate to have studies on different populations of the same species.

1. In the northern fur seal we can compare the colonies on the Pribilofs (at 57°N) and on San Miguel Island (34°N).
2. The South American fur seal was studied in Peru (15°S) and Chile (20–23°S), and the very similar Galapagos fur seal on the Galapagos (0°S).
3. For the California sea lion, data were obtained on the most varied aspects of behavior, physiology, and life history of any species in the book. These data originated from work on the California Islands, at the northernmost edge of the breeding range (33–34°N), and can be compared to information on the populations from Baja California (23°–28°N; the southernmost breeding range of *Z. c.*

californianus) and from the Galapagos Islands (at 0°N/S). The latter population is accorded subspecific status (*Z. c. wollebaeki*).

4. Lastly, we have detailed data on northern elephant seals from the center of their distribution on the California Channel Islands (33°–34°N) to the northern edge of their range at Año Nuevo Island (37°N) and on the Farallon Islands (38°N).

   Unfortunately, we have been unable to obtain substantial information on EN effects on the population of Steller sea lions (*Eumetopias jubatus*) in the north and obtained only limited information on the South American sea lion (*Otaria byronia*) in Chile and Peru (Guerra and Portflitt, Chap. 4; Majluf, Chap. 5). Data on the Harbor seal (*Phoca vitulina*) are also unavailable at present.

To structure this material for the following discussion (Sect. 26.2), we will first summarize our results (see Tables 1–3) with several general questions in mind:

1. How did effects on population parameters change with latitude?
2. What were the time courses of the various effects?
3. Were age groups and sexes differentially affected, and if so, how?
4. How did foraging behavior change during EN?

In Section 26.3 we then use this information on similarities and differences between species, and between sites within a species, to examine the stages in pinniped like cycles which suffer most from environmental disturbances. We then attempt to draw conclusions on the importance that environmental variance may have for the evolution of life history traits of phocids and otariids exposed to latitudinally different environmental regimes.

26.2 Summary

It is generally difficult to prove that changes in population parameters are directly caused by unusual environmental circumstances (Arntz et al., Chap. 2). This problem was demonstrated by Hatch (1987) in his report on seabirds in the Bering Sea: he observed substantial breeding failures during EN which were difficult to interpret since similar failures had been previously observed in non-EN years. Differential effects of EN among seabird species were weakly related to their foraging tactics, but showed marked geographical variation for a given species even within the area of the Bering and Chukchi Sea and the Gulf of Alaska. The variation in the response of these populations emphasizes how risky it is to ascribe changes in population parameters directly to EN. We are confronted with the same problem in the interpretation of our pinniped data.

Our data base, although in most cases encompassing several years, is not comprehensive enough to address all short- and long-term effects of this EN. Furthermore, most of our observations were made at the breeding sites and do not provide information on the nonbreeding season, during which extensive feeding occurs over wide areas (but see York, Chap. 9; Huber, Chap. 13). Since the migration period figures importantly in the life cycle of most pinniped species we are likely to underrate the importance of EN effects in our conclusions.

Table 1. Summary of the information on fur seals[a]

	C. ursinus Pribilof Islands	*C. ursinus* San Miguel Island	*A. galapagoensis* Galapagos	*A. australis* Peru
Latitude	57°N	34°N	0°	15°S
Pupping season	June-July	June-July	Sept.-Oct.	Oct.-Dec.
Pup mortality	No change	Increase	1982 100% Loss 1983 Normal	At least 41% increase
Juvenile mortality	Decreased by 15%	Presumably more than doubled in 1982 and 1983	100% Loss of 1980 and 1981 cohorts; partial loss of 1979 cohort	Increase
Adult female mortality	No change	50% Population decline mortality or decreased immigration? Increased emigration?	At least 30% lost in 1982–83	No exact data; most likely increase
Adult male mortality	No change	20% Less males; cause unclear	1982 100% Loss 1983 Back to normal	No quantitative data presumably increased
Number of pups born	No change	60% Decrease 1983 50% Decrease 1984	Normal 1982 89% Decrease 1983	No data
Pup growth rate	No data	1983 ca. 40% Decreased 1984 ca. 25% Decreased	1982 No growth 1983 Normal	1983 Decreased by 50%
Time to weaning	No data	No data	1982 All dead 1983 Normal	No data
Foraging trip duration	No change	Increased in 1983 Slight increase in 1984	About 3 times longer with larger variance	Increased
Prey composition	No change	Changed in 1983	Changed in 1983/84	Changed in 1983/84
Migration/ movement	No change on land	Local redistribution; more northward migration of juveniles	Local redistribution perhaps immigration into best habitats	Some movement into and establishment in northern Chile
Population recovery	No change	Still incomplete in 1987	Very slow	Established in Chile No date for Peru

Before we summarize our results, we need to define two terms for this summary. *Pups* are young animals which depend *entirely* on maternal milk for their energy input; juveniles are either recently weaned or living on a mixed diet of maternal milk and solid food for which they forage independently. Depending on the species, juveniles may be 4 weeks (*M. angustirostris*) to 3 years (*A. galapagoensis*) old. This definition stresses the change from total dependence on maternal care to partial or complete nutritional independence rather than absolute age (which is also used elsewhere in this book).

26.2.1 The Relationship Between the Impact of the EN Event on Population Parameters and Latitude

Because EN originates near the equator and is propagated poleward, the severity of EN-related impacts on pinniped populations declines from low to high latitudes

Table 2. The influence of El Niño on sea lions at various study sites (listed from N to S)[a]

	Channel Islands	Mexican Pacific	Sea of Cortez	Galapagos
Latitude	33–34°N	24°N	24–26°N	0°
Pupping season	June/July	June/July	June/July	Mostly Sept./Oct.
Adult female mortality	Slight?	No data	No data	Some?
Adult male mortality	No indication	No data	No data	Increased from 1982–83
Juvenile mortality	Increased on SC in 1984; no trend on SM	No data	No data	Strongly increased certainly >50%
Pup mortality	1982: 10% 1983 and 1984: 17% many abortions	No data	No change	ca. 100% in 1982 Normal (<10%) in 1983
Foraging trip duration	Increased in 1983 Almost normal in 1984	No data	No data	No data
Pup growth rate	Reduced in 1983 and 1984	No data	No data	No data
Number of pups born	30–70% Reduced in 1983	45% Reduced in 1983	No change	Reduced in 1983
Time to weaning	1984 More yearlings suckling	No data	No data	1982 Cohort lost
Migration/ movement	More animals migrate north	Less males on coast in 1982/83	No change	Unusual sightings on mainland Ecuador
Population recovery	Pup numbers in 1986 still < 1982	No data	No change	slow

[a]SC, San Clemente Island; SM, San Miguel Island, SN, San Nicholas Island.

Table 3. The influence of El Niño on elephant seals at three study sites (years refer to the time when young weaned)

	Farallons	Año Nuevo	San Nicolas Island
Latitude	38°N	37°N	33°N
Pupping season	Dec./Jan.	Dec./Jan.	Dec./Jan.
Adult female mortality	No change	No change	No data
Adult male mortality	No change for dominant males; increase for subadults	Slight increase	No data
Juvenile mortality	Increase for 1982, 1983 and 1984 cohorts	Increased for 1982, 1983 and 1984 cohorts	No data
Pup mortality	No change; more stillbirths 1984 and 1985	Increase to 70% by storms in 1983	Increase from 4 to 20% in 1982/83
Time at sea between lactation and molt	No data	Increase in 1983 and 1984	Increase in 1983
Pup mass	No data	No change 1982 and 1984 No data 1983	No data
Number of pups born	Decrease in 1984 and 1985	No change	No data
Lactation duration	No change	No change	Shorter in 1983 and 1984
Skipped breeding season	More males and females skipped the 1984/85 breeding season	No data	No data
Age at primiparity	No data	Increase in 1983	No data

(Fahrbach et al., Chap. 1; Arntz et al., Chap. 2). At low latitudes EN effects were strongest during late 1982 until June 1983, whereas at temperate latitudes effects were strongest during the winter of 1982–83, lasted through 1983, and lingered on with strong aftereffects in 1984. It is instructive to examine how population parameters varied with latitude because this instantaneous picture of all populations is analogous to the responses that a population at one location may show to environmental perturbations of various magnitudes over time. This is suggested by the study of the South American fur seal in Peru where the strong EN of 1982–83 almost exterminated the 1982 cohort, whereas the weak EN of 1986–87 caused lower pup body mass at 1 month of age than in previous years, but apparently had no major effect on pup survival to 1 year of age (Majluf, Chap. 5).

We first summarize the effects on otariids and then on the elephant seal, the only phocid species for which we have detailed information.

Otariids

The effects of EN on otariid pup survival followed a latitudinal gradient. Pups experienced reduced milk intake during the years of EN effects, which led to reduced growth rate and associated reduced body mass for a given age. Consequently, survival of pups during their first 3 months of life decreased (Tables 1 and 2). The most severe negative effect was observed in the Galapagos Islands, where 100% mortality of fur seal pups in 1982 and nearly 100% mortality in the sea lions occurred (Table 1; Trillmich and Dellinger, Chap. 6). In the near-tropical population of the South American fur seal in Peru, 42% of pups died within a 22-day observation period in February 1983 (Trillmich et al. 1986; Majluf, Chap. 5). However, survival of pups in the insulated population of California sea lions in the Sea of Cortez did not appear to be affected (Aurioles and Le Boeuf, Chap. 11). Further north in more temperate latitudes on the California Channel Islands (about 33°–35°N) the 1982–83 EN had a less drastic effect on pup survival. In 1983 and 1984, it approximately doubled normal early postpartum pup mortality among California sea lions (Francis and Heath, Chap. 12; DeLong et al., Chap. 17) and northern fur seals on San Miguel Island (DeLong and Antonelis, Chap. 7). The very high latitude Pribilof population of northern fur seals was also unaffected (Gentry, Chap. 8).

The loss of juveniles from these populations varied with latitude in the same way as pup mortality. The most equatorial species, the Galapagos fur seal, lost the 1980 and 1981 cohorts completely and part of the 1979 cohort. In the area of the California Channel Islands, high mortality of underweight weanlings during their first year at sea was documented in juvenile California sea lions and northern fur seals (Tables 1, 2). Although exact mortality estimates are missing, an increase in the number of beached yearling sea lions and fur seals was noted (J. Antrim, Sea World, San Diego, pers. comm.; DeLong and Antonelis, Chap. 7). At 52°N, the 1983 year class of northern fur seals was of normal size initially, but its subsequent survival to 2 years of age may have been greater than normal, because of enhanced food supplies available to juveniles after weaning as shown by York (Chap. 9) for previous EN events. This may have occurred despite the fact that the 1982–83 EN had a less beneficial effect on fish stocks in the north than previous EN events even though they were not as strong in the tropics (Niebauer 1988; Fahrbach et al., Chap. 1; Arntz et al., Chap. 2).

During most EN events, the mortality effects on pups and juveniles may not show such an extreme equatorial-temperate trend as in the 1982–83 event. Weaker events appear to affect near-equatorial populations much less severely than the exceptional 1982–83 EN (Majluf, Chap. 5). In fact, moderate and weak ENs may only cause reduction in growth rates of pups in tropical areas and may have no effects on temperate populations.

Mortality of adult females was also increased during EN and varied with lati-tude. The effect was largest in the tropics, where an estimated 30–50% of the female fur seals on Galapagos died (Table 1). The larger Galapagos sea lion females had a lower mortality rate than the fur seals (Trillmich and Dellinger, Chap. 6). We have no estimates of fur seal mortality in Peru, but it is clear that female mortality there was high in 1983. Some animals may have escaped starvation by migrating south into Chile (Guerra and Portflitt, Chap. 4). Movement to more productive areas was not an option for the Galapagos species.

The data for the Channel Island populations of otariid pinnipeds showed no increase of female mortality among sea lions (Table 2), but marked shifts in distribution (Francis and Heath, Chap. 12; DeLong and Antonelis, Chap 7; DeLong et al., Chap. 17). A dramatic decline in the number of northern fur seal females on San Miguel was noted (Table 1). Whether this decline was due to migration from San Miguel or mortality is unknown (DeLong and Antonelis, Chap. 7). Female survival of northern fur seals in the Bering Sea was unaffected by the EN (Gentry, Chap. 8).

Adult males of otariid species, which fast during territory tenure, went to sea with their fat stores seriously depleted. Again, the effect of this depletion differed by latitude. The virtually complete loss of reproductive male Galapagos fur seals indicates how risky the reproductive fast is for males of polygynous pinnipeds (Table 1; Trillmich and Dellinger, Chap. 6). There was less mortality among Galapagos sea lion males than among the fur seals, but the extent of the mortality was not quantified (Trillmich and Dellinger, Chap. 6). In contrast, censuses for the California sea lion on the Channel Islands indicated no measurable change in adult male mortality (Table 2; Ono et al. 1987; Francis and Heath, Chap. 12). Since these males migrate north after the breeding season they may have been able to avoid the worst EN effects.

Female fertility decreased in almost all otariid species during EN (Tables 1, 2). Reproductive rates of adult females decreased most for the Galapagos fur seal (70–90%). Pup production of California sea lions on the outer coast of Baja California and on the Channel islands was reduced by about 50% in the first breeding season following the onset of EN (Table 2). These lower natalities were probably a result of undernutrition of females (Francis and Heath, Chap. 12). Female condition was not directly measured but appeared to be poor in 1983 as reflected by decreased copulation rates and increased abortion, resulting in decreased female fertility in 1983 and 1984 (Francis and Heath, Chap. 12). Pup births in the northern fur seal in California fell by about 60%, but in this case we cannot discriminate between reduced female fertility, and massive emigration combined with some mortality of reproductive females (DeLong and Antonelis, Chap. 7).

Another possible factor leading to decreased numbers of pups late in or immediately after the EN year 1983 may have been an increase in the age at first reproduction in tropical and temperate species. When marine productivity is high, food for recently weaned juveniles is abundant and causes fast growth which leads to early maturation as seen in northern fur seals (York 1983). Conversely, reduced growth early in life may lead to a later onset of reproduction. Juvenile females of the cohorts, which were born, maturing, or reproducing for the first time during EN, may have delayed their first pupping due to reduced food availability. Evidence for this effect was observed in northern elephant seals (see below).

Phocids

The range of our study sites of northern elephant seals was about 400 km, too limited to reveal a latitudinal trend in pup mortality (Table 3), but site effects were important. Pups of this species were well buffered against EN effects while they were still suckling. Only direct storm effects increased the mortality rate of pups during early 1983 and the effect was site-specific. At the Farallon Islands (38°N) pup mortality rates were normal (Huber et al., Chap. 24). At Año Nuevo Island (37.5°N), pup mortality roughly doubled (Le Boeuf and Reiter, Chap. 23). At San Nicolas Island (33°N), the pup mortality rate was five times above normal (Stewart and Yochem, Chap. 25). At both of the latter locations mortality increased due to a coincidence of unusually strong winter storms and high sea levels which accompanied EN at temperate latitudes (Fahrbach et al., Chap. 1). The same storms struck the Farallon islands, but mortality did not differ from non-EN years because the animals there were on high ground, away from the sea.

The difference in storm effects among these islands was a function of population density and beach topography rather than of latitude. Population density was highest at Año Nuevo (cf. Table 1 in Le Boeuf and Reiter, Chap. 23). When the animals had to move during storms, high population density led to more trampling of pups and a greater probability of mother-pup separation. If, in addition, the animals were unable to reach higher ground because of cliffs at the back of the beach, pups drowned. Therefore, increased mortality of northern elephant seal pups occurred where high population density, limited beach space, and cliffs behind the beach exacerbated the storm effects (Le Boeuf and Reiter, Chap. 23; Stewart and Yochem, Chap. 25).

In conclusion, EN 1982–83 had a highly negative influence on several pinniped species at or near the equator, and a moderately negative influence in the temperate zone. The effect was negligible for the variables monitored in the subpolar northern fur seal.

26.2.2 Differential Effects of EN on Age Classes and Sex

The impact of EN on pinnipeds was strongly influenced by the age and sex of individuals. Age is a good measure of physical maturation and behavioral experience, and is therefore expected to influence the resistance of an individual to environmental perturbations. All the species studied show strong sexual dimorphism in size, and size influences the diving abilities of pinnipeds (Kooyman 1989). Furthermore, the reproductive strategies of males and females of these species differ widely and exert different constraints on the patterning of foraging time and the migratory behavior of the sexes. The way in which EN affects a pinniped individual is therefore also expected to be influenced by its sex.

Otariids

During the lactation period, otariid pups rely on the foraging success of their mothers. Pup growth rate will depend on foraging conditions, if, at lower food availabilities, females do not allocate increasing amounts of their resources to pups. During acute food shortages these pups may be more susceptible to EN impacts than are juveniles and adults since pups cannot contribute independently to their energy intake. Consequently, pups suffer more than any other age class wherever EN-related changes in the marine environment have negative effects on maternal foraging efficiency (Tables 1, 2).

Male otariid pups demand more milk from their mothers than female offspring, mainly because male pups are bigger and grow faster than female pups (Kerley 1985; Costa and Gentry 1986; Trillmich 1986b: Oftedal et al. 1987b). This suggests another way in which females may cope with reduced food resources. According to the Trivers and Willard (1973) hypothesis, mothers may produce more female offspring during times of food scarcity. Previously reported data for the Galapagos fur seal (Trillmich and Limberger 1985; Trillmich 1986b) and the California sea lion (Ono et al. 1987) showed no significant relationship between environmental quality and secondary sex ratio measured during the following EN. However, two studies in this volume have indicated instances of sex bias in California sea lion pups in relation to EN. When all 4 years of study on California sea lions (1982–1985) were considered simultaneously, the secondary sex ratio was shown to be female-biased during EN on San Nicolas Island (Ono and Boness, Chap. 20). Also, the sex ratio among pups 2 months of age on San Nicolas and Santa Barbara Islands was weakly female-biased during EN compared to the 2 years prior to EN (Francis and Heath, Chap. 21). Sex ratio changes in the predicted direction were mentioned for northern elephant seals on San Miguel Island (Stewart and Yochem, Chap. 25), but were not corroborated by more extensive observation on the same species on Año Nuevo and the Farallon Islands. The problem of adaptive sex ratio shifts clearly needs more investigation since it is hard to imagine how pinniped females could predict foraging conditions 1 year ahead of time in order to shift sex ratio in an adaptive way at the time of conception. They would have to make such a prediction because there is a 1-year delay between fertilization of the egg and birth of the pup. Alternatively, females may resorb embryos of the wrong sex previous to parturition, and thereby lose a major part of their lifetime reproductive effort. Such a mechanism has not been documented in pinnipeds and would a priori seem to be of questionable adaptive value.

Temperate and tropical otariid species are flexible in the timing of weaning (Gentry et al. 1986a; Trillmich 1986b; Francis and Heath, Chap. 21). Under poor conditions these species delay weaning and can thus reduce, or at least delay, mortality of juveniles during periods of low food availability. In fact, more California sea lion yearlings were suckling following the EN than in other years (Francis and Heath, Chap. 21). While the proportion of suckling male yearlings increased in

1984, there was no significant change across years in the proportion of juvenile females suckling. Differences between years in migratory behavior of yearling males may account for this finding: males in 1984 were more likely to stay on their island of birth and therefore remain in contact with their mothers. In addition, their mothers were less likely to have new pups to compete for milk than in years before EN. This difference in migratory behavior of male yearlings may have accounted for the lack of sex difference in juvenile mortality. Older sea lion juveniles migrated in greater than usual numbers from southern to central California during 1983 and 1984 and may thus have avoided the worst impact of EN (Huber, Chap. 13). Data from the South American fur seal in Peru also indicate that under EN conditions weaning is delayed (Majluf 1987).

Adults were most resistant to EN-related changes since they were the most efficient foragers (see Sect. 26.2.3). Moreover, they sometimes reacted to changes in the marine habitat by emigrating (Tables 1, 2).

In otariids, the extent of the increase in adult mortality during EN was related to sex-specific differences in foraging and migration. After the reproductive season, males have to recover from the territorial fast. Similarly, females must restore their body reserves after the initial 7-day fast following parturition. Recovery from poor body condition depends not only on food resources after this stressful period, but also relates to a sex difference in foraging options. California sea lion males migrate north, into areas less affected by EN, immediately after the end of the breeding season. Consequently, if they have enough reserves to get away from the breeding grounds after the fast, they are free to forage wherever they find food most plentiful. In contrast, female sea lions continue to forage near the breeding colonies and do so under the additional stress of lactation. They would therefore seem to be in greater danger than males when feeding conditions deteriorate since they cannot escape by migrating away without potentially losing the young of the year. In Galapagos fur seals and sea lions, on the other hand, both sexes appear to be resident and forage in the same upwelling areas near the islands. Under these conditions reproductive males clearly suffered higher mortality than females (Tables 1, 2). The combination of a more marked EN effect together with a more restricted foraging environment in the tropics may explain why sea lion males on Galapagos were more likely to die during EN than their conspecifics in California.

Females showed a marked reduction in fertility after the EN, whereas no such effect was noted for males. If males survived the stressful reproductive period they were able to forage continually during the nonbreeding season. This apparently provided them enough opportunities to recover and store reserves for the next breeding attempt. In contrast, females must spend a large percentage of their time ashore to suckle their pup during which they obviously cannot feed and must also incur the cost of lactation (Ono et al. 1987; Heath et al., Chap. 14; Costa et al., Chap. 16; Boness et al., Chap. 18). As a consequence, they were often unable to produce the extra energy necessary for a simultaneous pregnancy.

Phocids

Data on northern elephant seals do not suggest major differential sex effects of EN. Mortality of pups and juveniles appeared to be indiscriminate with respect to sex.

The studies showed no negative effect of EN on lactation in elephant seals, and presumably pup growth during nursing was also normal. No data on pup mass are available for pups born in December 1982 to January 1983. Mean weaning mass in 1982 and 1984 was no different from preceding and succeeding years.

However, the impact of EN varied strongly with the age of elephant seals. Mortality of juvenile elephant seals was elevated for the 1982 (i.e., born Dec. 81-Feb. 82), 1983 and 1984 cohorts compared to previous and later years (Table 3; Le Boeuf and Reiter, Chap. 23; Huber et al., Chap. 24), an effect that was most marked for the 1983 cohort. Juveniles, who enter the sea after a 2.5-month postweaning fast, appeared to suffer from reduced food availability during their first year(s) at sea. Compared to previous cohorts, less than half the normal number of females from the 1983 cohort appeared to reach sexual maturity.

Mortality among adult female elephant seals did not change on Año Nuevo or at the Farallon Islands, but mortality of adult males increased on Año Nuevo and, at the Farallons, subadult males seemed to suffer even more than adults (Huber et al., Chap. 24; Table 3). This finding is unexpected since the fully adult, dominant males presumably spend more energy during the reproductive season fighting and guarding females than subadults and hence have more to recover.

EN reduced the probability of breeding, and thus reduced lifetime fertility, of both sexes (Table 3; Huber et al., Chap. 24). More females than usual skipped 1 or 2 years of pupping and a few males also skipped a breeding season on the Farallons (Huber et al., Chap. 24). The data do not reveal whether males or females were more likely to skip a breeding season. In addition, age at first reproduction also increased significantly for the 1983 female cohort at Año Nuevo Island, an effect that may have been caused by reduced growth rates of that cohort during the first years at sea.

26.2.3 Effects of EN on Foraging and Energetics of Mothers and Pups

Otariids

EN was obviously a time of food shortage for temperate and tropical pinnipeds. In otariids, lactating females were ideal subjects for studying the effects of food shortage on foraging behavior because they forage between periods of pup nursing. This restricts them to a foraging site relatively close to the rookery. A pup's physical condition is a sensitive indicator of its mother's foraging success.

During EN, female otariids stayed away from their pups increasingly longer between nursing visits on land (Tables 1, 2; Trillmich and Limberger 1985; Ono et al. 1987; Majluf, Chap. 5; Trillmich and Dellinger, Chap. 6; Heath et al., Chap. 14; Feldkamp et al., Chap. 15). The effects were most marked in the tropics where females more than doubled the time away from their pups. It is important to note the distinction between *time away* from the pup and *time at sea*. This is best demonstrated by the data for the California sea lion (Feldkamp et al., Chap. 15). Here, and in Galapagos fur seals (Trillmich and Kooyman unpubl.), females sometimes rested elsewhere on land before returning to the pup. This observation supports the hypothesis by Costa and Gentry (1986) and Costa et al. (Chap. 16) that females attempt to restore their physical condition to an upper threshold of body mass before returning to their pups to lactate.

During EN, the time mothers spent away from their pups lengthened to the extent that the energy budget of many pups became negative; they lost mass and finally died. Reduced milk transfer (Oftedal et al. 1987b; Iverson et al., Chap. 19) was the main reason for mass loss and this reduction was reflected in decreased suckling times (Ono and Boness, Chap. 20). In the Galapagos, the mean absence duration of mothers was no longer than under normal warm season conditions. Nevertheless, pups starved to death because the variance associated with the mean was very high (Trillmich and Dellinger, Chap. 6). In California sea lions, pups with mothers gone for more than 9 days died (Heath et al., Chap. 14). Although pups decreased their activities, these compensatory mechanisms were insufficient to balance the reduced energy intake during the peak of EN effects (Ono et al. 1987; Boness et al., Chap. 18; Iverson et al., Chap. 19; Ono and Boness, Chap. 20).

As a consequence of EN, foraging females were also faced with changes in prey composition and abundance while at sea (Arntz et al., Chap. 2). Changes in prey availability were reflected to a certain degree in changes in the diet of several species (Majluf, Chap. 5; Trillmich and Dellinger, Chap. 6; DeLong and Antonelis, Chap. 7; DeLong et al., Chap. 17). The most obvious example is the increase of northern anchovy in the diet of northern fur seals and California sea lions in 1983. This was apparently caused by a northward shift in the distribution of this species (Arntz et al., Chap. 2). On the other hand, the proportion of scats with market squid changed little from 1982 to 1983 (DeLong and Antonelis, Chap. 7; DeLong et al., Chap. 17) despite the crash of the fishery for market squid in central California (Arntz et al., Chap. 2). Northern fur seals and California sea lions were apparently flexible enough in their diet choice to change to other food items where and when it was profitable, but also kept hunting for organisms that were decreasing in abundance.

The data on foraging behavior and energetics of female California sea lions (Feldkamp et al., Chap. 15; Costa et al., Chap. 16) suggest the mechanism by which foraging females attempted to compensate for reduced prey abundance. They appeared to reduce the time spent swimming and resting, and to increase the percentage of time spent diving (as in Antarctic fur seals, Costa et al. 1989). Dive bouts, which may reflect active foraging on individual prey patches (Gentry et al. 1986a; Feldkamp et al., Chap. 15), increased in length during 1983 when compared with the previous year.

The apparent decline in percent time swimming, while diving effort increased (Feldkamp et al., Chap. 15), suggests that sea lions do not initially travel further away to search for prey. Transit times to the first dive bout were not different between pre-EN and EN years. Prey were apparently found at similar distances from the islands in 1983 as in 1982, but occurred at lower density in 1983. Sea lions increased their search effort by spending a greater percentage of their time diving instead of swimming long distances in search for prey. Diving depths and durations showed a tendency to increase, perhaps reflecting an increased search effort for less available prey. Sea lions examined during the EN years appeared to make a greater percentage of their dives to greater depths or for longer durations (Feldkamp et al., Chap. 15). The South American fur seal (Majluf, Chap. 5) and the Antarctic fur seal (Costa unpubl. data) appear to compensate in a similar fashion for lowered prey availability.

All the above compensatory mechanisms were obviously not sufficient to bal-ance reduced prey availability. The animals found it increasingly difficult to find and capture enough prey to satisfy their energy demands. While the rate of food in-take (in biomass) appeared to be constant between years in California sea lions, more energy was expended to obtain this intake, compared with pre-EN years (Costa et al., Chap. 16). Furthermore, the energy content of prey may have been lower, resulting in a reduced energy intake per unit of biomass eaten. The decline in foraging efficiency manifested itself through increased time spent feeding at sea, and through a decline in the amount of milk provided to the offspring (Heath et al., Chap. 14; Boness et al., Chap. 18; Iverson et al., Chap. 19). It may be significant that under non-EN conditions, in 1982, all California sea lion females with time-depth recorders returned to their pups, while in 1983 three of five did not return. The ani-mals that did not return may have been unable to make up for the additional energy demand required by carrying the instrument. They gave up their pups and did not return to the site of capture. The animals from which we obtained dive data are therefore perhaps exceptionally efficient foragers.

Phocids

Similar effects must have influenced the elephant seals. Females spent more time at sea in 1983 between the end of lactation and molt than in any other year (Table 3; Le Boeuf and Reiter, Chap. 23; Stewart and Yochem, Chap. 25). It is not clear how EN affected their food resources since elephant seals forage off the continental shelf at great depth, 98.5% of their dives are deeper than 200 m (Le Boeuf et al. 1988). Perhaps, EN reduced prey abundance off the shelf. This is likely since we know that Pacific hake is prey of elephant seals and these fish seemed to be distributed more to the north and offshore during 1983 than in normal years (Arntz et al., Chap. 2). If the distribution of other prey species also changed, this would explain the reduced survival of elephant seal juveniles of the 1982 and 1983 cohorts (Le Boeuf and Reiter, Chap. 23; Huber et al., Chap. 24) which – as smaller and less experienced ani-mals – are presumably less able divers than adult females.

26.2.4 Time Lags of Effects on Pinnipeds Relative to the Onset of EN

The time scale of EN effects on pinnipeds ranged from the immediate to those with a lag of several years. "Immediate" effects were caused mainly by physical, oceano-graphic disturbances such as storms, shifts of the thermocline, and the accompany-ing changes in the distribution of prey species (Fahrbach et al., Chap. 1; Arntz et al., Chap. 2). In contrast, effects which showed up one to several years after the phe-nomenon and were potentially long-lasting were primarily caused by slower biolog-ical processes affecting mortality and fertility schedules of prey (Arntz et al., Chap. 2) and predator. Following the time course of effects provides clues to the causal chain through which EN influenced pinniped populations.

The most immediate effects of EN were created by the increase in sea level com-bined with heightened storm activity in the eastern Pacific. The best example of this was the large storm-related mortality of elephant seal pups described in Section 26.2.1.

The rapid change in SST accompanied by the drop in thermocline to greater depth during EN also had an immediate effect on pinnipeds by causing changes in the local distribution of prey. Deepening of the thermocline together with reduced upwelling in the shelf ecosystem, the otariids' foraging habitat, caused a decline in phyto- and zooplankton density near the surface. This in turn diminished the near-surface food resources of fish and cephalopod prey species of pinnipeds. This led to a decrease in body condition of fish, best documented for the Humboldt current system off Chile and Peru (Arntz et al., Chap. 2). It seems likely that changed vertical migration (i.e., fish stayed deeper) and increased dispersal of fish shoals made foraging increasingly difficult for pinnipeds (sea Sect. 26.2.3).

A second effect of EN was large-scale movement of prey organisms. This came about simultaneously with local redistribution and produced longer-lasting effects. Off Peru, benthic fish migrated to greater depths after the onset of EN, while fish with a strong cold preference initially gathered in remaining coastal pockets of cold water upwelling. The South American fur seal in Peru may have experienced this change as an initial increase in prey availability with a subsequent decline as local resource depletion and migration of fish to depth or south to cooler waters occurred. Similar migrations influenced the food resources of pinniped species in temperate zones (Arntz et al., Chap. 2; see Sect. 26.2.3). Redistribution and decrease in energy content of prey species thus had strong and almost immediate effects on pinnipeds.

The decline in prey availability associated with EN 1982–83 led to an observable reduction in body condition of adult animals of some species. Decreased physical condition of adult females then caused reduced milk production (Iverson et al., Chap. 19), decreases in the number of estrus females, or reduced pregnancy rates as indicated by the reduced number of copulations (Francis and Heath, Chap. 12). In gray seals (*Halichoerus grypus*), implantation is closely linked to an increase in food availability resulting in fast deposition of blubber (Boyd 1984). The reduction in pregnancy rates during EN may indicate that pinniped females have to increase fat reserves before reproduction (for similar effects in humans, see Frisch 1985, 1988).

The lowered growth rates of pups of many species led to poor condition of surviving young and to long-term reduced survival of cohorts born during the peak of EN (Tables 1, 2). For elephant seals, reduced juvenile survival occurred as a consequence of lower prey availability (see Sect. 26.2.3), even though initial growth of pups appeared normal during EN (Le Boeuf and Reiter, Chap. 23; Huber et al., Chap. 24; Stewart and Yochem, Chap. 25).

Most of these effects became evident while the meteorological and oceanographic signals of EN were still recognizable; but other effects of EN on pinniped food resources became obvious only when these signals had largely disappeared. The delay was apparently caused through loss (by mortality or migration) and reduced fertility of adult fish. Smaller fish stocks, with individuals of low body condition, reproduced little or not at all in the EN year or the year following it (Arntz et al., Chap. 2). Thus, whole cohorts of prey species were lost with the corollary that fish standing stocks were decreased for one to several years after EN. These effects on prey stocks may explain the lasting influence of EN on pinnipeds after the oceanographic signals had subsided. York (Chap. 9) discusses the ways in which a

4-year delay between an EN event and its positive effect on cohort survival can come about in subpolar areas. For northern fur seals, this is probably related to better juvenile survival of some fish like herring which in turn then provide more food to juvenile fur seals when they begin independent foraging.

Interestingly, the rebound of the marine ecosystem appeared to be more rapid in the tropical Pacific where initial effects were more extreme (Arntz et al., Chap. 2). At temperate latitudes, full EN effects took longer to develop and lasted longer than in the tropics. For example, in the Peruvian system the food chain is especially short and usually ends with a single species, the anchovy. Anchovy consume most of the primary production and produce most of the biomass that is harvested by top predators (Pauly and Tsukayama 1987). Fast recovery of this species can therefore mean a quick return of the ecosystem to an apparently normal state. Perhaps the less complex food web of tropical upwelling systems recovers faster from population crashes or the component species of tropical ecosystems are better adapted to swings from boom to bust conditions than those of temperate systems.

The longest lags in EN effects on pinnipeds are caused by changes in population structure due changes in mortality and fertility. Adult mortality coupled with the loss of several cohorts (Majluf, Chap. 5; Trillmich and Dellinger, Chap. 6) and partial loss of one or two cohorts as observed in temperate zone pinnipeds (DeLong and Antonelis, Chap. 7; Francis and Heath, Chap. 12; Le Boeuf and Reiter, Chap. 23; Huber et al., Chap. 24; Stewart and Yochem, Chap. 25) will change the population composition and its reproductive output for a long time. Full recovery to pre-EN population sizes will vary with latitude and with the number of year classes affected. Twenty years after an intentional kill of females, the Pribilof Island population of northern fur seals was still not fully recovered due to the loss of potential female recruits (York and Hartley 1981). The Galapagos fur seal population, which lost a large proportion of adult females and three full cohorts during EN 1982–83, will similarly show the effects for many years to come. More temperate populations can be expected to experience less dramatic, but nevertheless significant, long-term reductions in numbers due to a decrease in the number of female recruits. EN after-effects will show up as a decline in the number of pups born when females of the reduced cohorts are entering reproductive age.

In summary, the most important long-term population effect of EN, with delays on the order of one to several generation times, is increased adult female mortality. Next in importance (and for elephant seals perhaps the most important EN effect) is the loss or partial loss of cohorts of juveniles and pups. Another major long-term population change was brought about by large-scale emigration. In California sea lions on the Channel Islands, this may have led to a major population redistribution (DeLong et al., Chap. 17), and in northern fur seals on San Miguel Island, it temporarily reversed the long-term trend of a population increase (DeLong and Antonelis, Chap. 7). Emigration may also have caused a major loss to the Peruvian population of the South American fur seal (Guerra and Portflitt, Chap. 4; Majluf, Chap. 5).

26.3 Conclusions

26.3.1 Comparison of Pinnipeds and Seabirds

During the 1982–83 EN, ornithologists gathered extensive data on seabirds in the same area of the eastern Pacific as the pinniped studies reported in this book. A comparison between these two widely differing groups is of interest because pinnipeds and seabirds have to deal with similar ecological problems. Similarities and differences in the effects of EN on the two groups can therefore be used to expose, and increase understanding of, specific adaptations and constraints in the way in which species deal with environmental disturbances.

During the migration period, phocids and otariids behave like open ocean foraging birds, which are free to follow food-rich currents or stay in patches of high food abundance (Hunt and Schneider 1987). Nonmigratory and lactating otariids resemble inshore feeding seabirds during the breeding season when they are tied to the local food resources in their sometimes large "patch" of nearshore habitat. Although the monogamous mating system permits seabirds to share the burden of chick rearing, seabirds as well as otariid pinnipeds have to leave their young while hunting at sea and to return to them at intervals. The length of the feeding cycle partially depends on the distance to the foraging areas. Thus seabirds, like otariids, follow a central-place rearing strategy.

There are, of course, pronounced differences in the physiology and ecology of pinnipeds and seabirds. For allometric reasons (Lindstedt and Boyce 1985), as well as to maintain flying ability, even a maximally fat seabird will store less fat relative to metabolic demand than a much larger pinniped. Only the penguins, because of their size and flightlessness, can build up substantial nutrient deposits to cover their own metabolic needs while fasting for prolonged periods during chick rearing.

Since seabird parents cannot store nutrients in their own body tissues for later transfer to young, they must transport food to their chicks in their stomachs (Costa 1990). This method of provisioning young contrasts strongly with the ability to digest food and convert it into blubber and milk, which are easier to store and higher in energy content. Lactation thus allows pinnipeds to store large amounts of energy and nutrients over a relatively long period for later feeding of young. As a consequence of their lack of storage ability, adult seabirds and their chicks are less capable than pinnipeds of bridging long intervals of food shortage.

Differences in the foraging ecology of the two groups are also marked: most seabirds gather their food from the surface or from a thin upper layer of the ocean. Only penguins, alcids, and perhaps cormorants appear able to dive consistently to depths greater than 20 m (Kooyman and Davis 1987; Kooyman 1989). During warm water conditions, only these species may thus be expected to follow prey species to depths as pinnipeds seem able to do, whereas all other surface-feeding seabird species would be cut off from their food supply.

26.3.1.1 *Effects on Mortality and Breeding Success*

From these differences in size, provisioning strategy and foraging ecology between seabirds and pinnipeds one would expect stronger EN effects on seabirds than on pinnipeds. This is indeed the general pattern. In the tropical Pacific (Galapagos and Peru) boobies, cormorants, pelicans, and penguins abandoned their nests, eggs, and chicks leading to almost total breeding failure (Arntz 1986; Valle et al. 1987; Duffy et al. 1988; Schreiber and Schreiber 1989). In addition, adult mortality in seabirds was far more massive than in pinnipeds. Duffy et al. (1988) estimated adult mortality of Peruvian boobies (*Sula variegata*), Guanay cormorants (*Phalacrocorax bougainvillii*), and Peruvian pelicans (*Pelecanus occidentalis*) as near 85% (cf. our pinniped data in Tables 1, 2).

Data for the northern hemisphere show the same trends, but the responses of seabirds in this area were more varied (Ainley et al. 1988; Schreiber and Schreiber 1989). Along the coast of North America, EN effects on seabirds were delayed by 1 year, i.e., most pronounced in 1983 and 1984 as we noted for pinnipeds. All seabird species on the especially well-studied Farallon Islands were strongly affected during 1983. Adult mortality of cormorants and alcids was higher than normal, breeding success was reduced, and chick growth was slowed, leading to longer times to fledging and lower fledging weights (Ainley et al. 1988). Species feeding locally on benthic organisms (pigeon guillemots, *Cepphus columba*, and pelagic cormorants, *Phalacrocorax pelagicus*) were most strongly affected. Even further north, along the Oregon coast, researchers noted deleterious effects of EN on seabirds in 1983 (Hodder and Graybill 1985). Finally, in Alaskan waters, effects were noted in 1983 but were hard to interpret, as mentioned earlier (Hatch 1987).

In general, latitudinal trends in seabird mortality and the reduction in breeding success parallel effects found in pinnipeds, but EN effects and after effects seem to have caused stronger negative effects on chick growth and survival and perhaps adult mortality among seabirds than among pinnipeds. This may be related to the lower fasting tolerance of seabirds.

26.3.1.2 *The Role of Emigration*

The majority of seabirds can emigrate from areas of poor food conditions much faster than pinnipeds (or penguins). The ability to emigrate large distances from the core of the EN event may have saved many adult seabirds. In Peru, boobies, cormorants, and pelicans were observed to emigrate southward, and Galapagos blue-footed boobies (*Sula nebouxii*) apparently moved toward mainland South America. The return of thousands of adults of these species shortly after the end of EN in 1983 indicated that emigration had been successful for many. Similarly, the Humboldt penguin (*Spheniscus humboldti*) colony in Punta San Juan, Peru, was reduced to only 1% of the birds during EN (Duffy et al. 1988). Later on, many returned and by 1985 colony size was back to normal (Hays 1986).

In Galapagos, nearshore feeders (the Galapagos penguin, *Spheniscus mendiculus*, and the flightless cormorant, *Nannopterum harrisi*) showed the strongest adult mortality (77% and 49%, respectively; Valle et al. 1987). Given these species' limited traveling speeds and body reserves, they may have been unable to emigrate successfully to more suitable habitat, a problem which they apparently shared with the Galapagos pinnipeds.

Along the coast of North America major movements of seabirds were also noted. Cool water species migrated further offshore and many subtropical species followed warm water further north than their usual northern limit (Ainley et al. 1988). Increased northward migration of pinnipeds was also noted by Huber (Chap. 13), particularly for immature California sea lions. In summary, dispersal appears to have been a successful strategy to avoid the worst impact of EN-related changes in food resources for both seabirds and pinnipeds. But this strategy was not open to species feeding locally on benthic organisms, like many of the seabirds mentioned above, or to species which were stuck in the isolated food patches of near-coastal habitat.

Monitoring seabirds would therefore appear to provide a faster indication of changes in marine conditions than similar monitoring of pinnipeds. On the other hand, most seabirds more easily abandon breeding attempts and emigrate from disturbed oceanic areas than pinnipeds and therefore pinnipeds, and especially lactating females, may provide better indicators of local long-term effects.

26.3.1.3 Other Effects on Population Dynamics

An interesting phenomenon that deserves further attention in studies of pinniped populations was noted by ornithologists on the Farallon Islands during EN. The assumption of stability of some seabird populations over the period of EN proved to be erroneous since different individuals were breeding before and after EN (Ainley et al. 1988). Previously nonbreeding birds moved in to fill vacancies left by the mortality of breeding birds. This was demonstrated for populations of banded western gulls (*Larus occidentalis*) and Cassin's auklets (*Ptychoramphus aleuticus*) on the Farallons.

In some species of seabirds the breeding failure during EN was followed by an apparent population increase after the event ended. This was most likely caused by more synchronous breeding of the remaining birds, the majority of which responded with a breeding attempt when food resources returned to a more favorable state (Ainley et al. 1986, 1988). We have noted the same phenomenon in the Galapagos fur seal (Trillmich and Limberger 1985; Trillmich and Dellinger, Chap. 6). This effect will, in both groups, speed up recovery from environmentally caused population crashes.

Seabird observations also suggest another interpretation of the delayed onset of breeding seasons during and after EN, which was observed in two pinniped species (DeLong and Antonelis, Chap. 7; Stewart and Yochem, Chap. 25). Later onset of seabird breeding seasons on the Farallons could have been a secondary consequence of the changed age structure of the populations (Ainley et al. 1988). If floaters entered the population of breeders after EN, their lower efficiency may have caused later onset of breeding and reduced mean fertility in 1984 (Ainley et al. 1988). We cannot presently decide whether this effect might also have contributed to the later arrival of breeding female pinnipeds ashore.

26.3.2 *Phocid-Otariid Differences in Reproductive Strategy*

Pinnipeds face a foraging versus reproduction dilemma. They forage at sea and breed on land because their young are initially bound to land. Females of the two families of pinnipeds, the phocids and otariids, use two entirely different strategies in dealing with this dilemma. In most phocids, reproductive events are isolated from foraging in space and time while in otariids the two activities are intertwined.

Elephant seals forage most of the year without returning to the breeding grounds except to molt. Females travel as far as necessary and stay in good food patches as long as profitable while storing large amounts of nutrients. Costa et al. (1986) calculated that a female elephant seal need only increase her food intake by about 10% per day above her own needs to gather all the nutrient reserves needed for the 28-day period of lactation ashore. As in most phocids, lactation is then fueled entirely from body reserves. Therefore, the location of breeding areas is largely independent of the location of concentrated food resources. Fasting proves to be the key strategy phocids use for solving the food versus reproduction dilemma.

In phocids, large amounts of high energy milk are transferred to the young in a very short time, 4 days to about 6 weeks depending on the species. At the end of this period young are abruptly weaned and are left to their own devices. At this stage they have about 25% of the mother's body mass (Costa 1990). Much of the mass of recently weaned young is a large energy deposit stored as subcutaneous fat which enables them to live up to several months without feeding until they begin to feed on their own.

In contrast, otariids reproduce in a way that we described above as central-place rearing. They give birth to their young in places where high food-abundance ecosystems are nearby and permit fast gathering of food during short foraging trips. Mothers alternate between short 1–3 day long stays ashore and hunting sojourns at sea which may last from 1 to 8 days depending on the species. Trip length is constrained by the pups' ability to fast and to maintain a net mass gain over time. Given that otariid milk never contains much more than 50% fat, and that there is a limit to the quantity of milk a female can produce, otariid foraging trips appear limited to about 7 days. This constraint makes the location of otariid breeding grounds dependent on a combination of distance to and food density in foraging areas: they can afford to travel farther if resources are richer. Mothers must not only replenish their own body stores during the time at sea, they must also pay for the cost of travel between breeding and foraging site and store enough nutrients for lactation during the next stay ashore with the pup. The metabolic overhead of this strategy is much larger than of the phocid rearing strategy, since considerable energy is expended for shuttling between the breeding colony and the foraging grounds. In addition, otariids take more time before weaning their pups, usually between 4 and 12 months, but in exceptional cases up to 36 months. Pups also begin feeding on their own before weaning. They are weaned when they have attained between 40 and 55% of maternal body mass.

As expected from the differences in the lactation strategies of otariids and phocids, starvation of pups due to insufficient milk transfer did not play a role in elephant seal pup mortality. Even if our studies spanned a wider range of latitudes and species we would not expect phocids to show the same tropical-temperate gradient of EN effects on pup mortality as otariids.

Another difference between phocids and otariids that follows from the differences in rearing strategy is that phocids, and to some extent subpolar otariids, provide their young with fat stores and wean them when they are still naive about foraging. This results largely from the short time they spend at maternal care. This strategy can only be successful if food is relatively easy for an unexperienced animal to catch. Temperate and tropical otariids, on the other hand, wean young that are not very fat, but that have some foraging experience. If food is less abundant or difficult to catch, gaining more foraging experience while still supported by maternal milk may be critical for juvenile survival and the otariid strategy allows for this.

Both phocid and otariid males fast during the reproductive season because intrasexual competition does not permit them to forage intermittently at sea. If they did, they would lose opportunities for reproduction and may risk losing their territories or dominance positions to other males. The obvious solution to this problem is to forage more or less independently of a land base during most of the year gathering body stores. These reserves are then used for maintenance during the breeding season on shore. Males are thus shielded from fluctuations in prey abundance during the breeding season, but are fully impacted by whatever conditions they meet upon returning to the sea.

Knowing about the differences in the life history of female otariids and phocids and male vs female pinnipeds, we are now in a position to evaluate the differences found in the effects of EN on otariids and phocids. This can tell us which stages in their life cycles are most susceptible to environmental fluctuations.

26.3.3 The Influence of the Timing of EN Effects

Otariids

The onset of EN will have different effects depending upon which stage of the life cycle is affected. We observed EN effects at various stages of the otariid reproductive cycle, although our data set is confounded by the change in latitude and the corresponding change in the strength of the effects. We can adjust for some of these confounding variables by also observing the effects of weaker ENs on the same population as e.g. in Peru (Majluf, Chap. 5).

In Galapagos, EN began during the peak of the birth season of fur seals and sea lions. In both species growth rates of pups decreased immediately as mothers' time at sea increased (Trillmich and Dellinger, Chap. 6). Decreased provisioning rates led to almost complete pup mortality within a few months of the onset of the event. However, where EN is not as strong, flexibility in the duration of maternal care buffers the impact. Buffering is best demonstrated by Majluf's data for the South American fur seal in Peru. During the 1987 EN, growth rates of pups decreased but mortality rates did not increase. Even though pups were small at 4 months of age, they nevertheless survived as well as other cohorts which grew much faster (Majluf, Chap. 5).

Otariid mothers are continuously exposed to and informed about foraging conditions during lactation, and therefore, are immediately affected by deterioration of the food supply. This may actually work to their advantage. If pup survival becomes unlikely, the earliest possible cessation of investment will allow a female to attend to her own survival as well as garner energy for subsequent offspring. The mechanism by which maternal effort is ended may lie in maternal physiology. Since lactating females forage longer during poor food conditions (see Sect. 26.2.3) in order to obtain enough resources for lactation as well as their own metabolic demands, pups may starve to death during very protracted feeding trips, especially if such trips are consecutive. This effectively constitutes pup "abandonment".

Otariids in the California system were not affected by EN in 1982 before their young were 3–4 months old. The only clear change here was an increase in the period of maternal care (Francis and Heath, Chap. 21) thus reducing the impact of EN. However, increased maternal care entailed a great cost to mothers through reduced fertility. This effect was more pronounced in otariids than in phocids (Tables 1–3; cf. DeLong et al., Chap. 17, on sea lions; DeLong and Antonelis, Chap. 7, on fur seals; Aurioles and Le Boeuf, Chap. 11, on sea lions; Le Boeuf and Reiter, Chap. 23, and Huber et al., Chap. 24, on elephant seals). While phocid mothers showed only slight decreases in pup production, sea lions produced only half as many pups in 1983 as in 1982.

Reproductive output of otariid mothers was reduced even when lactation was not extended: in 1983, northern fur seals at San Miguel weaned their young as usual after about 4 months. Nevertheless, they produced fewer young in 1984. The mortality of recently independent northern fur seal young also appeared to increase during the subsequent months (DeLong and Antonelis, Chap. 7 see also Sect. 26.3.4). Even in the tropics, otariid populations began to recover immediately after the end of EN as soon as feeding conditions returned to normal. This is possible because mothers do not depend on body reserves for pup rearing but on food available at the time of rearing. In California, recovery in the otariid populations did not occur until 1985 since the long-term effects of the EN carried over into 1984 (see above).

Phocids

EN in 1982 first influenced temperate waters in Sept./Oct. when elephant seal mothers had already gathered the majority of the nutrient reserves needed for successful pup rearing. Consequently, the period of pup rearing showed few effects of EN (aside from the direct physical storm effects). However, the subsequent stages in the cycle, foraging to recover some of the resources lost in lactation prior to molting, and foraging until the next pupping season, were affected as shown by the longer foraging of females before returning to molt (Le Boeuf and Reiter, Chap. 23) and by reduced natality in the next season (Huber et al., Chap. 24).

Phocid pups are shielded from the effects of environmental fluctuations while still dependent on their mothers, but juveniles are adversely affected when they begin foraging at times of low food abundance. This was most noticeable in the reduced long-term survival of the 1983 cohort. Juveniles are expected to be less capable of sustained deep diving than their mothers because of lower body size. This has been shown for the Weddell seal, *Leptonychotes weddelli* (Kooyman et al. 1983), where the 13-min aerobic dive limit (i.e., the time they can spend under water without incurring an oxygen debt) of juveniles was about half the 20–25-min aerobic dive limit of adult animals (see also Gentry et al. 1986a). However, this explanation may not apply to elephant seals. By 14 months, juvenile elephant seals showed the same dive depth/duration pattern as adult females (Le Boeuf unpubl. data). Alternatively, less hunting experience of juveniles as compared to adults may cause the decreased survival of juvenile elephant seals when food is scarce.

The unusually long duration of the 1982–83 EN provided the opportunity to record the effects of long-term reductions in food supply on a phocid. The 1984 elephant seal reproductive season was preceded by a year of EN influence. This had no effect on adult female mortality or on the number of pups born on Año Nuevo island (Le Boeuf and Reiter, Chap. 23), but it increased the females' tendency to skip one reproductive season, and decreased the number of pups born on the Farallon Islands (Huber et al., Chap. 24). Apparently, the animals at the species' northern distribution limit were affected more by the changes than were Año Nuevo animals. The reasons for this difference are not known, but clearly food shortages can influence a female's ability to implant the fertilized egg or to pay the metabolic cost of gestation. Thus, the phocids are most sensitive to food shortages during the long foraging period at sea, the nonreproductive season. As a corollary of reduced food intake during their first year at sea, growth of juveniles is presumably retarded, and age at primiparity is delayed.

EN events of more "normal" duration, the composite ENs of Wyrtki (see Fahrbach et al., Chap. 1), may begin to affect the temperate areas off California in January. This would presumably lead to less pronounced changes in adult behavior and fertility, but may influence juvenile survival similarly as we found for the 1982–83 EN.

Unlike lactating otariids, breeding phocids of both sexes as well as breeding male otariids do not gather information on feeding conditions during the reproductive season. During very strong ENs they may not be able to replenish body reserves lost during the reproductive season quickly enough. Could such an effect be responsible for the absence of phocids in the eastern tropical Pacific?

26.3.4 *Potential Selective Effects of Environmental Fluctuations Like EN on Pinnipeds*

None of our results are sufficiently detailed to permit direct measurements of selection due to EN conditions. Nevertheless, on the basis of the documented effects, we consider it useful and of value for future studies to build hypotheses on the selective effects of environmental variance on pinniped life history.

26.3.4.1 Effects on Maternal Strategies

What effect could recurrent EN events have on the evolution of the patterning of maternal care in otariids? We noticed no effect of EN on the population of northern fur seals in the Bering Sea (Gentry, Chap. 8), but a severe one on the small population which recently colonized temperate San Miguel Island (DeLong and Antonelis, Chap. 7). This species invariably weans its young after about 4 months. It shares this characteristic with its southern counterpart, the Antarctic fur seal (*Arctocephalus gazella*) which also lives in a relatively predictable subpolar environment. Populations of both species suffered high pup mortality under adverse environmental conditions (Croxall et al. 1988; Costa et al. 1989; DeLong and Antonelis, Chap. 7). Apparently, the maternal strategy of these subpolar species is not flexible enough to compensate for periods of low food availability by increasing the lactation period, thus helping juveniles to ease into independent foraging and increasing their probability of survival.

These results are of great interest because they are our only example of the impact of EN on a population with a fixed time to weaning. Growth of the San Miguel colony presently still partially depends on continuing immigration. This apparently precludes genetic adaptation to local conditions in the California Current system. Northern fur seals on San Miguel followed a rigid, genetically programmed time course in their pup rearing. This strategy is probably adaptive in the subpolar areas where they normally breed, but has major disadvantages when animals live under temperate conditions. The same applies to Antarctic fur seals which breed on temperate Marion Island (Kerley 1985).

All other otariids, sea lions and fur seals alike, are much more flexible in the duration of lactation, which allows them to buffer their young against unpredictable reductions in food resources. It seems logical that a more flexible patterning of maternal effort will be selected for when environmental circumstances vary unpredictably, if the benefit of lengthening the lactation period (i.e., increased juvenile survival) outweighs any decrement in future offspring production (i.e., reduced female fertility).

At the same time, lengthening of the nursing period should allow juveniles to gather foraging experience before they become totally independent. In such a variable rearing system there needs to be communication between young and their mothers as to the degree of nutritional dependence or independence. While such a system is open to cheating from the side of the young (Trivers 1974), it allows females to adjust maternal effort in relation to environmental circumstances. EN could be one source of variation selecting for increased flexibility of the maternal strategy. This agrees with the degree to which this flexibility is observed at different latitudes: at high latitudes with highly predictable seasonality northern fur seals (and Antarctic fur seals) show no flexibility of weaning age, whereas at low latitudes Galapagos fur seal mothers wean their young at any time between 10–36 months. A similar though lower flexibility was also noted for the South American fur seal in Peru (Majluf, Chap. 5) and the California sea lion (Francis and Heath, Chap. 21).

Whereas very strong EN events like the one in 1982/83 destroy all offspring of a given cohort, as in the Galapagos fur seal, less violent fluctuations will lead to differential survival of young as observed in the California otariids and the fur seal in Peru. If such events recur frequently enough they could lead to selection on mothers for giving up the current reproductive effort in favor of a future one if the probability of less severe environmental conditions in the next reproductive cycle is sufficiently high. During EN in 1982/83 many female otariids were at sea for so long that their young starved to death. It has been postulated for the northern fur seal that mothers return to their pups when they have replenished their body stores up to a set threshold (Costa and Gentry 1986). They then transfer milk to the pup until their body stores reach a lower threshold whereupon they leave again to forage. Interindividual variation in the position of the upper and lower threshold could well become a target for selection. Mothers in a relatively predictable environment may face less of a survival cost if they allow body stores to fall to a lower setpoint than animals in a less predictable environment. This would apply if lowered fat stores imply more of a mortality risk to a mother in an unpredictable environment.

In phocid seals, it is more difficult to envisage how selection would act on maternal strategies since survival of both mother and pup depend on body stores during and at the end of lactation. Bigger females may have an advantage during periods of food shortage since they can carry and transfer more reserves to their young than smaller mothers (Lindstedt and Boyce 1985). To young, the advantage of being bigger (bigger in terms of more lean body mass) may consist of increased diving ability once they begin their independent foraging. In addition, pups with larger fat stores at weaning can spend more time gaining hunting experience once they enter the sea before their body stores fall to critically low levels (Reiter et al. 1978).

For mothers, being bigger could also be disadvantageous, since in absolute terms a bigger animal utilizes more food per unit time. This need for more food could become critical for a big female under conditions of food shortage, whereas at the same time the larger size of the young of a big mother (Costa et al. 1988) would confer an advantage to her due to a higher survival probability of her offspring. Without detailed modeling it is hard to guess where the optimum size of a female would fall under conditions of variable prey abundance, but it seems a worthwhile exercise to model this problem of a size-dependent shift in the trade-off between fertility and survival.

In the elephant seal, age at primiparity was delayed in cohorts grown up under EN conditions (Le Boeuf and Reiter, Chap. 23; Huber et al., Chap. 24). If large size of mother and pup alike confer a survival advantage, unpredictable variance in environmental conditions could select for delayed primiparity in both phocids and otariids. This would allow females to grow to a larger size before first reproduction. Costa et al. (1988) have shown that in the Antarctic fur seal female size is correlated positively with pup size. Thus, bigger mothers produce bigger pups and may increase the survival chances of their pup. If under fluctuating prey abundance, growth rates of subadults fluctuate strongly between cohorts, age at primiparity may come more under the control of body size rather than of age per se and this could result in greater variance in the age at primiparity in more tropical pinniped populations. We presently have no data to test this hypothesis.

26.3.4.2 *Effects on Male Size and Sexual Selection*

A decrease in male size reduces the absolute amount of food needed for maintenance metabolism. If food shortages occur frequently and cannot be evaded by migration to more productive areas, this could select for reduced male size. These arguments may explain why the Galapagos fur seal has the least pronounced sexual size dimorphism of all otariids. Similarly, fully adult male California sea lions on Galapagos (*Zalophus californianus wollebaeki*) appear smaller than males in California (*Z. c. californianus*), but exact morphometric data on California sea lion males from the Galapagos population are not available.

Environmental fluctuations, which decrease male adult or juvenile survival for certain cohorts, may decrease the average intensity of intrasexual selection. During EN, increased mortality or skipping of reproductive seasons among breeding males was observed in many species (Tables 1–3). Nearly 100% of the territorial male fur seals died in the Galapagos. If many breeding males die during EN, male-male competition among the survivors necessarily decreases. The extent of this decrease in male-male competition depends on the proportion of fully adult males that actually become territorial in a given season. Males, which are subadult at the time that breeding males experience increased mortality, also face less fierce competition by older males in the next breeding season. Similarly, males born in, just before, or just after the cohort(s) with reduced survival at the pup stage also face less competition when they become territorial.

The intensity of male-male competition is considered the selective process responsible for the major size difference between male and female polygynous pinnipeds (e.g. Bartholomew 1970). Environmental fluctuations, which cause increased male mortality, would therefore be expected to reduce intrasexual selection for large male size. This agrees with the latitudinal trend in size dimorphism found in otariids: at nearly equal female body mass, male northern fur seals have five to six times higher body mass than females, whereas male Galapagos fur seals weigh only about twice as much as females.

References

Adams P, Samiere W (1987) Juvenile rockfish (Sebastes spp.) in the Gulf of the Farallones. CalCoFi Invest Annu Meet, Nov 1987, Lake Arrowhead, California

Adolph EF, Heggeness FW (1971) Age changes in body water and fat in fetal and infant mammals. Growth 35:55–63

AGU (American Geophysical Union – ed) (1987) El Niño. An AGU Chapman conference. Reprinted from J Geophys Res 92 C: 14, 187–14, 479

Aguayo A, Maturana R (1973) Presencia del lobo marino común (Otaria flavescens) en el litoral chileno, Arica (18°27′S) a Punta Maiquillahue (39°27′S). Biol Pesq Chile 6:45–75

Ailey DG, Boekelheide RJ (1989) Farallon Islands seabirds: ecology, dynamics and community structure of an upwelling system. Univ Press, Standford

Ainley DG, Huber HP, Bailey KM (1982) Population fluctuations of California sea lions and the Pacific whiting fishery off central California. Fish Bull 80:253–258

Ainley DG, Spear LB, Boekelheide RJ (1986) Extended post-fledging parental care in the red-tailed tropic bird and sooty tern. Condor 88:101–102

Ainley DG, Carter HR, Anderson DW, Briggs KT, Coulter MC, Cruz F, Cruz JB, Valle CA, Fefer SI, Hatch SA, Schreiber EA, Schreiber RW, Smith NG (1988) Effects of the 1982–83 El Niño – Southern Oscillation on Pacific ocean bird populations. In: Ouellet H (ed) Acta 19th Congr Int Ornithol, vol 2. Univ Press, Ottawa, pp 1747–1758

Alamo AV, Bouchon M (1987) Changes in the food and feeding of the sardine (Sardinops sagax sagax) during the years 1980–1984 off the Peruvian coast. J Geophys Res 92 C:14, 411, 415

Alderdice DF, Velsen FPJ (1971) Some effects of salinity and temperature on early development of Pacific herring (Clupea pallasi). J Fish Res Board Can 28:1545–1562

Allen SG, Peaslee SC, Huper HR (in press) Northern elephant seals breeding on the Point Reyes peninsula. Mar Mammal Sci

Alvarez-Borrego S (1983) Gulf of California. In: Ketchum B (ed) Ecosystems of the world, vol 26: Estuaries and enclosed seas. Elsevier, Amsterdam, pp 427–450

Alvial A (1985) Programa de vigilancia del fenomeno El Niño en la zona norte de Chile. Resultados y perspectivas. Invest Pesq (Spec Issue) 32:69–77

Alvial A (1988) Impactos de “El Niño” sobre el manejo de las pesquerias pelagicas en el Pacifico oriental. Med Ambiente 9:35–41

American Society of Mammologists (ed) (1967) Standard measurements of seals. J Mammal 48:459–462

Antonelis GA Jr, DeLong RL (1985) Population and behavioral studies, San Miguel Island, California. In: Kozloff P (ed) Fur seal investigations, 1983. NOAA Tech Mem NMFS F/NWC-78, pp 32–41

Antonelis GA Jr, DeLong RL (1986) Population and behavioral studies, San Miguel Island, California. In: Kozloff P (ed) Fur seal investigations, 1984. NOAA Tech Mem NMFS F/NWC-97, pp 49–53

Antonelis GA Jr, Fiscus CH (1980) The pinnipeds of the California Current. CalCoFi Invest Rep 11:68–78

Antonelis GA Jr, Perez MA (1984) Estimated annual food consumption by northern fur seals in the California Current. CalCoFi Invest Rep 25:135–145

Antonelis GA Jr, Fiscus CH, DeLong RL (1984) Spring and summer prey of California sea lions (Zalophus californianus) at San Miguel Island, California, 1978–79. Fish Bull USA 82 (1):67–76

Antonelis GA Jr, Lowry MS, DeMaster DP, Fiscus CH (1987) Assessing northern elephant seal feeding habits by stomach lavage. Mar Mammal Sci 3:308–322

Arnold W, Trillmich F (1985) Time budget in Galapagos fur seal pups: the influence of the mother’s presence and absence on pup activity and play. Behaviour 92:302–321

Arntz WE (1984) El Niño and Perú: positive aspects. Oceanus 27:36–39

Arntz WE (1986) The two faces of El Niño 1982–83. Meeresforschung 31:1–46

Arntz WE, Fahrbach E (1991) El Niño – Klimaexperiment der Natur. Physikalische Ursachen und biologische Folgen. Birhäuser-Verlag, Basel: 264 pp

Arntz WE, Valdivia J (1985) Vision integral del problema "El Niño": introducción. In: Arntz WE, Landa A, Tarazona J (eds) "El Niño". Su impacto en la fauna marina. Bol Inst Mar Perú, Callao (Spec Issue), pp 5–10
Arntz WE, Flores LA, Maldonado M, Carbajal G (1985a) Cambios de los factores ambientales, macrobentos y bacterias filamentosas en la zona de mínimo de oxígeno frente al Perú. In: Arntz WE, Landa A, Tarazona J (eds) "El Niño". Su impacto en la fauna marina. Bol Inst Mar Perú, Callao (Spec Issue), pp 65–77
Arntz WE, Landa A, Tarazona J (eds) (1985b) "El Niño". Su impacto en la fauna marina. Bol Inst Mar Perú, Callao (Spec Issue), 224 pp
Ashton HJ, Haist V, Ware DM (1985) Observations on abundances and diet of pacific mackerel (Scomber japonicus) caught off the west coast of Vancouver Island, September 1984. Can Tech Rep Fish Aquat Sci 1394:11 pp
Aurioles D (1982) Contribución al conocimiento de la conducta migratoria del lobo marino de California, Zalophus californianus. Tésis de licentiatura en biología marina. Univ Auton Baja California Sur, Mexico, 75 pp
Aurioles D (in press) Dispersal and migration of California sea lions in the Gulf of California. J Mammal
Aurioles D, Sinsel F (1988) Mortality of California sea lion pups at Los Islotes Baja California Sur, México. J Mammal 69:180–183
Aurioles D, Sinsel F, Fox C, Alvaro E, Maravilla O (1983) Winter migration of the subadult male California sea lions in the southern part of Baja California. J Mammal 64:513–518
Austad SN, Sunquist ME (1986) Sex-ratio manipulation in the common opposum. Nature (London) 324:58–60
Avaria S (1985a) Variaciones en la composición y biomasa del fitoplancton marino del norte de Chile entre diciembre 1980 y junio 1984. Invest Pesq Chile (Spec Issue) 32:191–193
Avaria S (1985b) Efectos de El Niño en las pesquerias del Pacífico Sureste. Taller nacional fenómeno El Niño 1982–83. Invest Pesq Chile 32:101–116
Bailey KM, Ainley DG (1982) The dynamics of California sea lion predation on Pacific whiting. Fish Res 1:163–176
Bailey KM, Incze LS (1985) El Niño and the early life history and recruitment of fishes in temperate marine waters. In: Wooster WS, Fluharty DL (eds) El Niño north. Washington Sea Grant Program. Univ Press Washington, Seattle, WA, pp 143–165
Bailey KM, Francis RC, Stevens PR (1982) Life history and fishery of Pacific whiting. CalCoFi Rep 23:81–98
Baker RJ, Nelder JA (1978) Generalized linear interactive modelling. The GLIM system release 3.77 Manual. Numerical Algorhythms Group Ltd. Oxford, UK
Bakkala RG (1988) Condition of groundfish resources of the eastern Bering Sea and Aleutian Island region in 1987. US Dep Commerc NOAA Tech Mem NMFS F/NWC-139
Bakkala RG, Wespestad VG, Low LL (1987) Historical trends in abundance and current condition of walleye pollock in the eastern Bering sea. Fish Res 5:199–215
Barber RT (1986) The biology of El Niño. Abstract. In: Intergov Oceanogr Comm Worksh Rep 49. Guayaquil, Ecuador, p 3
Barber RT, Cháves FP (1983) Biological consequences of El Niño. Science 222:1203–1210
Barber RT, Chávez FP (1986) Ocean variability in relation to living resources during the 1982–83 El Niño. Nature (London) 319:279–285
Barber RT, Kogelschatz JE, Chávez FP (1985) Origin of productivity anomalies during the 1982–83 El Niño. CalCoFi Rep 26:65–71
Barnett T, Graham N, Cane M, Zebiak S, Dolan S, O'Brien J, Legler D (1988) On the prediction of El Niño 1986–1987. Science 241:192–196
Bartholomew GA (1967) Seal and sea lion populations of the California Islands. In: Proc Symp Biology of California Islands. Santa Barbara Botanic Garden, Santa Barbara, CA, pp 229–244
Bartholomew GA (1970) A model for the evolution of pinniped polygyny. Evolution 24:546–559
Bartholomew GA, Wilke F (1956) Body temperature in the northern fur seal, Callorhinus ursinus. J Mammal 37:327–337
Bauman DE, Elliot JM (1983) Control of nutrient partitioning in lactating ruminants. In: Mepham TB (ed) Biochemistry of lactation. Elsevier, Amsterdam, pp 437–469
Baumgartner TR, Robles JM, Ferreira V (1987) Factores que controlan la productividad primaria del golfo de California en condiciones de El Niño. Abstract. In: 7. Congr Nacl Oceanografía. Inst Nacl Pesc, Ensenada, Baja California, pp 261–262
Beach RJ, Geiger AC, Jefferies SJ, Treacy SD, Trootman BL (1985) Marine mammals and their interactions with fisheries of the Columbia river and adjacent waters, 1980–1982. NMAFC Proc Rep 85–04, Seattle, WA
Bedford D, Jow T, Klingbeil R, Read R, Spratt J, Warner R (1983) Review of some California fisheries for 1982. CalCoFi Rep 22:6–10

Behrens W-U (1964) The comparison of means of independent normal distributions with different variances. Biometrics 20:16–27
Berhage HP (1957) Fluctuations in the general atmospheric circulation of more than one year, their nature and prognostic value. Koningl Ned Meteor Inst Meded Verh 69:152 pp
Bernal PA, McGowan JA (1981) Advection and upwelling in the California current. In: Richards FA (ed) Coastal and estuarine science, vol 1. Am Geophys Un, Washington, DC, pp 381–395
Bernard HJ, Hedgepeth JB, Reilly SB (1985) Stomach contents of albacore skipjack, and bonito caught off California during summer 1983. CalCoFi Invest Rep 26:175–182
Bigg MA (1973) Census of California sea lions on southern Vancouver Island, British Columbia. J Mammal 54:285–287
Bigg MA (1985) Status of the Steller sea lion (Eumetopias jubatus) and California sea lion (Zalophus californianus) in British Columbia. Can Spec Publ Fish Aquat Sci 77:20 pp
Bindmann AG (1986) The 1985 spawning biomass of the northern anchovy. CalCoFi Invest Rep 27:16–24
Blackboum DJ, Tasaka MB (1989) Marine scale growth in Fraser River pink salmon: a comparison with sockeye salmon marine growth and other biological parameters. In: Proc 14th NE Pac Pink and Chum Worksh (in press)
Blanco JL, Díaz M (1985) Características oceanográficas y desarrollo de El Niño 1982–83 en la zona costera de Chile. Invest Pesq (Spec Issue) 32:53–60
Boness DJ, Dabek L, Ono KO, Oftedal OT (1985) Female attendance behavior in California sea lions. In: Abstracts 6th Bien Conf Biol Ma Mammal, Vancouver, BC, 15 pp
Bonnell ML, Le Boeuf BJ, Pierson MO, Dettman DH, Farrens GD, Heath CB, Gantt RF, Larsen DJ (1980) Pinnipeds of the Southern California Bight. In: Norris KS, Le Boeuf BJ, Hunt GL Jr (eds) Marine mammals and sea bird surveys of the Southern California Bight area 1975–1978, vol 3. Gov Print Off, Washington, DC, 531 pp
Bonnell ML, Pierson MO, Farrens GD (1983) Pinnipeds and sea otters of central and northern California, 1980–1983: status, abundance, and distribution. Final Rep Miner Manag Serv, Dep Int Aff, Washington, DC, 221 pp
Bonnot P (1937) California sea lion census for 1936. Cal Fish Game 23:108–112
Bonnot P, Clark GH, Hatton SR (1938) California sea lion census for 1938. Cal Fish Game 24:415–419
Boveng P (1988) Status of the California sea lion population on the US West Coast. SWFC Admin Rep LJ-88-07, SWF Center, 8604 La Jolla Shores Dr, La Jolla, CA, 92038, 26 pp
Boyd CM (1967) The benthic and pelagic habits of the red crab, Pleuroncodes planipes. Pac Sci 21(3):394–403
Boyd IL (1984) The relationship betwen body condition and the timing of implantation in pregnant grey seals (Halichoerus grypus). J Zool London 203:113–123
Brambell FWR (1948) Prenatal mortality in mammals. Biol Rev 23:370–407
Brinton E (1981) Euphausid distribution in the California Current during the warm winter-spring of 1977–78, in the context of a 1949–1966 time series. CalCoFi Invest Rep 22:135–154
Brodeur RD, Pearcy WG (1986) Distribution and relative abundance of pelagic nonsalmonid nekton off Oregon and Washington, 1979–84. NOAA Tech Rep NMFS 46, 85 pp
Brodeur RD, Gadomski DM, Pearcy WG, Batchelder HP, Miller CB (1985) Abundance and distribution of ichthyoplankton in the upwelling zone off Oregon during anomalous El Niño conditions. Estuar Coast Shelf Sci 21:365–378
Brody S (1945) Bioenergetics and growth. Haffner, London
Brown RF (1988) Assessment of pinniped populations in Oregon, Apr 1984 – Apr 1985. NWAFC Proc Rep 88-085. Seattle, WA, 44 pp
Burgner RL (1980) Some features of ocean migrations and timing of Pacific salmon. In: McNeil WJ, Himsworth DC (eds) Salmonid ecosystems of the North Pacific. State Univ Press, Corvallis, Oregon, pp 153–164
Butler J (1987) Comparison of the early life history parameters of Pacific sardine and northern anchovy and implications for species interactions. PhD Thesis, Univ Cal, San Diego, 242 pp
CalcoFi (California Cooperative Fisheries Investigations – ed) (1984) Review of some California fisheries for 1983. CalCoFi Rep 25:7–15
CalcoFi (California Cooperative Fisheries Investigations – ed) (1985) Review of some California fisheries for 1984. CalCoFi Rep 26:9–16
CalcoFi (California Cooperative Fisheries Investigations – ed) (1986) Review of some California fisheries for 1985. CalCoFi Rep 25:7–15
CalcoFi (California Cooperative Fisheries Investigations – ed) (1987) Review of some California fisheries for 1986. CalCoFi Rep 28:11–20
CalcoFi (California Cooperative Fisheries Investigations – ed) (1988) Review of some California fisheries for 1987. CalCoFi Rep 29:11–20
Cane MA (1983) Oceanographic events during El Niño. Science 222:1189–1195
Can MA (1986) El Niño. Annu Rev Earth Planet Sci 14:43–70

Cane MA, Zebiak SE, Dolan SC (1986) Experimental forecasts of El Niño. Nature (London) 321: 827–832
Cannon GA, Reed RK, Pullen PE (1985) Comparison of El Niño events off the Pacific Northwest. In: Wooster WS, Fluharty DL (eds) El Niño north. Washington Sea Grant Program. Univ Washington, Seattle, WA, pp 75–84
Cañon J (1985) La variabilidad ambiental en la zona norte de Chile y su influencia en la pesqueria pelágica durante El Niño 1982–83. Invest Pesq Chile 32:119–128
Carrasco S, Santander H (1987) The El Niño event and its influence on the zooplankton off Perú. J Geophys Res 92 C:14, 405–14, 410
Cass VL (1985) Exploitation of California sea lions (Zalophus californianus) prior to 1972. Mar Fish Rev 47 (1):36–38
Caughley G (1977) Analysis of vertebrate population. John Wiley & Sons, New York London, 234 pp
Caughley G, Caughley J (1974) Estimating median date of birth. J Wildlife Manag 38:552–556
Chapman DG (1961) Population dynamics of the Alaska fur seal herd. Trans N Am Wildlife Nat Resour Conf 26:356–369
Chamov EL (1976) Optimal foraging, the marginal value theorem. Theor Pop Biol 9:129–136
Chatfield C (1975) The analysis of time series: theory and practice. Chapman & Hall, London, 263 pp
Chávez F (1987) The annual cycle of SST along the coast of Peru. Trop Ocean Atmosph Newslett 37:4–6
Chelton DB, Bernal PA, McGowan JA (1982) Large-scale interannual physical and biological interaction in the California Current. J Mar Res 40:1095–1125
Clutton-Brock T (1988) Reproductive success. Univ Press, Chicago
Cole DA, McLain DR (1989) Interannual variability of temperature in the upper layer of the North Pacific eastern boundary region, 1971–1987. NOAA Tech Mem NMFS SW Fish Cent 125, 10 pp
Colin C, Henin C, Hisard P, Oudot C (1971) Le Courant Cromwell dans le Pacifique central en février. Cah ORSTROM; Ser Oceanogr 9:167–186
CONCYTEC (Consejo Nacional de Ciencia y Technologia – ed) (1985) Ciencia, tecnología y agresión ambiental: el fénomeno El Niño. CONCYTEC, Lima, Perú, 692 pp
Condit R, Le Boeuf BJ (1984) Feeding habits and feeding grounds of the northern elephant seal. J Mammal 65:281–290
Cooper CF, Stewart BS (1983) Demography of northern elephant seals, 1911–1982. Science 210:969–971
Costa DP (1987) Isotopic methods for quantifying material and energy balance of free-ranging marine mammals. In: Huntley AC, Costa DP, Worthy GAJ, Castellini MA (eds) Approaches to marine mammal energetics. Soc Mar Mammal Sci Spec Publ 1. Allen, Lawrence, KA, pp 43–66
Costa DP (1988a) Methods for studying the energetics of freely diving animals. Can J Zool 66:45–52
Costa DP (1988b) Assessment of the impact of the California sea lion and the northern elephant seal on commercial fisheries. In: California Sea Grant Biennial Report of Completed Projects 1984–86. Cal Sea Grant Coll Progr, Univ Cal, La Jolla, Publ R-CSGCP-024, pp 36–43
Costa DP (1990) Reproductive and foraging energetics of pinnipeds, penguins and albatross, implications for life history strategies. Am Zool (in press)
Costa DP, Gentry RL (1986) Free-ranging energetics of the northern fur seals. In: Gentry RL, Kooyman GL (eds) Fur seals: maternal strategies on land and at sea. Univ Press, Princeton, NJ, pp 79–101
Costa DP, Trillmich F (1988) Mass changes and metabolism during the perinatal fast: a comparison between Antarctic (Arctocephalus gazella) and Galapagos fur seals (A. galapagoensis). Physiol Zool 61(2):160–169
Costa DP, Le Boeuf BJ, Huntley AC, Ortiz CL (1986) The energetics of lactation in the northern elephant seal. J Zool London 209:21–33
Costa DP, Trillmich F, Croxall JP (1988) Intraspecific allometry of neonatal size in the Antarctic fur seal (Arctocephalus gazella). Behav Ecol Sociobiol 22:361–364
Costa DP, Croxall JP, Duck CD (1989) Foraging energetics of Antarctic fur seals in relation to changes in prey availability. Ecology 70:596–606
Croxall JP, McCann TS, Prince PA, Rothery P (1988) Reproductive performance of seabirds and seals at South Georgia and Signy Island, South Orkney Islands 1976–1986: implications for Southern Ocean monitoring studies. In: Sahrhage D (ed) Antarctic Ocean and resources variability. Springer, Berlin Heidelberg New York, pp 261–286
Cucalón E (1987) Oceanographic variability off Ecuador associated with an El Niño event in 1982–83. J Geophys Res 92 C: 14, 309–14, 322
Cushing DH (1982) The effect of El Niño upon the Peruvian Anchoveta stock. In: Cushing DH (ed) Climate and fisheries. Academic Press, New York, London, pp 267–296
Dandonneau Y (1986) Monitoring the sea surface chlorophyll concentration in the tropical Pacific: consequences of the 1982–83 El Niño. US Fish Bull 84:687–695
Darwin C (1871) The descent of man and selection in relation to sex. Murray, London
Dasmann RF, Taber RD (1956) Behavior of Columbian black-tailed deer with reference to population ecology. J Mammal 37:143–164

Day DE (1987) Changes in the natural mortality rate of the S E Strait of Georgia sacroe herring spawning stock 1976–1985. Washington Dep Fish Tech Rep 98:30 pp
Dellinger T (1987) Das Nahrungsspektrum der sympathischen Galapagos-Seebären (Arctocephalus galapagoensis) und Galapagos-Seelöwen (Zalophus californianus) mit Versuchen zur Methodik der Kotanalyse. Diplomarbeit, Univ Konstanz, FRG, 71 pp
Dellinger T, Trillmich F (1988) Estimating diet composition from scat analysis in otariid seals (Otariidae): is it reliable? Can J Zool 66:1865–1870
DeLong RL (1982) Population biology of northern fur seals at San Miguel Island, California. PhD Thesis, Univ Cal, Berkeley, CA, 185 pp
DeLong RL, Antonelis GA Jr (1985) Population growth and behavior. San Miguel Island. In: Kozloff P (ed) Fur seal investigations, 1982. NOAA Tech Mem NMFS F/NWC-71, pp 54–60
DeLong RL, Stewart BS (in press) Diving patterns of northern elephant seal bulls. Mar Mammal Sci 7: in press
DeLong RL, Antonelis GA Jr, Jameyson EC (1982) Population biology of northern fur seals at San Miguel Island. In: Kozloff P (ed) Fur seal investigations, 1981. NOAA Tech Mem NMFS F/NWC-37, pp 33–34
DeMaster DP, Miller DJ, Goodman D, DeLong RL, Stewart BS (1982) Assessment of California sea lion fishery interactions. In: Sabol K (ed) Trans 47th N Am Wildlife Nat Resour Conf. Wildlife Manag Inst, Washington, DC, pp 253–264
de Muizon C (1981) Les vertébrés fossiles de la formation Pisco (Pérou), pt 1: Deux nouveaux Monachine (Phocidae, Mammalia) due Pliocène de Sud-Sacaco. Travaux de l'Institut francais d'études andines 22. Rech Grandes Civil Mem 6:161 pp
Deser C, Wallace JM (1987) El Niño events and their relation to the Southern Oscillation: 1925–1986. J Geophys Res 92 C: 14, 189–14, 196
Dessier A, Donguy JR (1987) Response to El Niño signals of the epiplanktonic copepod populations in the eastern tropical Pacific. J Geophys Res 92 C:14, 393–14, 403
DeVries TJ (1987) A review of geological evidence for ancient El Niño activity in Peru. J Geophys Res 92 C: 14, 471–14, 479
Dierauf LA, Vandenbroek DJ, Roletto J, Koski M, Amaya L, Gage LJ (1985) An epizootic of leptospirosis in California sea lions. J Am Vet Med Assoc 187 (11):1145–1148
Dioses T (1985) Influencia del fenómeno "El Niño" 1982–83 en el peso total individual de los pesces pelágicos, sardina, jurel y caballa. In: Arntz WE, Landa A, Tarazona J (eds) "El Niño". Su impacto en la fauna marina. Bol Inst Mar Perú, Callao (Spec Issue), pp 129–134
Doidge DW, Croxall JP (1985) Diet and energy budget of the Antarctic fur seal, Arctocephalus gazella, at South Georgia. In: Siegfried WR, Condy PR, Laws RM (eds) Antarctic nutrient cycles and food webs. Springer, Berlin Heidelberg New York, pp 543–550
Doidge DW, Croxall JP, Baker JR (1984) Density-dependent mortality in the Antarctic fur seal Arctocephalus gazella at South Georgia. J Zool London 202:449–460
Duffy DC, Arntz WE, Tovar H, Boersma PD, Norton RL (1988) A comparison of the effects of El Niño and the Southern Oscillation on birds in Peru and the Atlantic ocean. In: Ouellet H (ed) Acta 19th Congr Int Ornithol, vol 2. Univ Press, Ottawa, pp 1740–1746
Dunnett CW (1980) Pairwise multiple comparisons in the homogeneous variance, unequal sample size case. J Am Statist Assoc 75:780–795
Dunstone N, O'Connor RJ (1979) Optimal foraging in an amphibious mammal. I. The aqualung effect. Anim Behav 27:1182–1194
Eggers DM (1986) Preliminary forecasts and projections for 1986 Alaska salmon fisheries. Alaska Dep Fish Game Inf Leaflet 253:55 pp
Eggers DM, Dean MR (1987) Alaska commercial salmon catches, 1978–1986. Alaska Dep. Fish Game Reg Inf Rep 5J87-01:69 pp
Enfield DB (1987) Progress in understanding El Niño. Endeavour New Ser 11:197–204
Everitt RD, Fiscus CH, DeLong RL (1980) Northern Puget Sound marine mammals. Doc EPA-600/7-80-139 EPA, Washington, DC, 134 pp
Favorite F, McLain DR (1973) Coherence in transpacific movements of positive and negative anomalies of sea surface temperature, 1953–60. Nature (London) 244:139–143
Feldkamp SD (1985) Swimming and diving in the California sea lion, Zalophus californianus. PhD Thesis, Univ Cal, San Diego
Feldkamp SD (1987) Swiming in the California sea lion: morphometrics, drag and energetics. J Exp Biol 131:117–135
Feldkamp SD, DeLong RL, Antonelis GA (1989) Diving patterns of California sea lions, Zalophus californianus. Can J Zool 67:872–883
Feldman GC (1984) Satellites, seabirds and seals. Trop Ocean Atmosph Newslett 28:4–5
Feldman GC, Clark D, Halpern H (1984) Satellite color observations of the phytoplankton distribution in the eastern equatorial Pacific during the 1982–1983 El Niño. Science 226:1069–1071

Fiedler PC (1984a) Sattelite observations of the 1982–1983 El Niño along the US Pacific Coast. Science 224:1251–1254
Fiedler PC (1984b) Some effects of El Niño 1983 on the northern anchovy. CalCoFi Rep 25:53–58
Fielder PC, Methot RD, Hewitt RP (1986) Effects of California El Niño 1982–1984 on the northern anchovy. J Mar Res 44:317–338
Firing E, Lukas R, Sadler J, Wyrtki K (1983) Equatorial undercurrent disappears during 1982–83 El Niño. Science 222:1121–1123
Fisher RA (1930) The genetic theory of natural selection. Clarendon, Oxford
Fisher RA (1954) Statistical methods for research workers, 12th edn. Oliver & Boyd, Edinburgh, 356 pp
Fisher RA (1958) The genetical theory of natural selection, 2nd edn. Dover, New York
Fisher JP, Pearcy WG (1988) Growth of juvenile coho salmon (Oncorhynchus kisutch) in the ocean off Oregon and Washington, USA, in years of differing coastal upwelling. Can J Fish Aquat Sci 45:1036–1044
Flowerdew JR (1987) Mammals: their reproductive biology and population ecology. Arnold, London
Fonseca TR (1985) Efectos fisicos del fenomeno El Niño 1982–83 en la costa Chilena. Invest Pesq (Spec Issue) 32:61–68
Foucher RP, Tyler AV (1988) Pacific cod. In: Fargo J, Saunders MW, Tyler AV (eds) Groundfish stock assessments for the west coast of Canada in 1987 and recommended yield options for 1988. Can Tech Rep Fish Aquat Sci 1617, pp 45–74
Francis JM, Heath CB (1983) Carcass loss and the accuracy of pup mortality estimates for the California sea lion. In: Mugu Lagoon, San Nicolas Island Ecol Res Symp, Oct 21–22, 1983. Pacific Missle Test Center, Pt. Mugu, CA, pp 137–144
Frisch RE (1985) Fatness, menarche, and female fertility. Perspect Biol Med 28:611–633
Frisch RE (1988) Fatness and fertility. Sci Am March (3) 1988:70–77
Frisman E, Skalestkaya EI, Kuzin AE (1982) A mathematical model of the population dynamics of a local northern fur seal herd. Ecol Modell 16:151–172
Frost BW (1983) Interannual variation of zooplankton standing stock in the open Gulf of Alaska. In: Wooster WS (ed) From year to year. Interannual variability of the environment and fisheries of the Gulf of Alaska and the Bering Sea. Univ Washington Sea Grant Publ, Seattle, WA, pp 146–157
Fry DH (1939) A winter influx of sea lions from lower California. Fish Game Bull 25:245–250
Fuenzalida R (1985) Aspectos oceanograficos y meteorologicos de El Niño 1982–83 en la zona costera de Iquique. Invest Pesq (Spec Issue) 32:47–52
Fulton JD, LeBrasseur RJ (1985) Interannual shifting of the subarctic boundary and some of the biotic effects on juvenile salmonids. In: Wooster WS, Fluharty DL (eds) El Niño north: Niño effects in the eastern subarctic Pacific. Washington Sea Grant Publ WSG-WO 85-3, Seattle, WA, pp 237–247
Funk F, Savikko H (eds) (1989) Preliminary forecasts and projections for 1989, Alaska herring fisheries. Alaska Dep Fish Game Reg Inf Rep 5J89–02, 98 pp
Galindo I, Otaola JA, Grivel F, Gallegos A (1986) Meteorological and physical oceanography aspects of El Niño on the Mexican Pacific coasts. In: Abstr Intergov Oceanogr Comm Worksh Rep 49. Guayaquil, Ecuador, p 56
Gallegos A, de La Lanza G, Ramos F (1986) The 1982–1983 El Niño in the coastal waters off Guerrero, Mexico. In: Abstr Intergov Oceanogr Comm Worksh Rep 49. Guayaquil, Ecuador, p 27
Gearin P, DeLong R, Antonelis GA (1986a) Population and behavioral studies, San Miguel Island, California. (Unpubl MS, available from National Marine Mammal Laboratory, Northwest and Alaska Fisheries Center, Bldg 4, 7600 Sand Point Way NE, Seattle, WA 98115, USA)
Gearin P, Pfeifer B, Jefferies S (1986b) Control of California sea lion predation of winter-run steelhead at the Hiram M. Chittenden Locks, Seattle, Dec 1985 – Apr 1986. Washington Dep Game Fish Manag Rep 86-20, pp 1–108
Gentry RL, Holt JR (1982) Equipment and techniques for handling northern fur seals. US Dep Commerc NOAA Tech Rep NMFS SSRF-758
Gentry RL, Holt JR (1986) Attendance behavior of northern fur seals. In: Gentry RL, Kooyman GL (eds) Fur seals: maternal strategies on land and at sea. Univ Press, Princeton, NJ, pp 41–60
Gentry RL, Kooyman GL (eds) (1986a) Fur seals: maternal strategies on land and at sea. Univ Press, Princeton, NJ
Gentry RL, Kooyman GL (eds) (1986b) Methods of dive analysis. In: Fur seals: maternal strategies on land and at sea. Univ press, Princeton, J, pp 28–40
Gentry RL, Costa DP, Croxall JP, David JHM, Davis RW, Kooyman GL, Majluf P, McCann TS, Trillmich F (1986a) Synthesis and conclusions. In: Gentry RL, Kooyman GL (eds) Fur seals: maternal strategies on land and at sea. Univ Press, Princeton, NJ, pp 220–264
Gentry RL, Goebel ME, Roberts WE (1986b) Behavior and biology, Pribilof Islands, Alaska. In: Kozloff P (ed) Fur seal investigations, 1984. US Dep commerc NOAA Tech Mem NMFS F/NWC-97, pp 29–40
Gentry RL, Kooyman GL, Goebel ME (1986c) Feeding and diving behavior of northern fur seals. In: Gentry RL, Kooyman GL (eds) Fur seals: maternal strategies on land and at sea. Univ Press, Princeton, NJ, pp 61–78

George-Nascimento M, Bustamante R, Oyarzun C (1985) Feeding ecology of the South American sea lion Otaria flavescens: food contents and food selectivity. Mar Ecol Prog Ser 21:135–143

Gisiner RC (1985) Male territorial and reproductive behavior in the Steller sea lion Eumetopias jubatus. PhD Thesis, Univ, Cal Santa Cruz, 146 pp

Gittleman JL, Oftedal OT (1987) Comparative growth and lactation energetics in carnivores. Symp Zool Soc London 57:41–77

Glantz MH, Thompson JD (eds) (1981) Resource management and environmental uncertainty: lessons from coastal upwelling fisheries. John Wiley & Sons, New York, 491 pp

Goebel ME (1988) Duration of feeding trips and age-related reproductive success of lactating females, St. Paul Island, Alaska. In: Kozloff P, Kajimura H (eds) Fur seal investigations, 1985. US Dep Commerc NOAA Tech Mem NMFS F/NWC-146, pp 28–33

Goebel ME, Bengtson JL, DeLong RL, Gentry RL, Loughlin TR (1991) Diving patterns and foraging locations of female northern fur seals. Fish Bull, US, 89:171–179

Gosling LM, Baker SJ, Wright KMH (1984) Differential investment by female coypus (Myocaster coypus) during lactation. Symp Zool Soc London 51:273–300

Graham NE, White WB (1988) The El Niño cycle: a natural oscillator of the Pacific ocean-atmosphere system. Science 240:1293–1302

Groot C, Quinn TP (1987) Homing migration of sockeye salmon, Oncorhynchus nerka, to the Fraser River. Fish Bull 85:455–469

Grover JJ, Olla BL (1987) Effects of an El Niño event on the food habits of larval sablefish, Anoplopoma fimbria, off Oregon and Washington. Fish Bull 85:71–79

Guerra C, Torres D (1987) Presence of the South American fur seal, Arctocephalus australis, in northern Chile. In: Croxall JP, Gentry RL (eds) Proc Int Symp Worksh, Status, biology, and ecology of fur seals. Cambridge, UK, 23–27 Apr 1984. NOAA Tech Rep NMFS 51, pp 169–175

Guerra C, Portflitt G, Gomez-Bonta JM (1987) Criterios cientificos y técnicos para el manejo del lobo marino común Otaria flavescens (Shaw) en el norte del Chile. In: Arana P (ed) Manejo y desarollo pesquero, Escuela Cienc Mar, UCV, Valparaiso, pp 215–232

Guillén O, Lostaunau N, Jacinto M (1985) Characteristicas del fenomeno El Niño 1982–83. In: Arntz WE, Landa A, Tarazona J (eds) "El Niño". Su impacto en la fauna marina. Bol Inst Mar Perú, Callao (Spec Issue), pp 11–21

Hacker ES (1986) Stomach contents analysis of short-finned pilot whales (Globicephala macrorhynchus) and northern elephant seals (Mirounga angustirostris) from the southern California Bight. NMFS SW Fish Cent Admin Rep LJ-86-086, 34 pp

Hacker ES, Antonelis GA Jr (1986) Pelagic food habits of northern fur seals. In: Summary of joint research on diets of northern fur seals and fish in the Bering Sea during 1985. NW Alaska Fish Cent Proc Rep 86–19, pp 5–22

Haegele CW, Schweigert JJ (1985) Distribution and characteristics of herring spawning grounds and description of spawning behavior. Can J Fish Aquat Sci 42 (Suppl 1): 39–55

Haist V, Stocker M (1985) Growth and maturation of Pacific herring (Clupea harengus pallasi) in the Strait of Georgia. Can J Fish Aquat Sci 42 (Suppl 1):138–146

Haist V, Schweigert JF, Fournier D (1988) Stock assessments for British Columbia herring in 1987 and forecasts of the potential catch in 1988. Can MS Rep Fish Aquat Sci 1990, 63 pp

Hamilton K, Emery WJ (1985) Regional atmospheric forcing of interannual surface temperature and sea level variability in the northeast Pacific. In: Wooster WS, Fluharty DL (eds) El Niño North. Washington Sea Grant Progr, Univ Washington, SA, pp 22–30

Hatch SA (1987) Did the 1982–83 El Niño – southern oscillation affect seabirds in Alaska? Wilson Bull 99:468–474

Hay DE (1985) Reproductive biology of Pacific herring (Clupea harengus pallasi). Can J Fish Aquat Sci 42 (Suppl 1): 111–126

Hayes S (1985) Sea level and near surface temperature variability at the Galapagos islands, 1979–1983. In: Robinson G, del Pino EM (eds) El Niño en las islas Galápagos. El evento de 1982–1983. Fund Charles Darwin Islas Galápagos, Quito, Ecuador, pp 49–82

Hays C (1986) Efectos de "El Niño" 1982–83 en las colonias del pinguino de Humboldt en el Perú. Bol Lima 45:39–47

Heath CB, Francis JM (1983) Population dynamics and feeding ecology of the California sea lion with applications for management. Results of 1981–1982 research on Santa Barbara and San Nicolas Islands. SW Fish Cent Admin Rep LJ-83-04C, NMFS, La Jolla, CA

Heath CB, Francis JM (1984) Results of research on California sea lions, San Nicolas Island 1983. SW Fish Cent Admin Rep LJ-84–41C, NMFS, La Jolla, CA

Herdson D (1984) Changes in the demersal fish stocks and other marine life in Ecuadorean coastal waters during the 1982–83 El Niño. Trop Ocean Atmosph Newslett 28:14–16

Hewitt RP (1985) The 1984 northern anchovy spawning biomass. CalCoFi Rep 26:17–25

Hodder J, Graybill MR (1985) Reproduction and survival of seabirds in Oregon during the 1982-83 El Niño. Condor 87:535–541

Hollander M, Wolfe D (1973) Nonparametric statistical methods. John Wiley & Sons, New York
Hollowed AB, Bailey KM (1989) New perspectives on the relationship between recruitment of Pacific hake (Merluccius productus) and the ocean environment. Can Spec Publ Fish Aquat Sci 108 (in press)
Hollowed AB, Methot R, Dorn M (1988) Status of the Pacific whiting resource in 1988 and recommendations to managements in 1989. Appendix A. In: Status of the Pacific coast ground fishery through 1988 and recommended acceptable biological catches for 1989. Pac Fish Manag Connec, Portland, OR, 49 pp and appendices
Holms S (1979) A simple sequentially rejective multiple tests procedure. Scan J Statist 6:65–70
Hourston AS, Haist V, Humphreys RD (1981) Regional and temporal variation in the fecundity of Pacific herring in British Columbia waters. Can Tech Rep Fish Aquat Sci 118:101
Houvenaghel GT (1984) Oceanographic setting of the Galápagos Islands. In: Perry R (ed) Key environments Galápagos. Pergamon, Oxford, pp 43–56
Huber HR (1987) Natality and weaning success in relation to age of first reproduction in the northern elephant seal. Can J Zool 65:1311–1316
Huber HR, Ailey DG, Boekelheide RJ, Henderson RP, Lewis TJ (in press) Annual and seasonal variation in numbers of pinnipeds on the Farallon Islands, California, 1971–1986. Mar Mammal Sci
Hunt GL Jr, Schneider DC (1987) Scale-dependent processes in the physical and biological environment of marine birds. In: Croxall JP (ed) Seabirds: feeding ecology and role in marine ecosystems. Cambridge Univ Press, Cambridge, pp 7–41
Huyer, Smith RL (1985) The signature of El Niño off Oregon, 1982–1983. J Geophys Res 90, C4:7133–7142
Icochea L (1989) Análisis de la pesquería de arrastre pelágica en la costa peruana durante el período 1983–1987 y su relación con los cambios oceanográficos. Flota Pesquera Peruana (FLOPESCA), 51 pp (mimeo)
Idyll CP (1973) The anchovy crisis. Sci Am 228, 6:22–29
IFOP (Instituto de Fomento Pesquero – ed) (1985) Taller nacional fenómeno el Niño. Invest Pesq 32 (Spec Issue): 242 pp
IMARPE (ed) (1969) Resultados preliminares del primer crucero de exploración pesquera del SNP-1 6901. Bol Inst Mar Perú, Callao. Ser Inf Espec IM-39
IOC (Intergovernmental Oceanographic Commission – ed) (1986) AGU-IOC-WMO-CPPS. Chapman Conf Int Symp El Niño. Guayaquil, Ecuador, 83 pp
Iverson SJ (1988) Composition, intake and gastric digestion of milk lipids in pinnipeds. PhD Thesis, Univ Maryland, College Park, MD
Jefferies S (1984) Marine mammals of the Columbia River estuary. Final Rep Washington Dep Game, Olympia, WA
Jiménez C, Guzmán J (1986) Distribution and breeding success of the brown pelican in the Bay of La Paz BCS, Mexico. In: Abstr 12th Annu Meet Pacific Seabird Group, Dec 1986, La Paz BCS, Mex
Jiménez R, Intriago P (1986) Equatorial upwelling processes in the Galápagos islands after the 1982–83 "El Niño". In: Abstr Intergov Oceanogr Comm Worksh Rep 49, Guayaquil, Ecuador, p 68
Johnson AM (1968) Annual mortality of territorial male fur seals and its management significance. J Wildlife Manag 32(1):94–99
Johnson SL (1988) The effects of the 1983 El Niño on Oregon's coho (Oncorhynchus kisutch) and chinook (O. tsawytcha) salmon. Fish Res 6:105–123
Jordán R (1971) Distribution of anchoveta (Engraulis ringes J) in relation to the environment. Invest Pesq 35:113–126
Jordán R, Chirinos de Vildoso A (1965) La anchoveta (Engraulis ringens J). Conocimiento actual sobre su biología, ecologia y pesquería. Bol Inst Mar Perú, Callao, Inf 6
Kajimura H (1979) Fur seal pup/yearling distribution in the eastern North Pacific. In: Kajimura H, Lander RH, Perez MA, York AE, Bigg MA (eds) Preliminary analysis of pelagic fur seal data collected by the United STates and Canada during 1958–74. 22nd Annu Meet Standing Scientific Committee, North Pacific Fur Seal Commission, 2–6 April 1979, Washington, DC, pp 9–50
Kajimura H (1980) Distribution and migration of northern fur seals (Callorhinus ursinus) in the Eastern Pacific. In: Kajimura H, Lander RA, Perez MA, York AE (eds) Further analysis of pelagic fur seal data collected by the United States and Canada during 1958–74. 23rd Annu Meet Standing Sci Committee, North Pacific Fur Seal Commission, 7–21 April 1980, Moscow, USSR, pp 4–43
Kajimura H (1984) Opportunistic feeding of the northern fur seal, Callorhinus ursinus, in the eastern North Pacific Ocean and eastern Bering Sea. US Dep Commerc NOAA Tech Rep NMFS SSRF-799, 49 pp
Kajimura H, Fowler CW (1984) Apex predators in the walleye pollock ecosystem in the eastern Bering Sea and Aleutian Island regions. In: Ito DH (ed) Proc Worksh Walleye Pollock and its ecosystem in the eastern Bering Sea. NOAA Tech Mem NMFS F/NWC-62, pp 193–233
Kajimura H, Loughlin TR (1988) Marine mammals in the oceanic food web of the eastern subarctic Pacific. Bull Ocean Res Inst Univ Tokyo 26 (pt 2):187–223

Keith EO, Condit RS, Le Boeuf BJ (1984) California sea lions breeding at Año Nuevo Island, California. J Mammal 65:695
Kelly R (1985) Aspectos generales El Niño 1982–83. Invest Pesq (Spec Issue) 32:5–7
Kelly R, Blanco JL (1984) Estudio oceanográfico de las aguas del Norte de Chile durante Febrero y Marzo de 1983. Rev Com Perm Pac Sur 15:179–201
Kendall M, Stuart A (1977) The advanced theory of statistics, 4th edn. vol 1. Distribution theory. MacMillan, New York, 49 pp
Kerley GIH (1985) Pup growth in the fur seals Arctocephalus tropicalis and A. gazella on Marion Island. J Zool 205:315–324
Kerr RA (1983) Fading El Niño broadening scientists view. Science 221:940–941
Kerr RA (1988) La Niña's big chill replaces El Niño. Science 241:1037–1038
Keyes MC (1965) Pathology of the northern fur seal. J Am Vet Med Assoc 147:1090–1095
Kimura DK, Ronholt LL (1990) Atka mackerel. In: Condition of groundfish resources of the eastern Bering Sea and Aleutian Islands region in 1988. NOAA Tech Mem NMFS F/NWC-178, pp 93–98
Kleiber M (1975) The fire of life: an introduction to animal energetics. Krieger, Huntington, NY
Knutson C, Lesh J (1984) Salmon in 1983 – blame it on El Niño. Outdoor Cal March-April: 1–4
Kogelschatz J, Solorzano L, Barber R, Mendoza P (1985) Oceanographic conditions in the Galapagos islands during the 1982/1983 El Niño. In: Robinson G, del Pino EM (eds) El Niño en las Islas Galapagos. Fund Charles Darwin Islas Galapagos. Quito, Ecuador, pp 91–123
Kooyman GL (1989) Diverse divers. Springer, Berlin Heidelberg New York, 200 pp
Kooyman GL, Gentry RL (1986) Diving behavior of South African fur seals. In: Gentry RL, Kooyman GL (eds) Fur seals: maternal strategies on land and at sea. Univ Press, Princeton, NJ, pp 142–152
Kooyman GL, Trillmich F (1986a) Diving behavior of Galapagos fur seals. In: Gentry RL, Kooyman GL (eds) Fur seals: maternal strategies on land and at sea. Univ Press, Princeton, NJ, pp 186–195
Kooyman GL, Trillmich F (1986b) Diving behavior of Galapagos sea lions. In: Gentry RL, Kooyman GL (eds) Fur seals: maternal strategies on land and at sea. Univ Press, Princeton, NJ, pp 209–219
Kooyman GL, Gentry RL, McAlister WB (1976) Physiological impact of oil on pinnipeds. US Dep Commerc NOAA, NMFS, NW Alaska Fish Cent, Seattle, Washington, 23 pp
Kooyman GL, Billups JO, Farwell WD (1983) Two recently developed recorders for monitoring diving activity of marine birds and mammals. In: MacDonald AG, Priede IG (eds) Experimental biology at sea. Academic Press, New York London, pp 197–214
Kramer D (1988) The behavioral ecology of air breathing aquatic animals. Can J Zool 66:89–94
Krebs CJ (1972) Ecology: the experimental analysis of distribution and abundance. Harper & Row, New York
Krebs JB (1978) Optimal foraging: decision rules for predators. In: Krebs JB, Davies NB (eds) Behavioral ecology, an evolutionary approach. Sinauer, Sunderland, MA, pp 23–63
Kundu PK, Allen JS (1976) Some three-dimensional characteristics of low-frequency current fluctuations near the Oregon coast. J Phys Oceanogr 6:181–199
Labov JB, Huck UW, Vaswani P, Lisk RD (1986) Sex ratio manipulation and decreased growth of male offspring of undernourished golden hamsters (Mesocricetus and auratus). Behav Ecol Sociobiol 18:241–249
Lack D (1966) Population studies of birds. Clarendon, Oxford
Lander RH (1975) Method of determining natural mortality in the northern fur seal (Callorhinus ursinus) from known pups and kill by age and sex. J Fish Res Board Can 32:2447–2452
Lander RH (1979) Role of land and ocean mortality in yield of Alaskan fur seals. Fish Bull 77:311–314
Lara JR, Valdez E, Bazan C, Lara JL (1986) Primary productivity in the Gulf of California during 1982–1985. In: Abstr Intergov Oceanogr Comm Worksh Rep 49. Guayaquil, Ecuador, p 63
Lara-Lara JR, Holguín-Valdéz JE, Jiménez-Pérez LC (1984) Plankton studies in the Gulf of California during the 1982–83 El Niño. Trop Ocean Atmosph Newslett 28:16–17
Lara-Osorio J, Lara-Lara JR (1987) Productividad y biomasa del fitoplancton por tamano de clases en la parte central del Golfo de California en condiciones de El Niño. In: Abstr 7. Congr Nacl Oceanogr, 27–31 July 1987. Ensenada, BC, Mexico, p 202
Layne JN (1968) Ontogeny. Spec Pub Am Soc Mammal 2:148–253
Le Boeuf BJ (1974) Male-male competition and reproductive success in elephant seals. Am Zool 14:163–176
Le Boeuf BJ, Briggs KT (1977) The cost of living in a seal harem. Mammalia 41:167–195
Le Boeuf BJ, Bonnell ML (1980) Pinnipeds of the California Channel Islands: abundance of distribution. In: Power D (ed) The California Islands: Proc Multidisc Symp, Santa Barbara Mus Nat Hist, Santa Barbara, CA, pp 475–493
Le Boeuf BJ, Condit RS (1983) The high cost of living on the beach. Pac Discovery, July-September: 12–14
Le Boeuf BJ, Reiter J (1988) Lifetime reproductive success in northern elephant seals. In: Clutton-Brock T (ed) Reproductive success. Univ Press, Chicago, IL, pp 344–362
Le Boeuf BJ, Whiting RJ, Gantt RF (1972) Perinatal behavior of northern elephant seal females and their young. Behavior 43:121–156

Le Boeuf BJ, Ainley DG, Lewis TJ (1974) Elephant seals on the Farallons: population structure of an incipient breeding colony. J Mammal 55:370–375

Le Boeuf BJ, Aurioles D, Condit R, Fox C, Gisiner R, Romero R, Sinsel F (1983) Size and distribution of the California sea lion (Zalophus californianus) population in México. Proc Cal Acad Sci 43:77–85

Le Boeuf BJ, Costa DP, Huntley AC, Kooyman GL, Davis RW (1986) Pattern and depth of dives in northern elephant seals, Mirounga angustirostris. J Zool London 208:1–7

Le Bouef BJ, Costa DP, Huntley AC, Feldkamp SD (1988) Continuous, deep diving in female northern elephant seals, Mirounga angustirostris. Can J Zool 66:446–458

Le Boeuf BJ, Condit R, Reiter J (1989a) Parental investment and the secondary sex ratio in northern elephant seals. Behav Ecol Sociobiol 25:109–117

Le Boeuf BJ, Naito Y, Huntley AC, Asaga T (1989b) Prolonged, continuous, deep diving by northern elephant seals. Can J Zool 67:2514–1519

Lee PC (1987) Nutrition, fertility and maternal investment in primates. J Zool London 213:409–422

Lee PC, Moss CJ (1986) Early maternal investment in male and female elephant calves. Behav Ecol Sociobiol 18:353–361

Lewontin RC (1965) Selection for colonizing ability. In: Baker HG, Stebbins GL (eds) The genetics of colonizing species. Academic Press, New York London, pp 77–94

Lifson N, McClintock R (1966) Theory and use of the turnover rates of body water for measuring energy and material balance. J Theor Biol 12:46–74

Limberger D (1985) El Niño on Fernandina. In: Robison G, del Pino EM (eds) El Niño en las islas Galapagos. El evento de 1982–1983. Fund Charles Darwin Islas Galapagos. Quito, Ecuador, pp 211–225

Limberger D, Trillmich F, Kooyman GL, Majluf P (1983) Reproductive failure of fur seals in Galapagos and Peru in 1982–83. Trop Ocean Atmosph Newslett 21:16–17

Lindstedt SL, Boyce MS (1985) Seasonality, fasting endurance, and body size in mammals. Am Nat 125:873–878

Lluch-Belda D (1969) El lobo marino de California Zalophus californianus californianus (Lesson 1828) Allen, 1880. Observaciones sobre su ecologia y explotacion. Inst Mex Recur Nat Renov, AC Mexico, DF, 69 pp

Loudon ASI, Kay RNB (1984) Lactational constraints on a seasonally breeding mammal: the red deer. Symp Zool Soc London 51:233–252

Loughlin TR, Livingston PA (1986) Summary of joint research on the diets of northern fur seals and fish in the Bering Sea during 1985. US Dep Commerc NWAFC Processes Rep, pp 85–91

Loughlin TR, Bengston JL, Merrick RL (1987) Characteristics of feeding trips of female northern fur seals. Can J Zool 65:2079–2084

Loveridge GG (1986) Bodyweight changes and energy intake of cats during gestation and lactation. Anim Technol 37:7–15

Lowry LF, Frost KJ, Louglin TR (1989) Importance of walleye pollock in the diets of marine mammals in the Gulf of Alaska and Bering Sea, and implications for fishery management. In: Proc Int Symp Biol Manag Walleye Pollock, pp 701–726

Loy L (1970) Behavioral responses of free-ranging rhesus monkeys to food shortage. Am J Phys Anthrop 33:263–272

MacCall AD, Klingbeil RA, Methot RD (1985) Recent increased abundance and potential productivity of Pacific mackerel (Scomber japonicus). CalCoFi Invest Rep 26:119–129

Macy PT, Wall JM, Lampsakis ND, Mason JE (1978) Resources of the non-salmonid pelagic fishers of the Gulf of Alaska and the eastern Bering Sea, pt 1. US Dep Commerc NOAA, NMFS, NW Alaska Fish Cent Seattle, Washington, 355 pp

Macy SK (1982) Mother-pup interactions in the northern fur seal. PhD Thesis, Univ Washington, Seattle, WA, 155 pp

Majluf MP (1985) Comportamiento del lobo fino de Sudamerica (Arctocephalus australis) en Punta San Juan, Peru, durante “El Niño” 1982–83. In: Arntz WE, Landa A, Tarazona J (eds) “El Niño”. Su impacto en la fauna marina. Bol Inst Mar Perú, Callao (Spec Issue), pp 187–193

Majluf MP (1987) Reproductive ecology of female south American fur seals at Punta San Juan, Peru. PhD Thesis, Univ Cambridge, UK

Majluf MP, Trillmich F (1981) Distribution and abundance of sea lions (Otaria byronia) and fur seals (Arctocephalus australis) in Peru. Z Säugetierkd 46:384–393

Maridueña LS (1986) Important changes in the Ecuadorian pelagic fisheries since El Niño 1982–1983. In: Abstr Intergov Oceanogr Comm Worksh Rep 49. Guayaquil, Ecuador, p 34

Martínez C, Salazar C, Böhm G (1984) La pesquería cerquera ejercida en el norte de Chile y su relación con los cambios biológicos-pesqueros asociados al fenómeno de El Niño 1982–83. Rev Com Perm Pac Sur 15:203–221

Martínez CF, Salazar CZ, Böhm GS, Mendieta JC, Estrada CM (1985) Efectos del fenómeno El Niño 1982–83 sobre los principales recursos pelágicos y su pesquería (Arica-Antofagasta). Invest Pesq Chile (Spec Issue) 32:129–139

Mate BR (1973) Population kinetics and related ecology of the northern sea lion, Eumetopias jubatus, and the California sea lion, Zalophus californianus, along the Oregon coast. PhD Thesis, Univ Oregon, Corvallis
Mate BR (1975) Annul migration of the sea lions Eumetopias jubatus and Zalophus californianus along the Oregon USA coast. Rapp P Reun Cons Inst Explor Mer 169:455–461
Maynard Smith J (1978) The evolution of sex. Cambridge Univ Press, Cambridge, UK
Maynard Smith J (1980) A new theory of sexual investment. Behav Ecol Sociobiol 7:247–251
McClure PA (1981) Sex baised litter reduction in food-restricted wood rats (Neotoma floridana). Science 211:1058–1060
McCullagh P, Nelder JA (1983) Generalized linear models. Chapman & Hall, London
McGowan JA (1984) The California El Niño 1983. Oceanus 27:48–81
McGowan JA (1985) El Niño 1983 in the southern California Bight. In: Wooster WS, Fluharty DL (eds) El Niño north. Washington Sea Grant Progr, Univ Washington, Seattle, pp 166–184
McLain DR (1983) Coastal ocean warning in the northeast Pacific, 1976–1983. In: Pearcy WG (ed) The influence of ocean conditions on the production of salmonids in the North Pacific. Worksh Sea Grant Coll Progr Rep ORESU-W-83-001. Oregon State Univ Newport, RI, pp 61–86
McClain DR (1984) Coastal ocean warning in the northeast Pacific, 1976–1983. In: Pearcy WG (ed) The influence of ocean conditions on the production of salmonids in the North Pacific. Oregon Sea Grant Publ ORESU-W-85-001, Oregon State Univ, Corvallis, pp 61–86
McLain DR, Brainard RE, Norton JG (1985) Anomalous warm events in eastern boundary current systems. CalCoFi Invest Rep 26:51–64
Mee LD, Ramírez-Flores A, Flores-Verdugo F, González-Farias F (1985) Coastal upwelling and fertility of the southern Gulf of California: impact of the 1982–83 ENSO event. Trop Ocean Atmosph Newslett 31:9–10
Megrey BA (1989) Gulf of Alaska walleye pollock: population assessments and status of the resource as estimated in 1988. In: Conditions of groundfish resources of the Gulf of Alaska in 1988. NOAA Tech Mem NMFS F/NWC-165, pp 1–54
Methot RD, Lo NCH (1987) Spawning biomass of the northern anchovy in 1987. SW Fish Cent Admin Rep LJ-87–14, 34 pp
Miller JS (1977) Adaptive features of mammalian reproduction. Evolution 31:370–386
Miller CB, Batchelder HP, Brodeur RD, Pearcy WG (1985) Response of the zooplankton and ichthyoplankton off Oregon to the El Niño event of 1983. In: Wooster WS, Fluharty DL (eds) El Niño north. Washington Sea Grant Progr, Univ Washington, Seattle, pp 185–187
Miller DJ, Herder MJ, Schol JP (1983) California marine mammal/fishery interaction study, 1979–1981. Admin Rep LJ-83-13C SWFC, La Jolla, CA, 233 pp
Moulton CR (1923) Age and chemical development in mammals. J Biol Chem 57:79–97
Muck P, Sandoval de Castillo O, Carrasco S (1987) Abundance of sardine, mackerel and horse mackerel eggs and larvae and their relationship to temperature, turbulence and anchoveta biomass off Perú. In: Pauly D, Tsukayama I (eds) The Peruvian anchoveta and its upwelling ecosystem: three decades of change. ICLARM Stud Rev 15:268–275
Muller-Schwarze D, Stagge B, Muller-Schwarze C (1982) Play behavior: persistence, decrease, and energetic compensation during food shortage in deer fawns. Science 215:85–87
Muñoz P (1985) Estructura y comportamiento de las communidades fitoplanctónicas en el norte de Chile durante el fenómeno de El Niño 1982–83. Invest Pesq Chile (Spec Issue) 32:195–197
Mysak LA (1986) El Niño, interannual variability and fisheries in the northeast Pacific Ocean. Can J Fish Aquat Sci 43:464–497
Nagy KA (1975) Water and energy budgets of free-living animals: measurement using isotopically labaled water. In: Hadley N (ed) Environmental physiology of desert organisms. Dowden, Hutchinson & Ross, Stroudsberg, PA, pp 227–245
Nagy KA (1980) CO2 production in animals: analysis of potential errors in doubly-labeled water technique. Am J Physiol 238:R466-473
Nagy KA, Costa DP (1980) Water flux in animals: analysis of potential errors in the tritiated water method. Am J Physiol 238:R454–R465
Naito Y, Le Boeuf BJ, Asaga T, Huntley AC (1989) Preliminary analysis of long term continuous diving in female elephant seals, Mirounga angustirostris. Ant Rec 33:1–9
Neter JW, Wasserman W, Kutner MH (1985) Applied linear statistical models, 2nd edn. Irwin, Homewood, IL
Nicholas JW, Hankin DG (1988) Chinook salmon populations in Oregon coastal river basins: description of life histories and assessments of recent trends in run strengths. Oregon Dep Fish Wildlife Inf Rep 88–1:359 pp
Nicholson KA (1986) The movement patterns of California sea lions at the Monterey Coast Guard breakwater. MS Thesis. San Francisco State Univ, 53 pp
Nickelson TE (1986) Influences of upwelling, ocean temperature, and smolt abundance on marine survival of coho salmon (Oncorhynchus kisutch) in the Oregon production area. Can J Aquat Sci 43:527–535

Nie NH, Hull CH, Jemkins JG, Steinbrenner K, Bent DH (1975) Statistical package for the social sciences. McGraw-Hill, New York
Niebauer HJ (1985) Southern Oscillation/El Niño effects in the eastern Bering Sea. In: Wooster WS, Fluharty DL (eds) El Niño North. Washington Sea Grant Progr, Univ Washington, Seattle, pp 116–120
Niebauer HJ (1988) Effects of El Niño-Southern Oscillation and North Pacific weather patterns on interannual variability in the subarctic Bering Sea. J Geophys Res 93 C:5051–5068
Niebauer HJ, Day RH (1989) Causes of interannual variability in the sea ice cover of the eastern Bering Sea. Geo Journal 18:45–59
NOAA (ed) (1982) NOAA Oceanographic monthly summary 2 (11):1–23
NOAA (ed) (1983) NOAA Oceanographic monthly summary 3 (1–12):1–23
NOAA (ed) (1984) NOAA Oceanographic monthly summary 4 (1–12):1–23
Norton J, McLain D, Brainard R, Husby D (1985) The 1982–83 El Niño event off Baja and Alta California and its ocean climate context. In: Wooster WS, Fluharty DL (eds) El Niño north. Washington Sea Grant Progr, Univ Washington, Seattle, pp 44–72
Nunallee EP, Williamson EP (1988) Results of acoustic/midwater trawl survey of walleye pollock in Shelikof Strait, Alaska, in 1988. NW Alaska Fish Cent
Odell DK (1971) Census of pinnipeds breeding on the California Channel Islands. J Mammal 52:187–190
Odell DK (1972) Studies on the biology of the California sea lion and the northern elephant seal on San Nicolas Island, California. PhD Thesis, Univ, Los Angeles
Odell KD (1981) California sea lion – Zalophus californianus. In: Ridgway SH, Harrison RJ (eds) Handbook of marine mammals, vol 1. The walrus, sea lions, fur seals and sea otter. Academic Press, New York London, pp 67–97
Oftedal OT (1981) Milk, protein and energy intakes of suckling mammalian young: a comparative study. PhD Thesis, Cornell Univ, Ithaca, NY
Oftedal OT (1984) Milk composition, milk yield and energy output at peak lactation: a comparative review. Symp Zool Soc London 51:33–85
Oftedal OT (1985) Pregnancy and lactation. In: Hudson RJ, White RD (eds) Bioenergetics of wild herbivores. CRC, Boca Raton, FL, pp 215–238
Oftedal OT, Gittelman JL (1989) Patterns of energy output during reproduction in carnivores. In: Gittleman JL (ed) Carnivore behavior, ecology and evolution. Cornstock, Ithaca, pp 355–378
Oftedal OT, Iverson SJ (1987) Hydrogen isotope methodology for measurement of milk intake and energetics of growth in suckling pinnipeds. In: Huntley AC, Costa DP, Worthy GAJ, Castellini MA (eds) Approaches to marine mammal energetics. Soc Mar Mammal Spec Publ 1. Allen, Lawrence, KA, pp 67–96
Oftedal OT, Boness DJ, Iverson SJ (1983) The effect of sampling method and stage of lactation on the composition of California sea lion milk. In: Abstr 5th Biennial Conf Biol Mar Mammal, Boston, MA, p 73
Oftedal OT, Boness DJ, Tedman RA (1987a) The behavior, physiology and anatomy of lactation in the pinnipedia. Curr Mammol 1:175–245
Oftedal OT, Iverson SJ, Boness DB (1987b) Milk and energy intakes of suckling California sea lion Zalophus californianus pups in relation to sex, growth, and predicted maintenance requirements. Physiol Zool 60:560–575
Oliver CW, Jackson TD (1987) Occurrence and distribution of marine mammals at sea from serial surveys conducted along the US West Coast between December 15, 1980 and December 17, 1980. SW Fish Cent, NMFS, Admin Rep LJ-87–19, 189 pp
Ono KA, Boness DJ, Oftedal OT (1987) The effect of a natural environmental disturbance on maternal investment and pup behavior in the California sea lion. Behav Ecol Sociobiol 21:109–118
Orians GH, Pearson NE (1979) On the theory of central place foraging. In: Horn DJ, Mitchell R, Stair GR (eds) Analysis of ecological systems. Ohio State Univ Press, Columbus, pp 155–177
Orr RT, Poulter TC (1965) The pinniped population of Año Nuevo Island, California. Proc Cal Acad Sci 32:377–404
Orr RT, Shonewald J, Kenyon KW (1970) The California sea lion: skull comparison of two populations. Proc Cal Acad Sci 37:381–394
Pace N, Rathbun EN (1949) Studies on body composition. II. The body water and chemically combined nitrogen content in relation to fat content. J Biol Chem 158:685–691
Pacific Salmon Commission (ed) (1988) Chinook technical committee annual report 88–2, 60 pp
Paine RT (1986) Benthic community-water column coupling during 1982–1983 El Niño, are community changes at high latitudes attributable to cause or coincidence? Limnol Oceanogr 31:351–360
Pauly D, Tsukayama I (1987) The Peruvian anchoveta and its upwelling ecosystem: three decades of change, ICLARM Stud Rev 15:351 pp
Pearcy WG (1983) Commentary: abiotic variations in regional environments. In: Wooster WS (ed) From year to year. Washington Sea Grant Progr, Univ Washington, Seattle, pp 30–34
Pearcy WG (1988) Factors affecting survival of coho salmon off Oregon and Washington. In: McNeil MJ (ed) Salmon production, management, and allocation. Oregon State Univ Press, Corvallis, pp 67–73

Pearcy WG, Fisher JP (1988) Migrations of coho salmon, Oncorhynchus kisutch during their first summer in the ocean. Fish Bull 86:173–195
Pearcy WG, Schoener A (1987) Changes in the marine biota coincident with the 1982–1983 El Niño in the northeastern subarctic Pacific ocean. J Geophys Res 92C: 14,417–14,428
Pearcy WG, Fisher J, Brodeur R, Johnson S (1985) Effects of the 1983 El Niño on coastal nekton off Oregon and Washington. In: Wooster WS, Fluharty DL (eds) El Niño North. Washington Sea Grant Progr, Univ Washington, Seattle, pp 188–204
Perez MA, Bigg MA (1986) Diet of northern fur seals, Callorhinus ursinus, off western North America. Fish Bull 84:957–971
Peterson RS (1965) Behavior of the northern fur seal. PhD Thesis, Johns Hopkins Univ, Baltimore, MD, p 214
Peterson RS (1968) Social behavior in pinnipeds with particular reference to the northern fur seal. In: Harrison RJ, Hubbard RC, Peterson RS, Rice CE, Schusterman RJ (eds) The behavior and physiology of pinnipeds. Appleton-Century-Crofts, New York, pp 3–53
Peterson RS, Bartholomew GA (1967) The natural history and behavior of the California sea lion. Am Soc Mammal Spec Publ 1:79 pp
Peterson RS, Le Boeuf BJ, DeLong RL (1968) Fur seals from the Bering Sea breeding in California. Nature (London) 219:899–901
PFMC (Pacific Fishery Management Council – ed) (1984) A review of the 1983 ocean salmon fisheries and status of stocks and management goals for the 1984 salmon season off the coasts of California, Oregon, PFMC, Portland, Oregon
PFMC (Pacific Fishery Management Council – ed) (1985) Ocean salmon fisheries review. PFMC, Portland, Oregon
PFMC (Pacific Fishery Management Council – ed) (1989) Review of 1988 ocean salmon fisheries. PFMC, Portland, Oregon
Pierotti RJ, Ainley DG, Lewis TJ, Coulter MC (1977) Birth of a California sea lion on southeast Farallon Island. Cal Fish Game 63:64–66
Philander SGH (1983) El Niño Southern Oscillation phenomena. Nature (London) 302:295–301
Pitcher K, Calkins D (1981) Reproductive biology of Steller sea lions in the Gulf of Alaska. J Mammal 62:599–605
Ponce de Leon A, Bianco J, Vaz-Ferreira R (1986) Interrelaciones entre Otaria flavescens y Arctocephalus australis en islas Uruguayas. (Pinnipedia, Otariidae). In: Resum 2 Reun Trabalho de especialistas em Mamiferos aquaticos da America do Sul, 4–8, Aug, Rio de Janeiro, Bras
Price T, Kirkpatrick M, Arnold SJ (1988) Directional selection and the evolution of breeding date in birds. Science 240:798–799
Prime JH, Hammond PS (1987) Quantitative assessments of grey seal diet from fecal analysis. In: Huntley AC, Costa DP, Worthy GAJ, Castellini MA (eds) Approaches to marine mammal energetics. Soc Mar Mammal Sci Spec Publ 1. Allen, Lawrence, KA, pp 165–181
Pyke GH (1984) Optimal foraging theory: a critical review. Annu Rev Ecol Syst 15:523–575
Quinn WH, Zopf DO, Short KS, Kuo Yang RTW (1977) Historical trends and statistics of the Southern Oscillation, El Niño and Indonesian droughts. Fish Bull 76:663–678
Quinn WH, Neal VT, Antuñez de Mayolo SE (1987) El Niño occurrences over the past four and a half centuries. J Geophys Res 92C: 14,449–14,461
Ramos F, Gallegos A, de la Lanza G (1986) Nutrient distribution in the shelf waters off Guerrero, Mexico during the 1982–83 ENSO episode. In: Abstr Intergov Oceanogr Comm Worksh Rep 49. Guayaquil, Ecuador, p 65
Rasmusson EM (1984) El Niño: the ocean/atmosphere connection. Oceanus 27:5–12
Rasmusson EM (1985) El Niño and variation in climate. Am Sci 73:168–177
Rasmusson EM, Wallace JM (1983) Meteorological aspects of the El Niño/Southern Oscillation. Science 222:1195–1202
Reed DK, Schumacher JD (1985) On the general circulation in the Subarctic Pacific. In: Shomura RS, Yoshida HO (eds) Proc Worksh Fate and impact of marine debris, 26–29 Nov 1984, Honolulu, Hawaii. US Dep Commerce NOAA Tech Mem NMFS-SWFC-54
Reid JL, Roden GI, Syllie JG (1958) Studies of the California current system. CalCoFi. Invest Progr Rep 7-1-56 to 1-1-58, Mar Res Comm, Dep Fish Game, Sacramento, CA, pp 27–56
Reiter J, Stinson NL, Le Boeuf BJ (1978) Northern elephant seal development: the transition from weaning to nutritional independence. Behav Ecol Sociobiol 3:337–367
Reiter J, Panken KJ, Le Boeuf BJ (1981) Female competition and reproductive success in northern elephant seals. Anim Behav 29:670–687
Repenning CA, Peterson RS, Hubbs CL (1971) Contribution to the sytematics of the southern fur seals, with particular reference to the Juan Fernandez and Guadalupe species. In: Burt WH (ed) Antarctic pinnipedia. Antarct Res Ser 18. Am Geophys Un, Washington, DC, pp 1–34
Retamales R, Gonzales L (1985) Incidencia del fenómeno El Niño 1982–83 en el desove de sardina española (Sardinops sagax). Invest Pesq Chile (Spec Issue) 32:161–165

Rice WR (1990) A consensus combined p-value test and the family-wide significance of component tests. Biometrics 46:303–308

Robinson G (1985) Influence of the 1982–83 El Niño on Galapagos marine life. In: Robinson G, del Pino EM (eds) El Niño en las Islas Galapagos: el evento de 1982–83. Fund Charles Darwin Islas Galapagos. Quito, Ecuador, pp 153–190

Robinson G, del Pino EM (eds) (1985) El Niño en las Islas Galápagos. El evento de 1982–1983. Fund Charles Darwin Islas Galápagos. Quito, Ecuador, 534 pp

Rogers DE (1984) Trends in abundance of northeastern Pacific stocks of salmon. In: Pearcy WG (ed) The influence of ocean conditions on the production of salmonids in the north Pacific. Oregon State Univ Sea Grant Coll Prog ORESU-W-83-001, pp 100–127

Rogers DE (1987) Pacific salmon. In: Hood DW, Zimmermann ST (eds) The Gulf of Alaska, physical and biological resources. NOAA, Ocean Assess Div Dep Interior, Min Manag Serv, Washington, DC, pp 461–476

Rojas de Mendiola B, Gómez O, Ochoa N (1985) Efectos del fenómeno "El Niño" sobre el fitoplancton. In: Arntz WE, Landa A, Tarazona J (eds) "El Niño". Su impacto en la fauna marina. Bol Inst Mar Perú, Callao (Spec Issue), pp 33–40

Rollins HB, Richardson JB, Sandweiss DH (1986) The birth of El Niño: geoarchaeological evidence and implications. Geoarchaeology 1:3–15

Romo D (1985) Composición química de la harina de pescado chilena durane el fenómeno El Niño 1982–83. Invest Pesq Chile (Spec Issue) 32:141–151

Ropelewski CF, Halpert MS (1986) North American precipitation patterns associated with El Niño/ Southern Oscillation (ENSO). M Weather Rev 114:2352–2362

Ropelewski CF, Halpert MS (1987) Global and regional scale precipitation and temperature patterns associated with the El Niño/Southern Oscillation. M Weather Rev 115:1606–1626

Royer TC (1985) Coastal temperature and salinity anomalies in the northern Gulf of Alaska, 1970–84. In: Wooster WS, Fluharty DL (eds) El Niño north. Washington Sea Grant Progr, Univ Washington, Seattle, pp 107–115

Saetersdal G, Tsukayama I, Alegre B (1965) Fluctuaciones en la abundancia aparente del stock de anchoveta en 1959–1962. Bol Inst Mar Perú, Callao 1:36–85

Samamé M, Castillo J, Mendieta A (1985) Situación de las pesquerías demersales y los cambios durante la presencia del fenómeno "El Niño". In: Arntz WE, Landa A, Tarazona J (eds) "El Niño". Su impacto en la fauna marina. Bol Inst Mar Perú, Callao (Spec Issue), pp 153–158

Sambrotto RN (1985) The dependence of phytoplankton nutrient utilization on physical processes in the eastern Bering Sea area: mechanisms for yearly variation. In: Wooster WS, Fluharty DL (eds) El Niño north: Niño effects in the subarctic Pacific Ocean. Univ Washington Sea Grant Progr, Seattle, pp 268–282

Samiere W, Adams P (1987) Distribution and abundance of adult anchovies (Engraulis mordax) in the Gulf of the Farallons. In: CalCoFi Invest Annu Meet, Nov 1987, Lake Arrowhead, California

Santander H (1987) Relationship between anchoveta egg standing stock and parent biomass off Peru. In: Pauly D, Tsukayama I (eds) The Peruvian anchoveta and its upwelling ecosystem: three decades of change. ICLARM Stud Rev 15:179–207

Santander H, Zuzunaga J (1984) Impact of the 1982–83 El Niño on the pelagic resources off Perú. Trop Ocean Atmosph Newslett 28:9–10

Scheffer VB (1950a) Water injury to young fur seals on the northwest coast. Cal Fish Game 36:378–379

Scheffer VB (1950b) Growth layers on the teeth of Pinnipedia as an indication of age. Science 112:309–311

Schoeller DA, van Santen E (1982) Measurement of energy expenditure in human by doubly labeled water method. J Appl Physiol 53:955–959

Schoener A, Fluharty DL (1985) Biological anomalies off Washington in 1982–83 and other major Niño periods. In: Woosters WS, Fluharty DL (eds) El Niño north: effects in the eastern subarctic Pacific Ocean. Washington Sea Grant Prog, Univ Washington, Seattle, pp 211–225

Schreiber EA, Schreiber RW (1989) Insights into seabird ecology from a global "natural experiment". Natl Geogr Res 5:64–81

Schusterman RJ, Gentry RL (1971) Development of a fatted male phenomenon in California sea lions. Dev Psychobiol 4:333–338

Seften MA, Mackas DL, Fulton J, Ashton H (1984) Zooplankton abundance and composition off British Columbia during the 1983 El Niño. Eos Trans Am Geophys Un 65:909

Secretaria Regional de Planificatión y Coordinación (SERPLAC – ed) (1981) Estudio del aprovechiamento del lobo marino en la VII regíon, Chile. Informe de avance de la II y III etapas. Intendencia Regíon del Maule. Talca, Chile

Sharp GD (ed) (1980) Colonization and modes of opportunism in the ocean. In: Worksh Effects of environmental variations on the survival of larval pelagic fishes. IOC Worksh Rep 28, pp 143–166

Shaw W, Saunders MW, Hollowed AB (1988) Pacific hake. In: Fargo J, Saunders MW, Tyler AV (eds) Groundfish stock assessments for the west coast of Canada in 1987 and recommended yield options for 1988. Can Tech Rep Fish Aquat Sci 1617:126–150

Shen GT, Boyle EA (1984) Lead and cadmium in corals: tracers of global industrial fallout and paleo-upwelling. Eos 65:964

Shen GT, Boyle EA (1986) Cadmium in corals: construction of surface ocean fertility at Galapagos. In: Abstr Intergov Oceanogr Comm Worksh Rep 49. Guayaquil, Ecuador, p 28

Shoemaker VH, Nagy KA, Costa WR (1976) Energy utilization and temperature regulation in Mojave Desert jackrabbits (Lepus californicus). Physiol Zool 49:125–142

Sidwell VH, Foncannon PF, Moore NS, Bonnet JC (1974) Composition of the edible portion of raw (fresh or frozen) crustaceans, finfish and mollusks. I. Protein, fat, moisture, ash, carbohydrate, energy value and cholesterol. Mar Fish Rev 36:21–35

Siegel S (1956) Nonparametric statistics for the behavioural sciences. McGraw-Hill, New York

Simpson JJ (1984a) El Niño-induced onshore transport in the California current during 1982–1983. Geophys Res Lett 11:233–236

Simpson JJ (1984b) A simple model of the 1982–83 Californian "El Niño". Geophys Res Lett 11:237–240

Sinclair M, Tremblay MJ, Bernal P (1985) El Niño events and variability in Pacific mackerel (Scomber japonicus) survival index: support for Hjort's second hypothesis. Can J Fish Aquat Sci 42:602–608

Smart JL (1981) Undernutrition and aggression. In: Brain PF, Benton D (eds) Multidisciplinary approaches to aggression research. Elsevier/North Holland Biomedical Press, Amsterdam, pp 179–191

Smart JL (1983) Undernutrition, maternal behavior and pup development. In: Elwood RW (ed) Parental behavior of rodents. John Wiley & Sons, New York, pp 205–224

Smith PE (1985) A case history of an anti-El Niño to El Niño transition on plankton and nekton distribution and abundances. In: Wooster WS, Fluharty DL (eds) El Niño north. Washington Sea Grant Progr, Univ Washington, Seattle, pp 121–142

Smith RL (1983) Peru coastal currents during El Niño: 1976 and 1982. Science 221:1397–1399

Smith RL (1984) Coastal currents off Peru during El Niño of 1982–83. Prenatal to senescent conditions. Trop Ocean Atmosph Newslett 28:7–9

Smith TD, Polacheck T (1981) Reexamination of the life table for northern fur seals with implications about population regulatory mechanisms. In: Fowler CW, Smith TD (eds) Dynamics of large mammal populations. John Wiley & Sons, New York, pp 99–120

Sokal RR, Rohlf FJ (1969, 1981) Biometry. Freeman, San Francisco

Spratt JD (1979) Age and growth of the market squid, Loligo opalescens Berry, from statoliths. CalCoFi Invest Rep 20:58–64

Spratt JD (1985) Biological characteristics of the catch from 1984–85 Pacific herring, Clupea harengus pallasi, roe fishery in California. Cal Dep Fish Game Mar Res Div Admin Rep 85–3, 20 pp

Spratt JD (1987a) Biomass estimates of Pacific herring, Clupea harengus pallasi, in California from the 1986–87 spawning-ground surveys. Cal Dep Fish Game Mar Res Div Admin Rep 87–12, 29 pp

Spratt JD (1987b) Biological characteristics of the catch from the 1986–87 Pacific herring, Clupea harengus pallasi, roe fishery in California. Cal Dep Fish Game Mar Res Div Admin Rep 87–13, 17 pp

Spratt JD (1987c) Variations in the growth rate of Pacific herring from San Francisco Bay, California. Cal Fish Game 73:132–138

Spray CM, Widdowson EM (1950) The effect of growth and development on the composition of mammals. Brit J Nutrit 4:332–353

Stearns SC (1976) Life-history tactics: a review of the ideas. Q Rev Biol 51:3–47

Stearns SC, Koella JC (1986) The evolution of phenotypic plasticity in life-history traits: predictions of reaction norms for age and size at maturity. Evolution 40:893–913

Stein Z, Susser M (1975) Fertility, fecundity, famine: food rations in the Dutch famine 1944/45 have a causal relation to fertility, and probably fecundity. Human Biol 47:131–154

Stewart BS (1985) Population biology of pinnipeds at San Miguel Island, 16 June through 29 August 1985. Rep Natl Mar Mammal Lab, Contract 43 ABNF-5-2468, 26 pp (Available from NW Alaska Fish Cent, 7600 Sand Point Way NE, Seattle, WA 98115, USA)

Stewart BS (1989) The ecology and population biology of the northern elephant seal, Mirounga angustirostris Gill 1866, on the Southern California Channel Islands. PhD Dissertation, Univ Cal, Los Angeles, 195 pp

Stewart BS, DeLong RL (1989) Sexual differences in migrations and foraging behavior of northern elephant seals. Am Zool 30:44A

Stewart BS, Yochem PK (1984) Seasonal abundance of pinnipeds at San Nicolas Island, California, 1980–1982. Bull S Cal Acad Sci 83:121–132

Stewart BS, Yochem PK (1985) Feeding habits of harbor seals (Phoca vitulina richardsi) at San Nicolas Island, California, 1980–1985. In: Proc 6th Biennal Conf Biol Mar Mammals, Vancouver, BC, p 76

Stewart BS, Yockem PK (1986) Northern elephant seals breeding at Santa Rosa Island, California. J Mammal 67:402–403
Stewart BS, Yochem PS, Schreiber RW (1984) Pelagic red crabs as food for gulls: a possible benefit of El Niño. Condor 86:341–342
Stewart BS, Yochem PK, DeLong RL, Antonelis GA (in press) Status and trends in abundance and status of pinnipeds on the southern California Channel Islands. In: Proc 3rd Cal Islands Symp, Santa Barbara, CA, April 2, 1987
Stirling I (1983) The evolution of mating systems in pinnipeds. In: Eisenberg JF, Kleiman DG (eds) Recent advances in the study of mammalian behavior. Am Soc Mammal Spec Publ 7:489–527
Sutcliffe WH Jr, Drinkwater K, Muir BS (1977) Correlations of fish catch and environmental factors in the Gulf of Maine. J Fish Res Board Can 34:19–30
Sverdrup HU, Johnson MW, Fleming RH (1942) The oceans: their physics, chemistry, and general biology. Prentice Hall, Englewood Cliffs, NJ, 1087 pp
Tabata S (1984) Anomalously warm water off the Pacific coast of Canada during the 1982–83 El Niño. Trop Ocean Atmosph Newslett 24:7–9
Tabata S (1985) El Niño effects along and off the Pacific coast of Canada during 1982–83. In: Wooster WS, Fluharty DL (eds) El Niño north. Washington Sea Grant Progr, Univ Washington, Seattle, pp 85–96
Taft BA (1985) El Niño of 1982–83 in the tropical Pacific. In: Wooster WS, Fluharty DL (eds) El Niño north. Washington Sea Grant Progr, Univ Washington, Seattle, pp 1–8
Taylor FHC, Wickett WP (1967) Recent changes in abundance of British Columbia herring, and future prospects. Fish Res Board Can, Pac Biol Stn Nanaimo Circ 80:17 pp
Thompson GG, Shimada AM (1988) Pacific cod. In: Condition of groundfish resources of the eastern Bering Sea and Aleutian Islands region in 1988. NOAA Tech Mem NMFS F/NWC-178, pp 44–66
Thompson LG, Mosley-Thompson E (1986) Assessment of 1500 years of climate variability and potential of long term record of ENSO events from Andean glaciers. In: Abstr Intergov Oceanogr Comm Worksh Rep 49. Guayaquil, Ecuador, p 19
Thoresen AC (1959) A biological evaluation of the Farallon Islands. Rep Div Wildlife Fish Wildlife Serv, Portland, OR
Torres D, Guerra C, Salaberry M (1983) El Lobo fino del Sur, Arctocephalus australis, en el norte de Chile. Bol Antarct Chil 3(1):23–24
Torres-Moye G, Alvarez-Borrego S (1987) Effects of the 1984 El Niño on the summer phytoplankton of a Baja California upwelling zone. J Geophys Res 92C:14, 383–14, 386
Tovar S, Fuentes H (1984) Magnitud poblacional de lobos marinos en el litoral Peruano en Marzo de 1984. Inf Inst Mar Peru 88:5–32
Tovar S, Cabrera D, del Pino M (1985) Impacto del fenómeno El Niño en la población de lobos marinos en Punta San Juan. In: Arntz WE, Landa A, Tarazona J (eds) "El Niño" su impacto en la fauna marina. Bol Inst Mar Perú, Callao (Spec Issue), pp 195–200
Townsend CH (1912) The northern elephant seal. Zoologica 1:159–173
Trillmich F (1984) Natural History of the Galapagos fur seal (Arctocephalus galapagoensis, Heller). In: Perry R (ed) Key environments – Galapagos. Pergamon, Oxford, pp 215–223
Trillmich F (1985) Effects of the 1982/83 El Niño on Galapagos fur seals and sea lions. Not Galapagos 42:22–23
Trillmich F (1986a) Attendance behavior of Galapagos fur seals. In: Gentry RL, Kooyman GL (eds) Fur seals: maternal strategies on land and at sea. Univ Press, Princeton, NJ, pp 168–185
Trillmich F (1986b) Maternal investment and sex-allocation in the Galapagos fur seal, Arctocephalus galapagoensis. Behav Ecol Sociobiol 19:157–164
Trillmich F (1986c) Attendance behavior of Galapagos sea lions: In: Gentry RL, Kooyman GL (eds) Fur seals: maternal strategies on land and at sea. Univ Press, Princeton, NJ, pp 196–208
Trillmich F (1987) Galapagos fur seal, Arctocephalus galapagoensis. In: Croxall JP, Gentry RL (eds) Status, biology, and ecology of fur seals. NOAA Tech Report NMFS 51:23–27
Trillmich F, Lechne (1986) Milk of the Galapagos fur seal and sea lion with a comparison of the milk of eared seals (Otariidae). J Zool London 209:271–277
Trillmich F, Limberger D (1985) Drastic effects of El Niño on Galapagos pinnipeds. Oecologia 67:19–22
Trillmich F, Majluf P (1981) First observations on colony structure, behavior, and vocal repertoire of the South American fur seal (Arctocephalus australis Zimmermann, 1783) in Peru. Z Säugetierkunde 46:310–322
Trillmich F, Kooyman GL, Majluf MP, Sanchez-Griñan M (1986) Attendance and diving behavior of South American fur seals during El Niño in 1983. In: Gentry RL, Kooyman GL (eds) Fur seals: maternal strategies on land and at sea. Univ Press, Princeton, NJ, pp 153–167
Trites AW (1984) Population dynamics of the Pribilof Islands North Pacific fur seal (Callorhinus ursinus). MS Thesis, Univ British Columbia, Vancouver, 123 pp
Trivers RL (1972) Parental investment and sexual selection. In: Campbell B (ed) Sexual selection and the descent of man 1871–1971. Aldine, Chicago, pp 136–179

Trivers RL (1974) Parent-offspring conflict. Am Zool 14:249–264
Trivers RL, Willard EE (1973) Natural selection of parental ability to vary the sex ratio of offspring. Science 179:90–92
US Navy (ed) (1977) Marine climatic atlas of the world, vol 2. North Pacific Ocean. NAVAIR 50-1C-529. Washington, DC
US Navy (ed) (1979) Marine climatic atlas of the world, vol 5. South Pacific Ocean. NAVAIR 50-1C-532. Washington, DC
Valdivia J (1978) The anchoveta and "El Niño". Rapp P Reun CIEM 173:196–202
Valle CA, Cruz F, Cruz JB, Merlen G, Coulter MC (1987) The impact of the 1982–1983 El Niño-Southern Oscillation on seabirds in the Galapagos Islands, Ecuador. J Geophys Res 92C:14,437–14,444
Vaz-Ferreira R, Ponce de Leon A (1984) Estudios sobre Arctocephalus australis (Zimmermann, 1783), lobo de dos pelos sudamericano, en el Uruguay. Contrib Dep Oceanogr (FHC) Montevideo 1(8):1-8
Valleman PF (1980) Definition and comparison of robust nonlinear data smoothing algorithms. J Am Statist Assoc 75:609–615
Vilchez R, Muck P, Gonzales A (1988) Variaciones en la biomass y en la distribucion de los principales recursos pelagicos del Peru entre 1983 y 1987. In: Salzwedel H, Landa A (eds) Recursos y dinamica del ecosistema de afloramiento Peruano. Bol Inst Mar Peru, Callao (Spec Issue), pp 255–264
Villanueva R, Jordan R, Burd A (1969) Informe sobre el estudio de comportamiento de cardumenes de anchoveta. Inst Mar Peru Ser Inf Espec IM-45:18 pp
Vladimirov VA (1987) Age-specific reproductive behavior in northern fur seals on the Commander Islands. In: Croxall JP, Gentry RL (eds) Status, biology and ecology of fur seals. US Dep Commerc NOAA Tech Rep 51:113–120
Walker PL, Craig S (1979) Archaelogical evidence concerning the prehistoric occurrence of sea mammals at Point Bennett, San Miguel Island. Cal Fish Game 65(1):50–54
Wallace JM (1985) Atmospheric response to equatorial sea surface temperature anomalies. In: Wooster WS, Fluharty DL (eds) El Niño north. Washington Sea Grant Progr, Univ Washington, Seattle, pp 9–21
Walsh JJ (1981) A carbon budget for overfishing off Peru. Nature (London) 290:300–304
Ware DM, McFarlane GA (1986) Relative impact of Pacific hake, sablefish and Pacific cod on west coast of Vancouver Island herring stocks. Int N Pac Fish Comm Bull 47:67–77
Wasser SK, Barash DP (1983) Reproductive suppression among female mammals: implications for bio-medicine and sexual selection theory. Q Rev Biol 58:513–538
Watt BK, Merrill AL (1963) Composition of foods: raw, processed, prepared. Agricultural handbook, vol 8. Agric Res Serv, US Dep Agric. US Gov Print Off, Washington, DC
Wells LE (1987) An alluvial record of El Niño events from northern coastal Peru. J Geophys Res 92 C13:14, 463–470
Wespestad VG, Traynor JJ (1990) Walleye pollock. In: Condition of groundfish resources of the eastern Bering Sea and Aleutian Islands region in 1988. NOAA Tech Mem NMFS F/NWC-178, pp 19–43
Wolf P (1986) Jack mackerel. In: Klingbeil R (ed) Review for some California fisheries for 1985. CalCoFi Invest Rep 27
Wooster WS, Fluharty DL (eds) (1985) El Niño north. Washington Sea Grant Progr, Univ Washington, Seattle, 312 pp
Worcester K (1987) Market squid. In: Grant J (ed) Review of some California fisheries for 1985. CalCoFi Invest Rep 28
Wright SL, Crawford CB, Anderson JL (1988) Allocation of reprodutive effort in Mus domesticus: responses of offspring sex ratio and quality to social density and food availability. Behav Ecol Sociobiol 23:357–365
Wyrtki K (1975) El Niño – the dynamic response of the equatorial Pacific ocean to atmospheric forcing. J Phys Oceanogr 5:572–584
Wyrtki K (1982) The Southern Oscillation, ocean-atmosphere interaction and El Niño. Mar Technol Soc J 16:3–10
Wyrtki K (1985) Pacific-wide sea level fluctuations during the 1982–83 El Niño. In: Robinson G, del Pino EM (eds) El Niño en las islas Galápagos. El evento de 1982–1983. Fund Charles Darwin Islas Galápagos. Quito, Ecuador, pp 29–48
Wyrtki K (1986) Predicting El Niño: failures and hopes. In: Abstr Intergov Oceanogr Comm Worksh Rep 49. Guayaquil, Ecuador, p 5
Yochem PK (1987) Haul-out patterns and site fidelity of harbor seals at San Nicholas and San Miguel islands, California. MS Thesis, State Univ San Diego, Ca, 80 pp
Yochem PK, Stewart BS (1985) Radiotelemetry studies of hauling patterns, movements, and site fidelity of harbor seals (Phoca vitulina richardsi) at San Nicolas Island, California, 1983. HMRI Tech Rep 86–189
York AE (1983) Average age at first reproduction of the northern fur seal (Callorhinus ursinus). Can J Fish Aquat Sci 40:121–127

York AE (1985a) Forecast of the 1985 harvest on St. Paul Island, Alaska. In: Background paper, 28th Annu Meet Standing Scientific Committee, North Pacific Fur Seal Commission, April 1985. Tokyo, Jpn, 13 pp

York AE (1985b) Juvenile survival of fur seals. In: Kozloff P (ed) Fur seal investigations, 1982. NOAA Tech Mem NMFS-F/NWC,-71, pp 34–45

York AE (1987) Northern fur seal, Callorhinus ursinus, eastern Pacific population (Pribilof Islands, Alaska, and San Miguel Island, California). In: Croxall JP, Gentry RL (eds) Status, biology, and ecology of fur seals. NOAA Tech Rep NMFS 51, pp 23–27

York AE, Hartley JR (1981) Pup production following harvest of female northern fur seals. Can J Fish Aquat Sci 38:84–90

York AE, Kozloff P (1987) On the estimation of numbers of northern fur seal, Callorhinus ursinus, pups born on St. Paul Island, 1980–86. Fish Bull US 85:367–375

Zar JH (1974) Biostatistical analysis. Prentice-Hall, Englewood Cliffs, NJ, 620 pp

Zenger HH (1989) Pacific cod. In: Condition of groundfish resources of the Gulf of Alaska in 1988. NOAA Tech Mem NMFS F/NWC-165, pp 55–76

Zimmermann RR, Geist CR, Wise LA (1974) Behavioral development, environmental deprivation and malnutrition. In: Newton G, Riesen AH (eds) Advances in psychobiology. John Wiley & Sons, New York, pp 133–192

Zimmermann RR, Geist CR, Ackles PK (1975) Changes in the social behavior of rhesus monkeys during rehabilitation from prolonged protein-calorie malnutrition. Behav Biol 13:325–333

Zuta S, Rivera T, Bustamante A (1978) Hydrologic aspects of the main upwelling areas off Peru. In: Boje R, Tomczak M (eds) Upwelling ecosystems. Springer, Berlin Heidelberg New York, pp 235–257

Subject Index

Printing: Mercedesdruck, Berlin
Binding: Buchbinderei Lüderitz & Bauer, Berlin